ASSESSMENT OF MAMMALIAN EMBRYO QUALITY

INVASIVE AND NON-INVASIVE TECHNIQUES

ASSESSMENT OF MAMMALIAN EMBRYO QUALITY

Invasive and non-invasive techniques

edited by

Ann Van Soom

Department of Reproduction, Obstetrics and Herd Health,
Faculty of Veterinary Medicine, Merelbeke, Belgium

and

Marleen Boerjan

Research and Development, Pas Reform Hatchery Technologies,
Zeddam, The Netherlands

Springer Science+Business Media, B.V.

A C.I.P. Catalogue record for this book is available from the Library of Congress.

ISBN 978-94-010-3909-3 ISBN 978-94-010-0343-8 (eBook)
DOI 10.1007/978-94-010-0343-8

Printed on acid-free paper

TABLE OF CONTENTS

LIST OF CONTRIBUTORS

Alikani M

Institute of Reproductive Medicine and Science, Saint Barnabas Medical Center, 101 Old Short Hills road, Suite 501, West Orange, NJ 07052, USA

Bavister BD

Audubon Institute, Center for Research of Endangered Species, University of New Orléans, 14001 River Road, New Orleans, LA 70131, USA

Boerjan ML

PasReform, Hatchery technologies, PO Box 2, NL-7038 ZG Zeddam, the Netherlands

Cohen J

Institute of Reproductive Medicine and Science, Saint Barnabas Medical Center, 101 Old Short Hills road, Suite 501, West Orange, NJ 07052, USA

Donnay I

Université Catholique de Louvain, Unité des Sciences Vétérinaires, Place Croix du Sud 3, B-1348 Louvain-la-Neuve, Belgium

Greve T

Department of Clinical studies, Reproduction, The Royal Veterinary and Agricultural University, Dyrlægevej 68, 1870 Frederiksberg C, Copenhagen, Denmark

Hardy K

Department for Reproductive Science and Medicine, Institute of Reproductive and Developmental Biology, Imperial college, Hammersmith Hospital, Du Cane Road, London, W12 ONN, UK

Johnson MH

Department of Anatomy, University of Cambridge, Downing Street, Cambridge CB2 3DY, UK

Leibo SP

Audubon Institute, Center for Research of Endangered Species, University of New Orléans, 14001 River Road, New Orléans, LA 70131, USA

Maddox-Hyttel P

Department of Anatomy and Physiology, The Royal Veterinary and Agricultural University, Groennegaardsvej 7, 1870 Frederiksberg C, Copenhagen, Denmark

Massip A

Université Catholique de Louvain, Unité des Sciences Vétérinaires, Place Croix du Sud 3, B-1348 Louvain-la-Neuve, Belgium

Merton S

Holland Genetics, ET-IVP Laboratory, Broekstraat 5b, 7217 SL Harfsen, the Netherlands

Mullaart E

Holland Genetics, PO Box 5073, 6802 Arnhem, the Netherlands

Niemann H

Department of Biotechnology, Institute of Animal Science, Federal Agricultural Research Center (FAL), Höltystr. 10, D-31535 Neustadt, Germany

Sadowy S

Institute of Reproductive Medicine and Science, Saint Barnabas Medical Center, 101 Old Short Hills road, Suite 501, West Orange, NJ 07052, USA

Spanos S

Department for Reproductive Science and Medicine, Institute of Reproductive and Developmental biology, Imperial college, Hammersmith Hospital, Du Cane Road, London, W12 ONN, UK

Squirrell JM

Laboratory of Moleuclar Biology, University of Wisconsin, 1525 Linden Dr., Madison, WI 53717, USA

Tachataki M

Institute of Reproductive Biology, Faculty of Veterinary

Medicine, Imperial College, Hammersmith Hospital, Du Cane road , London W11 0NN, UK

Taylor D

Institute of Reproductive Biology, Faculty of Veterinary Medicine, Imperial College, Hammersmith Hospital, Du Cane road , London W11 0NN, UK

Thomsen PD

Department of Anatomy and Physiology, The Royal Veterinary and Agricultural University, Groennegaardsvej 7, 1870 Frederiksberg C, Copenhagen, Denmark

Van der Elst J

Infertility Centre, Ghent University Hospital, De Pintelaan 185, B-9000 Ghent, Belgium

Van Soom A

Department of Reproduction, Obstetrics and Herd Health, Faculty of Veterinary Medicine, Salisburylaan 133, B-9820 Merelbeke, Belgium

Viuff D

Molecular pharmacology, Novo Nordisk A/S, Novo Nordisk Park, 2760 Måløv, Denmark

Wrenzycki C

Department of Biotechnology, Institute of Animal Science, Federal Agricultural Research Center (FAL), Höltystr. 10, D-31535 Neustadt, Germany

Preface

The conceivement of this book happened more than 5 years ago, while we where discussing several criteria which could be useful to select the best embryo for transfer. Marleen Boerjan first mentioned the need for a book on embryo quality. Despite the availability of different techniques we still found it very difficult to estimate what a good embryo should be like. And through ongoing discussions in conferences and in embryo-mail, we got the impression that other people also found this issue very difficult.

In recent years, more and more techniques have become available to try and estimate the quality of a mammalian embryo. The organization of the chapters is designed to differentiate between invasive techniques, (which per definition destroy the embryo) and non-invasive techniques (which leave the embryo intact and suitable for transfer).

All contributions have been made by experts in the field. We would like to thank everyone for his or her contribution, because they were willing to spent some of their precious and limited time in writing a chapter and because they were able to respond to the deadlines and the sometimes tedious editorial demands. We would like to express our gratitude to our colleague Dr. Marc Coryn, who assisted us in reading the drafts and last but not least we would like to thank Mr Roger Verbeke who has spent numerous hours correcting and editing the many chapters of this book.

October, 2001

Ann Van Soom, Merelbeke, Belgium

Marleen Boerjan, Wageningen, The Netherlands

Introduction

ONLY THE BEST CONCEPTUS WILL DO! BUT WHAT DOES BEST MEAN?

Martin H. Johnson
Department of Anatomy
Downing Street
Cambridge CB2 3DY, UK

The history of mammalian pre-implantation embryology is both long and short. A number of classical studies on eggs and early embryogenic stages have been reported over the past 200 or so years. However, the major breakthrough came with the discovery in the 1950s of three technological advances: induced ovulation (Fowler and Edwards, 1957), in vitro spermatozoal capacitation leading to in vitro fertilization (IVF; Austin, 1951; Chang, 1951), and in vitro culture conditions that supported development of the conceptus (Whitten, 1956; Whitten, 1957; McLaren and Biggers, 1958; Whitten and Biggers, 1968; Whittingham, 1971; Brinster, 1972). These technical achievements permitted larger numbers of oocytes to be recovered at controlled and therefore predictable times, to be fertilized, and then to be held in vitro for varying periods under conditions that did not preclude further developmental progression upon transfer into the female genital tract. Immediately elements of this trilogy of technological innovation were in place, our knowledge about the reproductive biology of gametes and developmental mechanisms in mammals surged ahead. It was possible to supplement single time point observation and description with serial analysis and controlled experimentation. This possibility lead to the positing of testable hypotheses about early developmental mechanisms that, in a refined form, serve as the intellectual basis for our research, both basic and applied, today. It also laid the essential groundwork for the practical application of knowledge. Thus, although most of the basic research was initiated on the mouse, it has led subsequently to the development of more efficient and novel breeding programmes for live stock, the treatment of human infertility, conservation programmes for rare mammals, and the genetic manipulation of mammals in general. From these small but critical beginnings, whole new branches of science, medicine, veterinary practice, ethics and law have developed.

However, fantastic though our knowledge is today compared with 35 years ago, we must acknowledge that much remains to be understood. For example, despite over 23 years of human IVF, we are still not very good at it. Data taken from the Human Fertilization and Embryology Authority Annual Reports show that the proportions of live births for all cycles of IVF treatment initiated have not improved substantially with time across the UK as a whole. They also show that still only a minority of treatment cycles results in a live birth. These data are averages from all clinics, and therefore do mask considerable variation among clinics. However, the fact remains that this technique, more than 23 years after it was first used successfully (Edwards et al., 1969; Steptoe and Edwards, 1978), is still not sufficiently routine to yield comparable results across clinics, and even the most effective clinics only achieve live birth rates in the high 20% range. This relative failure is made more problematic, and indeed may in part be caused by a growing demand for infertility treatment from several quarters (Johnson, 2001b), in particular, from an increasing tendency to defer fertility or to experience serial relationships in each of which children are wanted. Moreover, the exposure of the population by the media to the new reproductive technologies encourages previously 'hopeless' patients to expect a solution to their infertility, and new sources of demand are also appearing, for example same sex couples want children in which both partners have a genetic interest (Kubiak and Johnson, 2001). Thus, there is pressure both to improve the outcome rates of existing treatments and to provide novel therapies to meet new expectations. Clearly, a basic requirement, if these needs are to be met, is for the clinical embryologist to be able to recognize and select those conceptuses (or better still in humans if multiple births are to be reduced that conceptus) with the maximum potential for development to term. The same requirement takes on a strong economic imperative when we consider the use of assisted reproduction in farm animals, in which in vitro production is mainly used to increase the number of offspring from high yield animals and/or to reduce generation intervals. Additionally, in the future, genetic manipulation of large farm animals will often accompany in vitro production and transfer, so the pressure to get it right becomes even more challenging. Why is it then, despite success, that we still have so far to go?

Why do reproductive technologies fail?

There are broadly two sorts of explanation. One explanation is that the recipient female is for some reason unable to carry the implantation and pregnancy successfully. This might be because the conditions used to recover the eggs affect adversely the female reproductive tract, or even, in cases of infertility, that the tract is inherently incompetent. This explanation is beyond the scope of this book, although very relevant for it. The second explanation focuses on the inadequacy of the conceptus itself, and this category of explanation lies at the heart of this book. Usually, the basis for conceptus inadequacy is divided broadly into inherent errors and those induced during the therapeutic intervention itself, although as we will see this is a fuzzy rather than clearly defined classification system. Inherent errors are thought to be those that exist because of developmental, endocrine or genetic pathology and/or of some earlier environmental insult such as infection, radiation or chemical exposure. Induced errors are those specifically arising from the treatment schedule itself.

The nature of the mammalian embryogenic phase

An early conclusion from the experimental study of the mammalian conceptus was that it differed fundamentally from the then established models of developmental biology, in particular sea urchin and Xenopus. Thus, the primary function of early mammalian development is the elaboration of a complex membrane support system, within which only a small minority of cells (<5%) becomes identified as "the embryo", which then embarks on embryonic and foetal development. This phase is therefore called the embryogenic phase of development, and the product of fertilization is called variously the conceptus, the pre-embryo, the embryogen or even, incorrectly, the embryo (after which the "embryo proper" is often said to emerge: Johnson and Selwood, 1996). Second, the developmental process underlying this embryogenic phase was found to be indeterminate and regulatory. If, during embryogenesis, embryogenic cells (or cytoplasm) were removed, added or rearranged, a perfectly normal embryo/foetus nonetheless resulted. These experiments suggested that there was no mosaic of developmental information encoded non-genetically in the gametes or zygote to be inherited and/or expressed differentially during cleavage. The results were interpreted to favour a developmental role for cell interaction, with the monitoring of relative cell position being critical for both morphogenesis and cytodifferentiation (Johnson, 2001a). Thus, an early hypothesis to explain the allocation of cells to the trophoblast and pluriblast lineages was the "inside-out" theory of Tarkowski and Wroblewska, (1967) which proposed that cells on the inside and the outside of the developing embryogenic cell mass were in some way able to recognize their relative position and respond accordingly. Later studies reinterpreted this proposal in more mechanistic cell biological terms in the polarisation hypothesis (Johnson et al., 1986; Johnson, 1996).

One practical consequence of this view of early mammalian development was a belief in the strong resilience of the cleaving and cavitating conceptus, and indeed the early mammalian conceptus is remarkably adaptive. However, since those early studies, a more thorough understanding of early development in many species has revealed that developmental plasticity is perfectly compatible with the mosaic organisation of developmental information, and it is now clear that the earliest stages of mammalian embryogenesis do make use of mosaic information (Johnson, 2001a). Thus, the delineation of the founder lineage populations that mark cytodifferentiation seems to occur as early as the oocyte in some mammals and even in the mouse there is evidence of influences from the early post-fertilization period (Pratt, 1989; Johnson, 1996; Selwood, 1996). More recently an influence of sperm entry and first cleavage plane on embryonic axis formation has been reported for the mouse (Gardner, 1997; Gardner, 2001; Johnson, 2001a; Piotrowska and Zernicka-Goetz, 2001).

These conclusions, in turn, require us to rethink our views about the adaptability of the early conceptus and what it might mean for manipulations in vitro. There are three questions of concern. First, is the way in which we produce the gametes for in vitro production procedures altering or even damaging them in some way? Second, once produced, are we handling the conceptuses in the best way for their long term survival and normality? And crucially for this volume, how can we know the answers to these questions short of placing the conceptuses in the female tract - what reliable and validated surrogate markers of real developmental potential are available to us? It is these markers that will allow us to measure the impact of our procedures and to discriminate between induced and inherent errors, and it is these markers, especially those that are non-invasive, which will in the longer term allow us practically to select the "best" conceptus(es) for transfer. In addressing these questions, it is important to be aware that the answers may differ in detail for different species. Much of the pioneering work has been undertaken on the mouse and to a lesser extent the hamster. However, these species differ from most other mammalian species in, for example, the orientation of the second meiotic spindle, the inheritance of centrosomal material from the sperm, the timing of key developmental transitions including the onset of zygotic transcription, the duration of the preimplantation period and the nature of attachment events that terminate it (Maro et al., 1985; Maro et al., 1986; Pickering et al., 1988; Simerly et al., 1995; Johnson, 1996; Johnson and Everitt, 2000 and see later). These differences are mainly of detail rather than principle, but may none the less mean that the types of problems encountered and the ways of identifying them may not be simply transferable between species. It is important to bear this in mind when reading the chapters in this book, which range across several species, but mostly bovine and human.

Induction of gamete production may induce abnormalities

The ability to recover immature ovarian oocytes and then to culture them to maturation in vitro (Edwards, 1965) is arguably a fourth major technological break through that has facilitated the study of early development and the practical application of IVF. These studies have identified clearly the complex nature of the maturation process and its importance for early development (Eppig and O'Brien, 1998; Moor et al., 1998), indicating the distinctive roles of chromosomal maturation from first meiotic prophase to second meiotic metaphase, cytoplasmic maturation, and cumulus:oocyte interactions. This knowledge permits a more refined analysis of oocytes maturing in vivo under exogenous hormonal stimulation so as to assess the relationship between measured parameters, such as ATP levels, follicular blood flow or mitochondrial activity, to developmental outcome (Van Blerkom et al., 1995; Van Blerkom., 2000; Wilding et al., 2001) and so to improve induction regimes. Induced ovulation has been shown to have an enduring impact on development (e.g. Van der Auwera and D'Hooghe, 2001).

Such studies have also revealed the extent to which, and by what mechanisms, the early development of the conceptus depends on the ovulated oocyte. The oocyte's mRNA population serves the conceptus exclusively until transcriptional activation, in some species for several days. Its proteins and organelles continue to be essential for weeks or, in the case of mitochondria, throughout life. These studies have made clear how errors of maturation (induced or natural) impact on development and has provided markers for practically assessing error incidence. This later point is important because errors are not necessarily lethal for oocyte or conceptus. Thus, Van der Elst (Chapter 4) shows that hormonally induced ovulation can lead to more diverse responses to post-fertilization freezing, which is thus acting as an indicator stressor. Most chromosomal abnormalities are compatible with survival in vitro (Taylor, Chapter 7 and Viuff et al., Chapter 10). Quite simple preventative measures during oocyte stimulation and recovery can lead to improvements in outcome. For example, the second meiotic spindle is known to be very sensitive to temperature (Pickering et al., 1990) and maintaining temperature during oocyte recovery in humans has been shown to improve IVF outcomes.

Culture of the conceptus requires its adaptation to in vitro conditions

It is now clear that the results from many of our studies may be compromised by the fact that the conceptus in vitro may differ significantly from the conceptus in vivo, and this is the case, whether the in vitro conceptus is produced by IVF or after recovery from the female tract after in vivo fertilization. Thus, although the mouse is undoubtedly the most successfully studied mammalian model, improvements to its preimplantation culture media continue to be reported even today (Summers et al., 2000). For many years, it has simply not been possible to culture preimplantation mouse conceptuses of most strains in vitro from oocyte to blastocyst. Even in those strains (or strain combinations) in which successful culture was possible, the development was slower, less cells were present at the blastocyst stage, and/or the in vitro developed conceptuses showed a number of molecular and biochemical differences from their counterparts that had developed in vivo. Thus, the conditions used to culture the conceptus had themselves induced changes. This knowledge has lead to the concept that even those in vitro conditions, which permit successful development both outside and then within the body after transfer to the female reproductive tract, involve the adaptation of conceptuses: the in vitro conditions do not truly replicate in vivo conditions (Johnson and Nasr-Esfahani, 1994; Leese, 1995; Biggers,

1998). For example, conceptuses developed in vitro can show altered morphology (Alikani, Chapter 1, and Merton, Chapter 2), altered metabolic properties (Edirisinghe et al., 1984;Tokura et al., 1993; Barnett and Bavister, 1996; Peippo et al., 2001) (Donnay, Chapter 3 and Bavister, Chapter 6), elevated free radical production (Nasr-Esfahani et al., 1990b), and altered patterns of both gene expression (Vernet et al., 1993; Ho et al., 1994; Ho et al., 1995; Lee et al., 2001; Lequarre et al., 2001; Minami et al., 2001; Stojanov and O'Neill, 2001; Wrenzycki et al., 2001 and Chapter 14) and genomic imprinting (Reik et al., 1993; Doherty et al., 2000; Khosla et al., 2001). In addition, media differing in composition can exert distinctive effects on each of these parameters. Moreover, these effects can be compatible with survival but then may have long term consequences that only express themselves at the foetal, neonatal or even later stages. Thus it is known that quite short periods in vitro can have enduring phenotypic effects on large farm animals, resulting in abnormally large and disproportionate foetuses with physiological anomalies and higher perinatal mortality (Walker et al., 1996; Young et al., 1998).

Recently, Fleming and his colleagues (Kwong et al., 2000) have further emphasized the environmental vulnerability of the preimplantation conceptus even in vivo, by showing that simple dietary protein deprivation of maternal rats over the first four days of pregnancy affects blastocyst organisation and, even when the pregnancy is allowed to continue on a full protein diet, the male offspring born develop high blood pressure and show disproportionate organ sizes. These results have ominous implications given that human epidemiological studies led Barker and his colleagues to conclude that small-for-dates babies correlated with both maternal undernutrition during pregnancy and a raised incidence of organ disproportional, cardiovascular disease and diabetes in the adults developing from such babies (Barker, 1997; O'Brien et al., 1999). Thus we have clear evidence that the early conceptus is rather sensitive to its environment, which can exert long term effects on both the anatomy and physiology of the offspring produced. Simply surviving the experience of a stressed preimplantation environment does not necessarily indicate that a normal foetus, neonate and adult will develop (see also: Boerjan et al., 2000).

Given this knowledge, it is important to know whether there is a particularly sensitive period during preimplantation development and/or whether simple duration of time in vitro leads to an accumulation of adaptive and potentially deleterious responses. The trend towards use of more prolonged culture in vitro of human conceptuses makes answers to this question especially important (Langley et al., 2001). There is clear evidence of sensitive periods during early development. Early on, the mouse was found to block at the 2-cell stage, and subsequently humans at the 4- to 8-cell stage (Braude et al., 1988), and sheep, cattle and goats at the 8- to 16-cell stage (Wright and Bondioli, 1981). In these species, the time of sensitivity to block coincides with the onset of major transcriptional activity of the zygotic genes. In the mouse, the period of block is also associated with elevated levels of reactive oxygen species (Nasr-Esfahani et al., 1990b). Exposure to antioxidants limited to the sensitive period alone can overcome or ameliorate the developmental block (Toyoda et al., 1989; Nasr-Esfahani et al., 1990a; Lee et al., 2000), suggesting that an acute and temporally defined period of free radical damage in vitro may be in part responsible for developmental arrest (Johnson and Nasr-Esfahani, 1994). However, arrest is in a way reassuring, because at least the conceptus is less likely to be transferred to the female tract. Much more serious are chronic and cumulative cultural stresses that permit survival but that may also lead to either non-lethal defects or to adaptations to the in vitro conditions that themselves have as a consequence later inappropriate and damaging foetal and neonatal responses to stress.

Conclusions

This consideration of induced deficiencies is important for a book which considers how to select the best conceptuses for transfer whether in the bovine or the human. First, the use of techniques for conceptus selection that rely on some aspect of in vitro survival must be scrutinized thoroughly, because in trying to sort current problems they may be setting up later ones. Second, as our methods for analysing the early conceptus become increasingly refined, we will have more and more markers of difference between in vivo and in vitro conceptuses. How will we know which differences are important and have enduring effects and which are trivial? For example, mRNA expression studies are now reporting large numbers of transcript differences between in vitro and in vivo conceptuses or between conceptuses cultured in different media (see Wrenzycki et al., 2001 and Chapter 14 for an example in the bovine): what does this mean exactly, and how significant is it for full developmental potential? Third, although there is a traditional division of embryogenic defects into those that are inherent and those induced by culture, this classification may be too simple. The inherent defects may include some which reflect simply an inability to respond to in vitro conditions adequately rather than representing any major block to development in vivo. A possible example is provided in a recent study on bovine conceptuses derived from in vitro matured oocytes cultured in glucose rich or poor media (Hashimoto et al., 2000). Excess glucose resulted in conceptuses with lower glutathione levels, higher free radical levels and impaired development. It is hypothesized that the elevated glucose inhibited glutathione synthesis by impairing expression of γ-glutamylcysteine synthetase (Nasr-Esfahani and Johnson, 1992) and accelerated reactive oxygen species production by stimulating xanthine oxidase activity (Nasr-Esfahani and Johnson, 1991). Such a developmental defect would not have been apparent in vivo, but is revealed in vitro after exposure to the stress of higher ambient oxygen tension. Moreover, since the glucose was almost certainly metabolized not by the maturing oocyte but by the associated follicular cells (Hashimoto et al., 2000), this model may not be entirely irrelevant to in vivo matured oocytes during superovulation during which disturbances in granulosa:oocyte communication may sometimes occur.

We are confronted therefore by a dilemma. We need to devise markers for the best developmental potential. As we become more sophisticated and increase the battery of markers available for study, how will we know which are useful and which are not? In particular, how can we best look for markers that predict the best developmental outcome for the in vitro period (well adapted conceptuses?), as well as for those predicting the best outcome after transfer for the remaining in vivo period. The two sets of markers need not necessarily be the same, indeed are unlikely to be so. What seems at the outset to be a relatively simple task, begins to take on a daunting complexity. The authors in this book rise to these challenges. However, I suspect that this will not be the last word on this subject. I already look forward to the second edition!

References

Austin CR. 1951. Observations on the penetration of the sperm into the mammalian egg. Aust J Sci Res 4:581-596.

Barker DJP. 1997. The fetal origins of coronary heart disease. Acta Paediatrica, Suppl 422:78-82.

Barnett DK, Bavister BD. 1996. What is the relationship between the metabolism of preimplantation embryos and their developmental competence? Mol Reprod Dev 43:105-133.

Biggers JD.1998. Reflections on the culture of the preimplantation embryo. Int J Dev Biol 42:879-884.

Boerjan ML, den Daas JHG, Dieleman SJ. 2000. Embryonic origins of health: long term effects of IVF in Human and Livestock. Theriogenology 53:537-547.

Braude P, Bolton V, Moore S. 1988. Human gene expression first occurs between the four- and eight-cell stages of preimplantation development. Nature. London 332:459-461.

Brinster RL 1972. Cultivation of the mammalian embryo. In Rothblatt G, Cristofalo V. .eds., Growth, Nutrition and Metabolism of Cells in Culture. Academic press, New York, Vol. II, pp. 251-286.

Chang MC. 1951. Fertilizing capacity of sperm deposited into the fallopian tubes. Nature 168:697-698.

Doherty AS, Mann MRW, Tremblay KD, Bartolomei MS, Schultz RM. 2000. Differential effects of culture on imprinted H19 expression in the preimplantation mouse embryo. Biol Reprod 62:1526-1535.

Edirisinghe WR, Wales RG, Pike IL. 1984. Degradation of biochemical pools labelled with [14C]glucose during culture of 8-cell and morula/early blastocyst-stage mouse embryos in vitro and in vivo. J Reprod Fertil 72:59-65.

Edwards R, Bavister B, Steptoe P. 1969. Early stages of fertilization in vitro of human oocytes matured in vitro. Nature 221:632-635.

Edwards RG. 1965. Maturation in vitro of mouse, sheep, cow, pig, rhesus monkey and human ovarian oocytes. Nature 208:349-351.

Eppig JJ, O'Brien MJ. 1998. Comparison of preimplantation developmental competence after mouse oocyte growth and development in vitro and in vivo. Theriogenology 49:415-422.

Fowler RE, Edwards, RG. 1957. Induction of superovulation and pregnancy in mature mice by gonadotrophins. J Endocrinology 15:374-384.

Gardner RL. 1997. The early blastocyst is bilaterally symmetrical and its axis of symmetry is aligned with the animal-vegetal axis of the zygote in the mouse. Development 124:289-301.

Gardner RL. 2001. Specification of embryonic axes begins before cleavage in normal mouse development. Development 128:839-847.

Hashimoto S, Minami N, Yamada M, Imai H. 2000. Excessive concentration of glucose during in vitro maturation impairs the developmental competence of ovine oocytes after in vitro fertilization: relevance to intracellular reactive oxygen species and glutathione contents. Mol Reprod Dev 56:520-526.

Ho Y, Doherty AS, Schultz RM. 1994. Mouse preimplantation embryo development in vitro: effect of sodium concentration in culture media on RNA synthesis and accumulation and gene expression. Mol Reprod Dev 38:131-141.

Ho Y, Wigglesworth K, Eppig JJ, Schultz RM. 1995. Preimplantation development of mouse embryos in KSOM: augmentation by amino acids and analysis of gene expression. Mol Reprod Dev 41:232-238.

Johnson M, Everitt B. 2000. Essential Reproduction. Blackwell Science, Oxford.

Johnson MH. 1996. The origins of pluriblast and trophoblast in the eutherian conceptus. Reprod Fert Dev 8:699-709.

Johnson MH. 2001a. Axes in the egg? Current Biology 11:R281-R284.

Johnson MH. 2001b. Reproduction in the Noughties: will the scientists have all the fun? J Anatomy 198:385-398.

Johnson MH, Maro B, Takeichi M 1986. The role of cell adhesion in the synchronization and orientation of polarization in 8-cell mouse blastomeres. J Embryol Exp Morphol 93:239-255.

Johnson MH, Nasr-Esfahani MH. 1994. Radical solutions and cultural problems: could free oxygen radicals be responsible for the impaired development of preimplantation mammalian embryos in vitro? BioEssays 16:31-38.

Johnson MH, Selwood L. 1996. Nomenclature of early development in mammals. Reprod Fertil Dev 8:759-764.

Khosla S, Dean W, Brown D, Reik W, Feil R. 2001. Culture of preimplantation mouse embryos affects fetal development and the expression of imprinted genes. Biol Reprod 64:918-926.

Kubiak J, Johnson MH. 2001. Human infertility, reproductive cloning and nuclear transfer: a confusion of meanings. BioEssays 23:359-364.

Kwong WY, Wild AE, Roberts P, Willis AC, Fleming TP. 2000. Maternal undernutrition during the preimplantation period of rat development causes blastocyst abnormalities and programming of postnatal hypertension. Development 127:4195-4202.

Langley MT, Marek DM, Gardner DK, Doody KM, Doody KJ. 2001. Extended embryo culture in human assisted reproduction treatments. Hum Reprod 16:902-908.

Lee CS, Koo DB, Fang N, Lee Y, Shin ST, Park CS, Lee KK. 2000. Potent and stage-specific action of glutathione on the development of goat early embryos in vitro. Mol Reprod Dev 57:48-54.

Lee KF, Chow JFC, Xu JS, Chan STH, Ip SM, Yeung WSB. 2001. A comparative study of gene expression in murine embryos developed in vivo, cultured in vitro, and cocultured with human oviductal cells using messenger ribonucleic acid differential display. Biol Reprod 64:910-917.

Leese HJ. 1995. Metabolic control during preimplantation mammalian development. Hum Reprod Update 1:63-72.

Lequarre AS, Feugang JM, Malhomme O, Donnay I, Massip A, Dessy F, Vanlangendonckt A. 2001. Expression of Cu/Zn and Mn superoxide dismutases during bovine embryo development: influence of in vitro culture. Mol Reprod Dev 58:45-53.

Maro B, Howlett SK, Webb M. 1985. Non-spindle microtubule organizing centers in metaphase II-arrested mouse oocytes. J Cell Biol 101:1665-1672.

Maro B, Johnson MH, Webb M, Flach G. 1986. Mechanism of polar body formation in the mouse oocyte: an interaction between the chromosomes, the cytoskeleton and the plasma membrane. J Embryol Exp Morphol 92:11-32.

McLaren A, Biggers JD. 1958. Successful development and birth of mice cultivated in vitro as early embryos. Nature 182:877-878.

Minami N, Sasaki K, Aizawa A, Miyamoto M, Imai H. 2001. Analysis of gene expression in mouse 2-cell embryos using fluorescein differential display: comparison of culture environments. Biol Reprod 64:30-35.

Moor RM, Dai Y, Lee C, Fulka J. 1998. Oocyte maturation and embryonic failure. Hum Reprod Update 4:223-236.

Nasr-Esfahani MH, Johnson MH, Aitken RJ. 1990a. The effect of iron and iron chelators on the in vitro block to development of the mouse preimplantation embryo: BAT6 a new medium for improved culture of mouse embryos in vitro. Hum Reprod 5: 997-1003.

Nasr-Esfahani MH, Aitken JR, Johnson MH. 1990b. Hydrogen peroxide levels in mouse oocytes and early cleavage stage embryos developed in vitro or in vivo. Development 109:501-507.

Nasr-Esfahani MH, Johnson MH. 1992. Quantitative analysis of cellular glutathione in early preimplantation mouse embryos developing in vivo and in vitro. Hum Reprod 7:1281-1290.

Nasr-Esfahani MH, Johnson MH. 1991. The origin of reactive oxygen species in mouse embryos cultured in vitro. Development 113:551-560.

O'Brien P, Wheeler T, Barker D. 1999. Fetal programming: influences on development and disease in later life. RCOG Press, London.

Peippo J, Kurkilahti M, Bredbacka P. 2001. Developmental kinetics of in vitro produced bovine embryos: the effect of sex, glucose and exposure to time-lapse environment. Zygote 9:105-114.

Pickering SJ, Braude PR, Johnson, MH, Cant A, Currie J. 1990. Transient cooling to room temperature can cause irreversible disruption of the meiotic spindle in the human oocyte. Fertil Steril 54:102-108.

Pickering SJ, Johnson MH, Braude PR, Houliston E. 1988. Cytoskeletal organization in fresh, aged and spontaneously activated human oocytes. Hum Reprod 3:978-989.

Piotrowska K, Zernicka-Goetz M. 2001. Role for sperm in spatial patterning of the early mouse embryo. Nature 409:517-521.

Pratt HP. 1989. Marking time and making space: chronology and topography in the early mouse embryo. Int Rev Cyt 117:99-130.

Reik W, Romer I, Barton SC, Surani MA, Howlett SK, Klose J. 1993. Adult phenotype in the mouse can be affected by epigenetic events in the early embryo. Development 119:933-942.

Selwood L. 1996. The blastocyst epithelium is not a protoderm in dasyurid marsupials: a review of the evidence. Reprod Fertil Dev 8:11-23.

Simerly C, Wu G, Ord T, Rawlins R, Jones J, Navara C, Gerrity M, Rinehart J, Binor Z. 1995. The paternal inheritance of the centrosome, the cell's microtubule-organising centre, in humans, and the implications for infertility. Nature Med 1:47-52.

Steptoe PC, Edwards RG. 1978. Birth after the reimplantation of a human embryo. Lancet 2:366.

Stojanov T, O'Neill C. 2001. In vitro fertilization causes epigenetic modifications to the onset of gene expression from the zygotic genome in mice. Biol Reprod 64:696-705.

Summers M, McGinnis L, Lawitts J, Raffin M, Biggers J. 2000. IVF of mouse ova in a simplex optimized medium supplemented with amino acids. Hum Reprod 15:1791-1801.

Tarkowski AK, Wroblewska J. 1967. Development of blastomeres of mouse eggs isolated at the 4- and 8-cell stage. J Embryol Exp Morph 18:155-180.

Tokura T, Noda Y, Goto Y, Mori T. 1993. Sequential observation of mitochondrial distribution in mouse oocytes and embryos. J Assist Reprod Gen 10:417-426.

Toyoda Y, Azuma A, Takeda S. 1989. Effects of chelating agents on preimplantation development of mouse embryos fertilized in vitro. In Yoshinaga K and Mori T. eds., Development of preimplantation embryos and their environment. Alan R. Liss, New York, pp. 171-179.

Van Blerkom J, Davis PW, Lee J. 1995. ATP content of human oocytes and developmental potential and outcome after in vitro fertilization and embryo transfer. Hum Reprod 10:415-424.

Van Blerkom J. 2000. Intrafollicular influences on human oocyte developmental competence: perifollicular vascularity, oocyte metabolism and mitochondrial function. Hum Reprod, 15 (Suppl. 2):173-188.

Van der Auwera I, D'Hooghe T. 2001. Superovulation of female mice delays embryonic and fetal development. Hum Reprod 16:1237-1243.

Vernet M, Cavard C, Zider A, Fergelot P, Grimber G and Briand P. 1993. In vitro manipulation of early mouse embryos induces HIV1-LTRlacZ transgene expression. Development 119:1293-1300.

Walker SK, Hartwich KM and Seamark RF. 1996. The production of unusually large offspring following embryo manipulation: concepts and challenges. Theriogenology, 45:111-120.

Whitten WK. 1956. Culture of tubal ova. Nature 177:96.

Whitten WK. 1957. Culture of tubal ova. Nature 179:1081-1082.

Whitten WK, Biggers JD. 1968. Complete development in vitro of the pre-implantation stages of the mouse in a simple chemically defined medium. J Reprod Fertil 17:399-401.

Whittingham DG. 1971. Culture of mouse ova. J Reprod Fertil 14:7-21.

Wilding M, Dale B, Marino M, di Matteo L, Alviggi C, Pisaturo ML, Lombardi L, De Placido G. 2001. Mitochondrial aggregation patterns and activity in human oocytes and preimplantation embryos. Hum Reprod 16:909-917.

Wrenzycki C, Herrmann D, Keskintepe L, Martins Jr A, Sirisathien S, Brackett B, Niemann H. 2001. Effects of culture system and protein supplementation on mRNA expression in pre-implantation bovine embryos. Hum Reprod 16:893-901.

Wright RW, Bondioli KR. 1981. Aspects of in vitro fertilization and embryo culture in domestic animals. J Anim Sci 53:702-729.

Young LE, Sinclair KD, Wilmut I. 1998. Large offspring syndrome in cattle and sheep. Rev Reprod 3:155-163.

Chapter 1

HUMAN EMBRYO MORPHOLOGY AND DEVELOPMENTAL CAPACITY

Mina Alikani, Sasha Sadowy and Jacques Cohen
Institute for Reproductive Medicine and Science
Saint Barnabas Medical Center
West Orange, New Jersey, USA

1. Introduction

Human embryos obtained during therapeutic in vitro fertilization (IVF) are morphologically diverse and display features not common among other mammalian embryos in vitro. Logic dictates and experience concurs that embryo quality assessment must entail more than phenotype alone, but so far and at least in the human, no other evaluation scheme has proven to be more practical or clinically useful. This chapter presents current opinions on the relationship between gross morphology and developmental capacity. An overview of data from a clinical embryology database (EggCyte™) with large numbers of records has also been incorporated.

2. EggCyte™ Embryology Database

Data presented in this chapter were collected and analysed using the EggCyte database (EggCyte™, ART Institute of NY and NJ, Livingston, NJ, USA 1995-2001) (Tomkin and Cohen, 2001). Our institute maintains a computer network with both paper and on-line database record-keeping with terminals in the laboratory and elsewhere. Using MS-Access (Microsoft

Corp., Seattle, USA), we have created a suite of tables, forms, programme functions and reports tailored to the requirements of clinical embryology for individual oocyte and embryo tracking. Like any relational database, the tables link the relevant records by indexed numeric fields common to each.

After retrieval, oocytes are numbered and tracked individually with separate pages or tables for each day of development, just as in a daily journal. When embryos are replaced, separate records for each are again created and linked to the record of the replacement procedure. As pregnancy results become available, they are entered in a further table, always linked by procedure number. Data for the analyses in this study were extracted by creating selection lists or queries from relevant tables in the database. EggCyte has over 40 data tables that can be linked for study. Queries of varying complexity with Structured Query Language (SQL) and Visual Basic programming functions (Microsoft Corp., Seattle, USA) are then designed to compile the results.

3. Zygote Morphology

Fertilization encompasses a cascade of spatial and temporal events that culminate in the appearance of two closely apposed pronuclei (PN) in the egg (Reviewed by Yanagimachi, 1988), each containing a number of nucleolar precursor bodies (NPB) or nucleoli. In the human, the number, size, and position of pronuclei, as well as the nucleoli have been shown to correlate with developmental potential of zygotes, the latter perhaps related to the manner of polarisation within the oocyte and the zygote. In a comprehensive examination of the subject, Edwards and Beard (1997) pointed to and provided evidence for ooplasmic rotation after fertilization to modify existing oocyte polarity, chromatin polarisation and rotation in maternal pronuclei, and establishment of polar axes by the sperm centrosome at syngamy.

3.1. Single Pronucleate Eggs

Following both conventional insemination and intracytoplasmic sperm injection (ICSI), 3-4% of eggs display only a single pronucleus (1PN). Karyotype studies indicate that many 1PN eggs are diploid and some suggest that this condition results from asynchronous development of male and female pronuclei (Staessen et al., 1993a). However, the work of Levron et al. (1995) suggests an alternative mechanism. They examined 1PN eggs obtained after conventional IVF after separation into a karyoplast (the pronucleus surrounded by scant cytoplasm and the plasma membrane) and a cytoplast (the remaining cytoplasm surrounded by a membrane); fluorescent in situ hybridization (FISH) was then applied to each part. Of the 16 karyoplasts examined, 6 were diploid, and 4 of the 6 contained both X and Y-chromosomes. The cytoplasts never contained any chromosomal material, demonstrating that the male and female nuclei may associate together to form a single diploid pronucleus following fertilization.

In a follow-up study, Sultan et al. (1995) examined 42 day 3 embryos that developed from 1PN zygotes. Sixty-two percent of those from conventional insemination were diploid, and nearly 50% contained a Y chromosome, indicating fertilization by a sperm. In contrast, only 14% of 1PN eggs obtained after ICSI were diploid, and 19% contained a Y chromosome. Based on these findings, in our clinical practice, eggs displaying a single (often) large pronucleus after conventional insemination are observed and considered for transfer if they cleave normally. On the other hand, 1PN eggs after intracytoplasmic sperm injection (ICSI) are not used clinically since a majority appears to be parthenogenetically activated.

3.2. Pronuclear and Nucleolar Morphology

Uneven pronuclear size (Figure 1) has been observed after routine clinical sperm injection in meiotically mature oocytes as well as insemination of entire cohorts of oocytes at mixed stages of nuclear maturation (Schmiady and Kentenich, 1993; Sadowy et al., 1998). Sperm injection in mature oocytes fused experimentally with immature cytoplasm (J. Cohen et al., unpublished results), and immature sperm injection into mature oocytes (Tesarik and Mendoza, 1996) have also led to this morphology.

Sadowy et al. (1998) evaluated the in vitro progression and developmental potential of zygotes with uneven pronuclei. They found that i) pronuclear size differences were maintained until the onset of syngamy, ii) arrest at the one-cell stage occurred significantly more frequently when the diameter of one pronucleus differed by at least 4 μm from the other, and iii) cleavage of these zygotes led to significantly higher frequencies of day-2 blastomere multinucleation and day-3 mosaicism. Clinically, cryopreservation or transfer of zygotes with distinctly uneven pronuclear size and embryos resulting from such zygotes is generally avoided when others with normal pronuclear morphology are available.

Figure 1: A human zygote displaying unevenly sized pronuclei. Pronuclear size differences are maintained until the onset of syngamy. The incidence of arrest, multinucleation, and mosaicism is higher among embryos developing from such zygotes.

Pronuclear alignment and nucleolar polarisation in zygotes have also been suggested (Van Blerkom, 1989) and shown to impact development post thaw (Wright et al., 1990) or after fresh transfer (Scott and Smith, 1998). In the study by Wright et al. (1990), zygotes with strongly adhered pronuclei and equatorially aligned nucleoli showed the highest implantation potential post thaw:11/39 or 28% foetal heart beat (FHB) per embryo replaced (ER). Scott and Smith (1998) described a series of successful fresh zygote transfers after careful selection based on a zygote morphology score. Zygotes were scored for alignment of pronuclei and nucleoli as well as cytoplasmic appearance: the highest score was assigned to zygotes with closely apposed pronuclei, aligned nucleoli, and a cytoplasmic "halo" or organelle-depleted sub-plasmalemmal area (Sathananthan et al., 1993; Payne et al., 1997). High scoring zygotes were associated with high implantation (49/175 or 28% FHB/ER) and clinical pregnancy rates (34/48 or 71%).

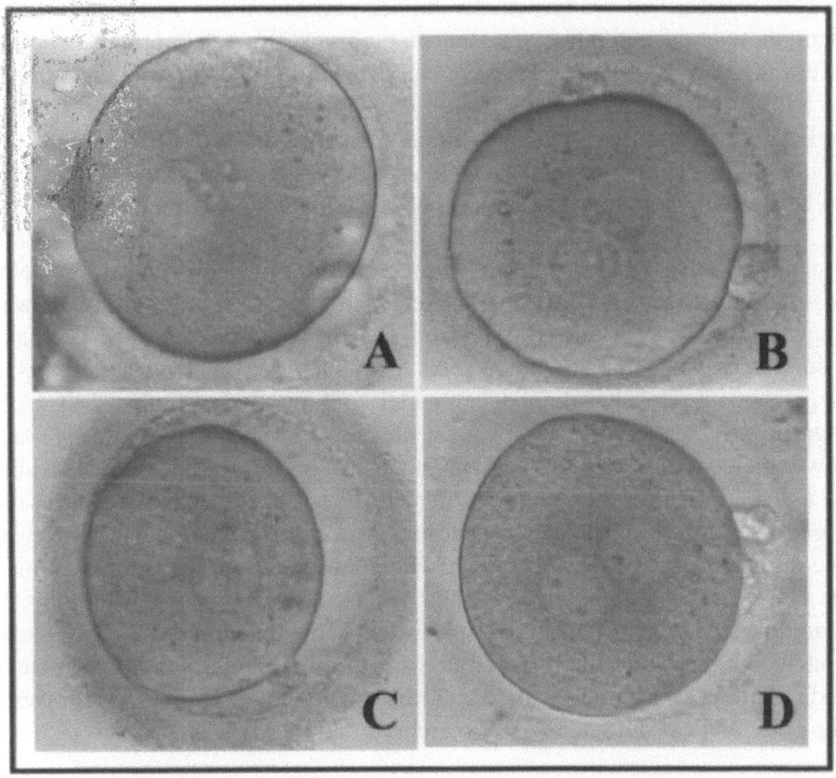

Figure 2: *Four distinct patterns of nucleolar distribution in pronuclei. (A) Closely aligned nucleolar apposition, (B) Nucleolar apposition, (C) Scattered nucleoli, (D) Scattered nucleoli in one pronucleus and distinctly polarized in the other pronucleus. Embryos derived from pattern (A) had higher cell numbers and lower incidence of multinucleation (overall better morphology), but this pattern occurred infrequently. Also note overt cytoplasmic contraction and polar body separation in (B), and large perivitelline space in (C).*

In a more recent study, Scott et al. (2000) scored zygotes according to distribution and size of nucleoli and correlated this score with the frequency of blastocyst formation on day 5; zygotes displaying inter-pronuclear uniformity had 49.5% blastocyst formation while those with unequal pronuclear size, nuclear number or distribution had 28% blastocyst formation.

Tesarik and Greco (1999) have suggested a role for interpronuclear synchrony as a determinant of developmental competence. These investigators later found the transfer of embryos that developed from "normal" zygotes established a significantly higher pregnancy and implantation rate than those achieved after transfer of embryos from "abnormal" zygotes (Tesarik et al., 2000). Zygotes were classified as normal when 1) the number of NPB in the two pronuclei did not differ by more than three, 2) NPB were polarized if fewer than seven and never polarized if more than seven, 3) the number of NPB in a pronucleus was not below three, and 4) the distribution of NPB was either polarized or non-polarized but not discordant.

Our own studies (Sadowy et al., in preparation) have shown that nucleolar distribution (Figure 2A-D) is associated with the duration of culture, as was suggested by Wright et al. (1990) and Tesarik and Greco (1999). We have also found that compared to zygotes with scattered nucleoli (Figure 2C), zygotes with closely aligned pronuclei and nucleoli (Figure 2A) have significantly higher cell numbers on day 3 (6.9 vs. 5.8, respectively) and lower incidences of multinucleation and arrest. So far, we have not been able to associate cytoplasmic contraction with development potential; nor have we seen an impact of polar body/pronuclear axis on development. However, the distance between the first and second polar bodies and development may be correlated. The latter two findings apparently contradict the findings of Garello et al. (1999) who examined the same phenomena. Their study concluded that while the distance between the first and second polar bodies did not affect embryo grade on day 2, the orientation of pronuclei, relative to the polar bodies could be related to embryo morphology.

Clearly, nucleolar patterns are linked to embryo viability, reinforcing the relevance of polarity in human oocytes and embryos (Edwards and Beard, 1997). It may be possible to determine the fastest and chromosomally most normal embryo by selecting zygotes in which central migration of the male and female pronuclei is complete and nucleoli are accumulated and aligned. This is particularly useful if transfer is to be performed at the pronucleate stage. However, it is important to remember that 1) embryos representing the "ideal" nucleolar pattern are relatively infrequently observed (approximately 9% of all zygotes in our studies), 2) there is considerable overlap with conclusions drawn from assessing cleaved embryos, and 3) the additional observation time necessary for detailed evaluations may lead to fluctuating culture conditions and can compromise zygote development.

4. Cleavage Stage embryos

A multitude of studies over the past two decades have identified cleavage rate, cell number, cleavage symmetry, cytoplasmic fragmentation, and multinucleation as key markers of viability in early human embryos. Zona pellucida structure, cell-cell contact, and cytoplasmic organisation are not as frequently noted, but are of obvious importance as predictors of developmental competence. In the next few sections, the clinical value of these markers is discussed.

4.1. Cleavage Timing and Rate

Several studies have suggested that timing of the first cleavage division is a strong marker of embryo quality and viability, early cleaving zygotes showing an advantage over those with delayed cleavage (Edwards et al., 1984; Shoukir et al., 1997; Sakkas et al., 1998). Early cleavage is defined as that occurring 25-27 hours post-insemination. In a recently published study (Sakkas, 2001), 45% of patients undergoing IVF or ICSI had early cleaving zygotes. Both implantation and pregnancy rates increased significantly as the number of early cleaving embryos in the transfer cohort increased from zero to two (mean number of embryos replaced was 2.2). Since the detection of early cleavage requires a second check of zygotes after fertilization has been confirmed, the benefits of embryo selection based on cleavage timing must be weighed against possible disadvantages of increased embryo handling.

Cleavage rate is also an important predictor of developmental potential. On day 2 of development, 4-cell embryos have been repeatedly shown to implant more frequently and establish more pregnancies compared to embryos with fewer or more cells at that time-point (e.g., Ziebe et al., 1997; Giorgetti et al., 1995). Moreover, a retrospective analysis of ongoing twin pregnancies after 2-embryo transfers showed that these embryos typically had 4 to 5 blastomeres on day 2 and at least 7 blastomeres on day 3 (Gerris et al., 1999; Van Royen et al., 1999; 2001).

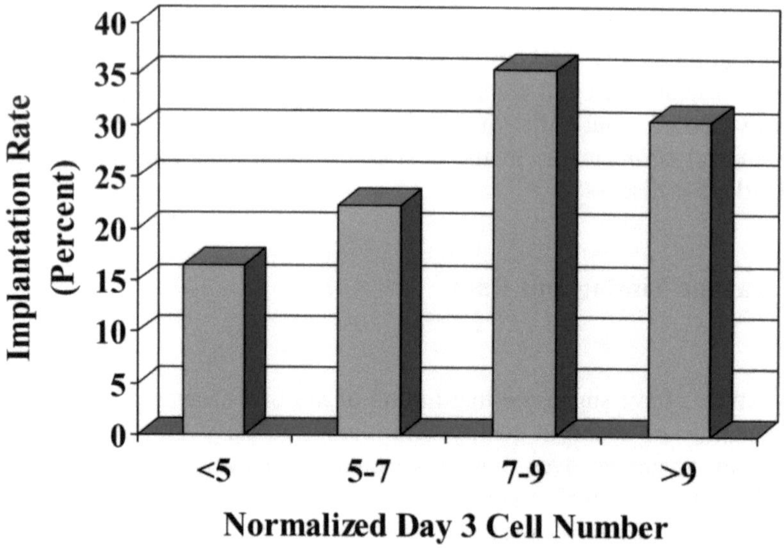

Figure 3: *The impact of cell number on implantation rate (Foetal hearts per embryo replaced). This analysis considers normalized cell number of embryos on day 3 (see text for formula) in homogeneous replacements only. The number of procedures in each plotted group was 85 (<5), 190 (5-7), 1266 (7-9), and 52 (>9). Implantation rate of embryos with fewer than 7 cells is significantly lower than those with more than 7 cells ($p<0.001$).*

Unusual acceleration or slow-down in cleavage rate has consequences as well. Fast cleavage has been linked to high levels of mosaicism and polyspermic fertilization (Harper et al., 1994). Magli et al. (1998) found as many as 76% chromosomal abnormality among embryos with 9 or more cells at 62 hours post-insemination.

Data on slow cleaving embryos is more rare. In the mouse model, lack of a specific protein product of the preimplantation development gene (*ped*), consistently leads to slowly developing embryos that implant as frequently as their fast developing counterparts, but are lost between mid-gestation and birth (Warner et al., 1998a). A homologue of this gene in the human probably exists (Jurisicova et al., 1996a; Cao et al., 1999). However, in the absence of concrete data and in view of the fact that the experimental data were obtained using highly inbred strains, non-arrested slow developing embryos are considered for transfer during IVF when no other embryos are available.

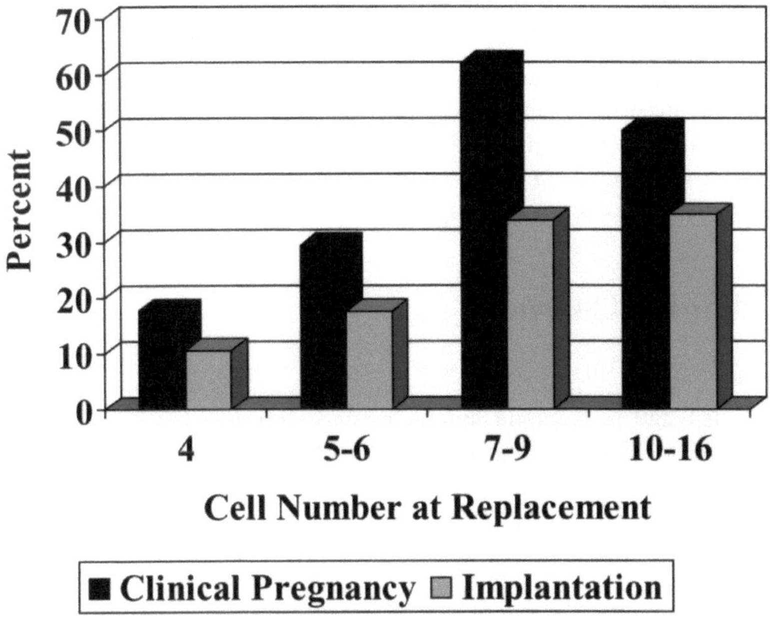

Figure 4: *The impact of cell number on clinical pregnancy (foetal heart detected) and implantation rates (Foetal hearts per embryo replaced). This analysis considers observed cell number at the time of replacement in homogeneous replacements only. The lowest pregnancy and implantation rates were obtained following exclusive replacement of 4-cell day 3 embryos (51 procedures with 85 embryos included; 17.7% pregnancy and 10.6% implantation rate; mean maternal age was 35.6). The highest rates were obtained after exclusive replacement of embryos with 7-9 cells (2099 procedures with 6548 embryos included; 61.9% pregnancy and 33.8% implantation rates; mean maternal age was 35.0).*

We have used two methods to evaluate the impact of cleavage rate on pregnancy and implantation outcome. Only replacements homogeneous with respect to cell number were considered in these analyses. In the first analysis (Figure 3), cell numbers on day-3 were adjusted according to time in culture after oocyte retrieval using the following formula: *[(Cell number/Hours in culture) x Average hours in culture for all embryos in the database]*. This was to minimize the effect of the time of evaluation. The average number of hours in culture from oocyte retrieval to day 3 was 71.68 hours. In the second analysis, only actual (observed) cell numbers at the time of replacement were considered (Figure 4).

Regardless of the method of analysis, the highest pregnancy and implantation rates were obtained after replacement of embryos with at least 7 cells. The lowest rates resulted from exclusive replacement of day 3 embryos at the 4-cell stage. Slow embryos also perform poorly in extended culture, forming few blastocysts and often showing abnormal morphology on day 5 of development (Alikani et al., 2000). In the clinical setting, we preferentially replace 7- to 9-cell embryos and cryopreserve those with fewer or more cells if all other morphological parameters are normal.

4.2. Cell Size and Symmetry

Uneven cleavage is common among human embryos developing in vitro (Figure 5A-C) and there is general agreement that transfer of embryos with this characteristic results in lowered pregnancy and implantation rates (e.g., Ziebe et al., 1997; Hardarson et al., 2001).

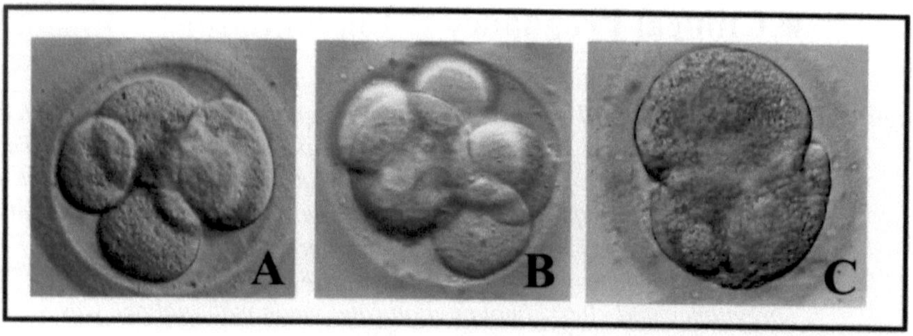

Figure 5: *Uneven cleavage among human embryos developing in vitro. (A) Loss of large volumes of cytoplasm through fragmentation leading to uneven appearance of groups of cells; fragments have been removed. (B) Uneven cells may appear as a result of cytokinesis arresting in one or more cells; in these instances, if karyokinesis continues, the arrested cells may appear multinucleated. (C) An overt form of blastomere size asymmetry is the presence of a single dominant cell among other smaller cells; such embryos are typically polyploid.*

This may occur for a number of reasons. In an interesting study by Goyanes et al. (1990), quantitative evaluation of human embryos revealed that those with uneven blastomeres had a total blastomere volume that was 8% lower than normal embryos. Since total blastomere volume during cleavage stages did not differ from ooplasmic volume at the pronucleate stage, the reduced volume of the uneven embryos implies a reduction in oocyte volume, possibly pointing to aberrant follicular development.

Asymmetric cell division can also result from asymmetric placement of the spindle, which is normal in some lower animals but is generally not expected in the cleaving human embryo. Both genetic and cytoplasmic equity (or polarity?) may be compromised as a result of unequal division. Loss of large volumes of cytoplasm through fragmentation is another mechanism leading to uneven appearance of groups of cells. Figure 5A depicts a fragmented embryo after fragment removal, with obvious discrepancy in blastomere size. These embryos have a poor development potential (Alikani et al., 1999) as will be discussed in a later section in this chapter. Alternatively, uneven cells may appear as a result of cytokinesis arresting in one or more cells; in these instances, if karyokinesis continues, the arrested cells may appear multinucleated (Figure 5B). Hardarson et al. (2001) found a significantly higher degree of both multinucleation and aneuploidy associated with blastomeres from uneven embryos. Finally, an overt form of blastomere size asymmetry is the presence of a single dominant cell among other smaller cells (Figure 5C). The dominant cell is often multinucleated, but regardless of nuclear morphology, these embryos are typically polyploid (Munné et al., 1994a).

Based on these observations, embryos with uneven cells, in particular those with overt forms, are excluded from transfer.

4.3. Fragmentation

Edwards et al. (1970) described fragmentation in human embryos as "cell division without nuclear division." Fragmentation is partial or total loss of cellular integrity and, according to our database, can affect nearly 90% of cleavage stage human embryos in vitro (50,974/57,761 or 88%). Its prevalence alone is an indication that fragmentation in the cleaving human embryo in vitro is not necessarily associated with embryonic demise, or even limited cell death as suggested previously (Jurisicova et al., 1996b). On the contrary, fragmentation in some instances may be integral to normal in vitro development.

Most fragments probably form sometime around the first cleavage division. Fragmentation before syngamy also occurs, albeit very rarely. A great majority of such fragmenting pronucleate zygotes in our database either show arrest by day 2 after one or two divisions, or become excessively fragmented by day 3 and are not suitable for transfer or cryopreservation. Likewise, if fragmentation is pervasive during syngamy, the developmental potential of the resulting embryo is usually severely compromised.

That fragmentation can limit the potential of human embryos for development became evident from early clinical data such as those presented by Puissant et al. (1987). Many studies have since been published on this topic, but the lack of sufficient data, low overall implantation rates, and ambiguous categorisation of embryos made it difficult to draw definitive conclusions. Three interesting studies on the impact of the degree of fragmentation on the outcome of IVF are those of Staessen et al. (1993b), Giorgetti et al. (1995), and Ziebe et al. (1997). These investigators reported an implantation rate of approximately 5% after transfer of embryos with 10-50% fragmentation on day 2 of development.

Although the prevalence of chromosomal abnormalities among severely fragmented embryos has been repeatedly demonstrated (e.g., Pellestor et al., 1994), the reduced development and implantation potential among such embryos should not be presumed to reflect chromosomal incompetence. It is noteworthy that nearly 50% of embryos with less than 35% fragmentation have the correct complement of those chromosomes tested during PGD; by contrast, the proportion is significantly lower among embryos with >35% fragmentation (Munné et al., 2001). Interestingly, although the direct correlation between aneuploidy and maternal age is well established (Márquez et al., 2000), fragmentation and maternal age seem to be inversely related (Alikani et al., 2001). Human embryos not only fragment to various degrees, they also fragment in distinct patterns (Alikani and Cohen, 1995; Warner et al., 1998b; Alikani et al., 1999). The pattern of fragmentation has a profound effect on the developmental fate of the embryo, thus should be considered in its evaluation. Five distinct patterns of fragmentation have been defined, four of which occur commonly (Figure 6A-D).

Fragmentation pattern I (Type I) is minimal in volume and the fragments are typically, but not always, associated with only one blastomere. Pattern II is characterized by highly localized fragments which often appear in a cluster or column. The most common pattern of fragmentation is Type III; fragments in this pattern are small and scattered and may be positioned between blastomeres, peripherally, or both. Large fragments represent fragmentation pattern IV; these often appear to be randomly distributed and

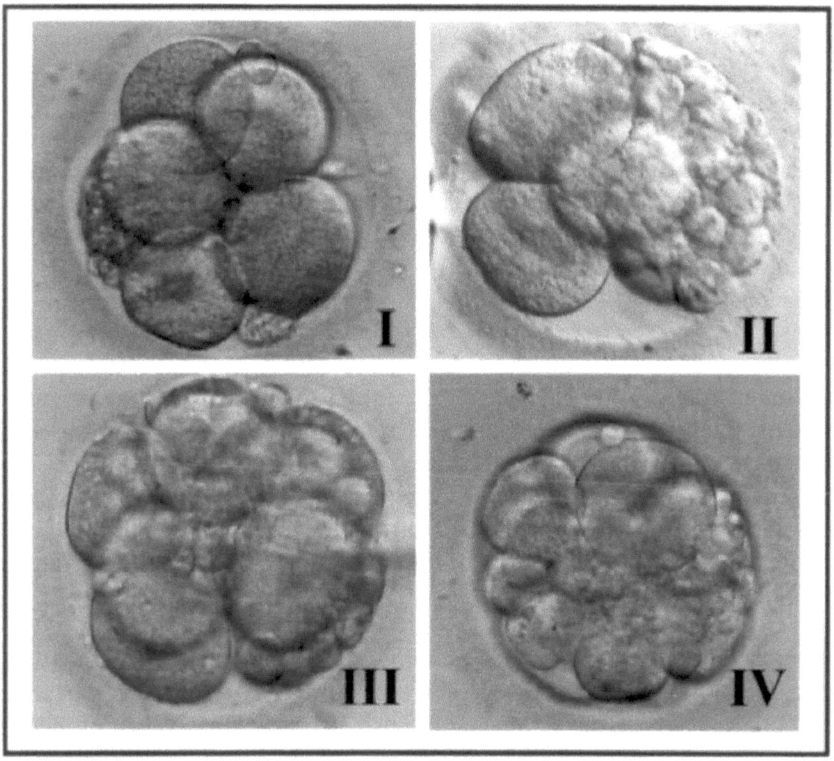

Figure 6: *Four of the five distinct patterns of fragmentation that are well established by the third day in culture. Fragmentation pattern I (I) is minimal in volume and the fragments are typically, but not always, associated with only one blastomere. Pattern II (II) is characterized by highly localized fragments which often appear in a cluster or column. Type III fragments (III) are small and scattered and may be positioned between blastomeres, peripherally, or both. Fragmentation pattern IV (IV) is represented by large fragments; these often appear to be randomly distributed and are always associated with uneven cells.*

are always associated with uneven cells.

Type V fragments appear necrotic, with a characteristic granularity and cytoplasmic contraction within the associating blastomeres. Some embryos do not show a distinct pattern of fragmentation and are classified as such. The patterns of fragmentation were not found to be correlated to maternal age (Alikani et al., 1999).

Type IV fragmentation appears detrimental to the embryo. Exclusive (and near-exclusive) replacement of embryos with large fragments produced an implantation rate significantly lower than that of other types. There was also a significant decrease in clinical pregnancy rate following transfers in which Type IV embryos predominated (40% compared to an overall pregnancy rate of 58.5%). In extended culture, fragmentation interfered with the development of embryos to the blastocyst stage: embryos with >15% fragmentation or Type IV fragmentation formed significantly fewer normal blastocysts than those with ≤15% fragmentation or Types I, II, or III fragmentation (Alikani et al., 2000).

A recent study by Antczak and Van Blerkom (1999) also suggested distinct patterns in fragmenting embryos. They further indicated that fragmentation resulted in partial or total loss of regulatory proteins from affected blastomeres. If at 1- or 2-cell stages, they proposed, the loss severely compromised the developmental potential of the embryo. By contrast, fragmentation from portions of the blastomeres in a 2-cell embryo without protein domains or complete loss of the domains after the 4-cell stage did not preclude further development.

Embryos without fragmentation or minor fragmentation are preferentially selected for transfer and cryopreservation, provided that they are at an appropriate stage of development. But when only extensively fragmented embryos are available, selection should not only be based on the degree, but also on the pattern of fragmentation. In these instances, our experience indicates that application of assisted hatching and fragment removal can restore developmental potential of some fragmented embryos, and improve the overall pregnancy and implantation rate in patients with an otherwise very poor prognosis for pregnancy (Reviewed by Alikani, 2001).

4.4. Multinucleation

Following the first cleavage division, some blastomeres in human embryos can display multiple nuclei (Figure 7A). One proposed cause of multinucleation is follicular under-oxygenation (Van Blerkom et al., 1997). Rapid follicular response has also been implicated in multinucleation (Jackson et al., 1998). The mechanisms of multinucleation include karyokinesis without cytokinesis, fragmentation of nuclei, and abnormal migration of chromosomes during anaphase (Tesarik et al., 1987; Winston et al., 1991; Hardy et al., 1993). Blastomere multinucleation has been shown to occur significantly more frequently in 2-cell embryos than 4-cell embryos (Balakier and Cadesky, 1997) on day 2; evaluation of day 2 embryos therefore is best done early and prior to the second cleavage division.

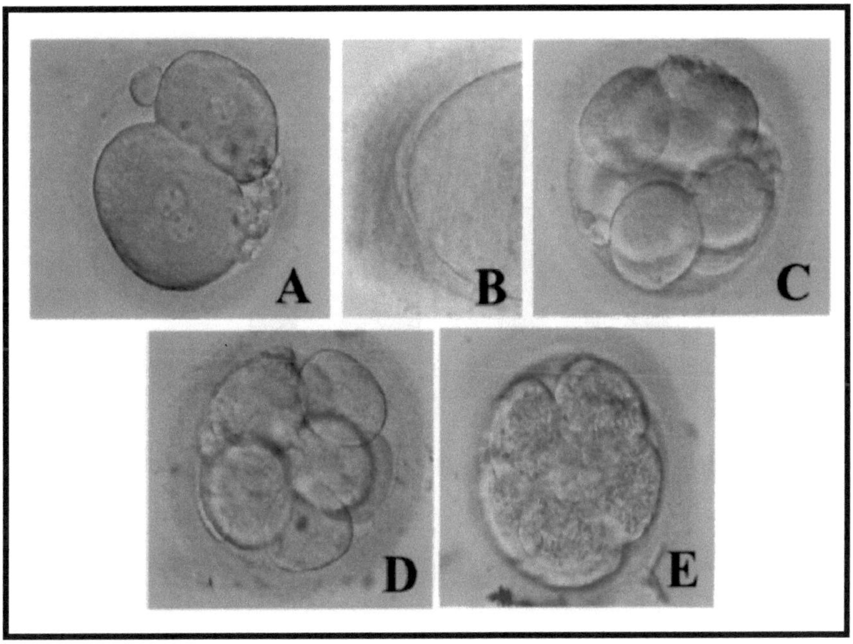

Figure 7: Day 3 embryos with multinucleation (A), distorted and multi-layered zona pellucida (B), poor cell adhesion (C), vacuolation (D) and vesiculation (E).

In a study by Kligman et al. (1996), the presence of multinucleated blastomeres (MNB) in developing day 2 and day 3 embryos was correlated to extensive mosaicism and/or polyploidy as assessed by FISH. Similar results were obtained by Laverge et al. (1997). By contrast, when MNB were detected on day 4 of development after biopsy of otherwise morphologically normal embryos, they were mostly chromosomally normal (Munné and Cohen, 1993; Munné et al., 1994b). Further studies in our institute have revealed a significant decrease in implantation and pregnancy rates as the proportion of embryos with multinucleated cells among replaced embryos increased (Figure 8). Staessen and Van Steirteghem (1998) found that a majority of 2-cell embryos with bi- and multinucleation cleaved further and gave rise to embryos containing mononucleated diploid blastomeres, mononucleated diploid and non-diploid blastomeres, or only non-diploid blastomeres (indicating chaotic division).

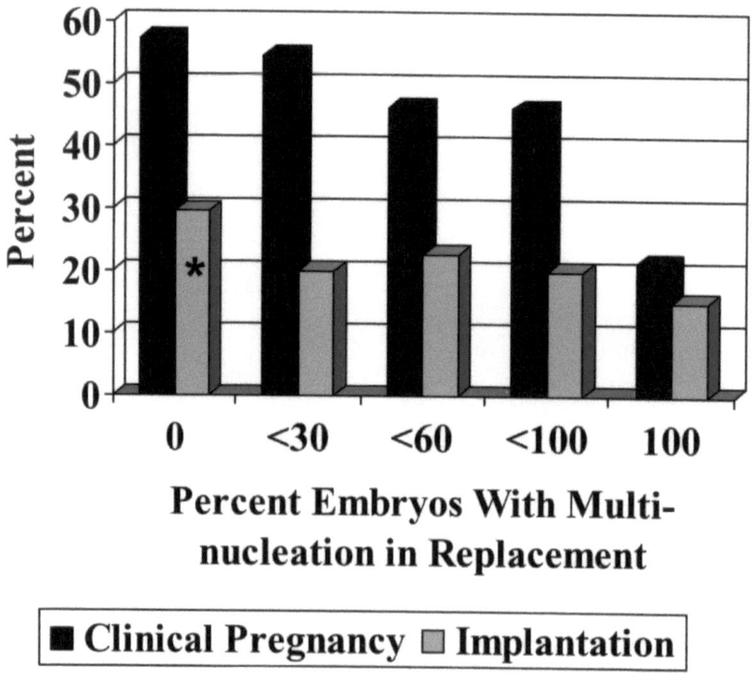

Figure 8: The relationship between multinucleation and embryo implantation. As the proportion of embryos with multinucleate blastomeres increases in the transferred cohort, the rate of implantation decreases significantly (none replaced with MNB (*) compared to all others, $p<0.001$).

When embryos were analysed at the 2-cell stage, results were split almost evenly between diploid and abnormal blastomeres. These investigators concluded that multinucleation may be temporary and a reversible phenomenon.

Generally, the transfer of embryos with MNB at the 2-cell stage is not advisable, while embryos with binucleate cells at the 8-cell stage or beyond may be transferred if no other morphologically normal embryos are available. Moreover, consistent with the findings of Balakier and Cadesky

(1997), we have found that embryos with MNB have a considerably reduced ability to reach the blastocyst stage in extended culture (Alikani et al., 2000).

4.5. Zona Pellucida Architecture

It is well established that the zona pellucida (ZP) plays an important role in normal mammalian embryo development (reviewed in Wassarman and Mortillo, 1991). A unique feature of this protein-carbohydrate moiety in human embryos was discovered by Cohen et al. (1989; 1990) who observed that zona pellucida thickness (ZPT) and variation were predictive of embryo hatching ability, hence viability following in vitro culture.

Garside and colleagues (1997) studied the normal progression of zona thinning during in vitro culture and the variables that correlate with zona thickness. They confirmed earlier findings of Cohen (1993) that ZPT decreased during in vitro culture, and that these changes were more evident in embryos that resulted in successful pregnancies. They also found a significant correlation between ZPT and cell number, embryo grade, fragmentation, and maternal age. Increasing cell number and grade as well as increasing maternal age (most notably >35 years) correlated with decreased ZPT, while increased fragmentation led to decreased thinning rate and thicker zonae. The application of selective assisted hatching to embryos with thick or abnormal zonae (e.g., multi-layered and distorted; Figure 7B) but otherwise good morphology, embryos with poor morphology and thick zonae, and embryos of patients >38 years of age with thick or thin zonae appears to overcome potential problems arising from these conditions (Cohen et al., 1992; Alikani et al., 1999).

4.6. Cell-Cell Contact

Experimental evidence unequivocally shows that cell adhesion plays a pivotal role in morphogenesis (e.g., Johnson et al., 1986; Fleming and Johnson, 1988). The expression of cell adhesion molecules is partly dependent on cell-cell contact. For instance, aggregation of developmentally blocked cells from 2-cell mouse embryos with cells from a non-blocking genotype fully restored the developmental capacity of the former, which resumed expression of cell adhesion molecules on their surfaces (Neganova et al., 2000). The degree to which human embryos show intercellular contact is therefore likely to impact their development, including allocation of cells to the inner cell mass. Disorganisation and reduced cell adhesion post thaw led to lowered implantation rates of cleavage stage embryos (Cohen et al., 1988). Fresh embryos can exhibit the same feature (Figure 7C). Cell-cell contact is also reduced by cytoplasmic fragmentation (Figure 6IV) when intercellular spaces are occupied by fragments of various sizes. It is conceivable that this morphology impacts a number of key events that precede differentiation, including cell polarisation, cell adhesion, tight and gap junction formation, and compaction.

4.7. Cytoplasmic Integrity

The presence of numerous intracytoplasmic vacuoles (Figure 7D) and severe cytoplasmic contraction in blastomeres signals possible degenerative processes (Sathananthan et al., 1990). Conversely, the appearance of vesicular (sometimes referred to as "pitted") cytoplasm (Figure 7E), commonly observed in day 3 embryos, may be a positive sign for genomic activation: fractional volumes of at least two cytoplasmic organelles, tubular endoplasmic reticulum and secondary lysosomes were found to significantly increase with transcriptional activity (Tesarik et al., 1988). In one scoring system devised by Desai et al. (2000), vesiculation was found to correlate well with pregnancy and implantation; embryos were given positive points in the absence of vacuoles and the presence of a “pitted” cytoplasm. Other cytoplasmic inclusions, such as "pyknotic nuclei" or areas of intracellular necrosis, observed and studied in oocytes (Van Blerkom, 1991; Alikani et al., 1995 b), are often isolated in one or more blastomeres of the cleaving embryo. Their specific impact on development is undetermined but appears to be insignificant.

5. Morulae and Blastocysts

Nuances of human morula and blastocyst morphology have not yet been fully explored and our understanding of the relationship between gross morphology at this stage and implantation potential is limited. This problem is compounded by the observation that blastocyst development (and morphology) is dramatically influenced by the type of culture medium and system employed (Conaghan et al., 1998).

A scoring system was devised by Cohen et al. (1985) to distinguish among different stages of blastocyst development: During Stage A, the blastocoel appeared as one or more small vacuoles; Stage B was represented by a distinct cavity one quarter to one half the volume of the embryo (unchanged embryonic diameter); At Stage C, a single cavity occupying most of the volume of the embryo appeared and the zona pellucida was thinned; By the final stage, Stage D, the blastocyst was fully expanded for 24 hours or more. A relationship between blastocyst stage and survival after cryopreservation and thaw was described by these investigators. The best results were obtained from Stage C blastocysts.

In studies by Dokras et al. (1991, 1993), a grading system based on development rate as well as morphology was used. Blastocyst Grade 1 (BG1) referred to early cavitation resulting in the formation of an eccentric and then expanded cavity lined by a distinct inner cell mass (ICM) region and trophectoderm (TE) layer on day 5. BG2 was assigned to "late" or "slow" developers with a one- or two-day delay in the visible differentiation of the two cell types. Vacuolated embryos as well as blastocysts with several degenerative foci in the ICM and collapse of the cavity within 24 hours without re-expansion were graded as BG3. Abnormal blastocyst morphology was correlated to reduced total cell count and low or absent hCG secretion in vitro. Cell number has also been correlated to the ability of blastocysts to hatch (Van Blerkom, 1993): none of the embryos with fewer than 43 nuclei on day 5.5 post-insemination hatched after 36 additional hours in culture.

In the scoring system devised by Gardner (Gardner and Schoolcraft, 1999), blastocysts were initially given a numerical score of 1-6 based on their degree of expansion and hatching status, similar to that described by Cohen et al. (1985): Score 1, blastocoel less than half the volume of the embryo; Score 2, blastocoel greater than half the volume of the embryo; Score 3, blastocoel completely filling the embryo, Score 4, blastocoel larger than the embryo, with zona thinning; Score 5, hatching blastocyst; and Score 6, hatched blastocyst. Gardner further scored the blastocysts with a score based on the development of the ICM and the TE. These were assigned a letter: for the ICM, Score A, tightly packed, many cells; Score B, loosely grouped, several cells; and score C, very few cells. For TE, Score A, many cells forming a cohesive layer; Score B, few cells forming a loose epithelium; Score C, very few large cells.

Some of these blastocysts that resulted in the birth of infants (designated male or female on the figure) have been depicted in Figure 9.

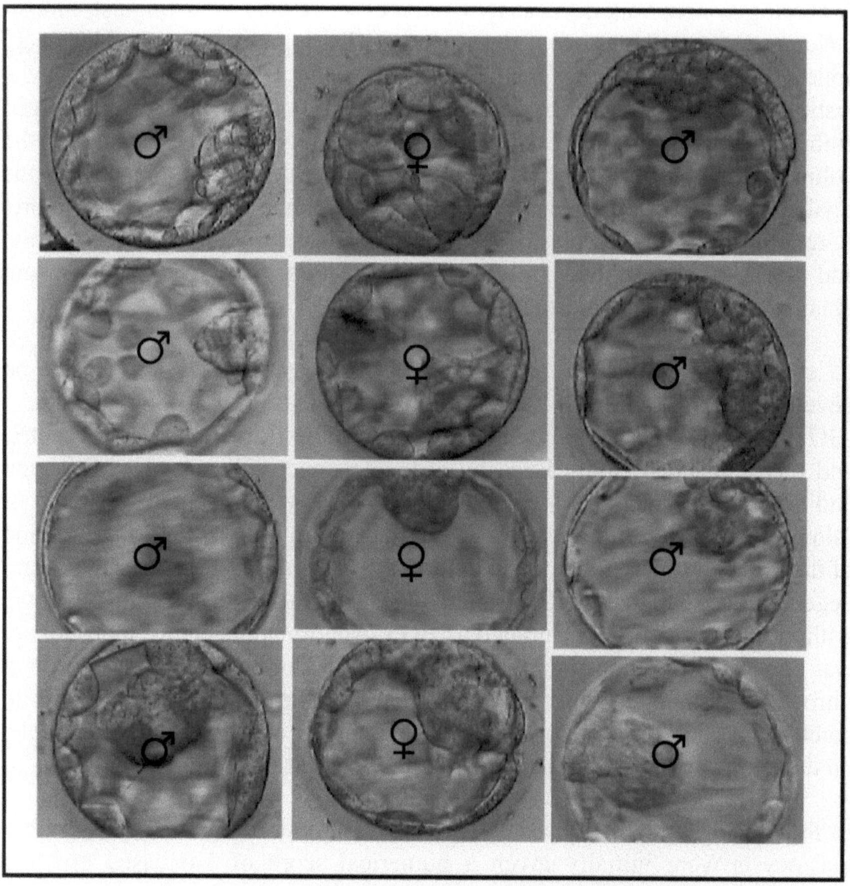

Figure 9: *Human blastocysts with apparently normal morphology. The sex of infants born after replacement of these embryos is designated on each micrograph. Note that at least one embryo (top row, middle) is early in the morula-blastocyst transition phase. ICM morphology is variable, however, all are distinctly visible. Trophectoderm cells are cohesive and well organized.*

While an overall implantation rate of 50% was achieved with blastocysts obtained after sequential culture in G1/G2 media (Gardner et al., 1998 a, b), transfer of blastocysts with a score of 3AA or better, that is, full blastocysts with a distinct ICM and a cohesive layer of TE-Cells, resulted in an implantation rate of 70% (Gardner et al., 2000).

We have evaluated the development of embryos in extended culture, with particular focus on different forms of compaction, cavitation, and ICM/TE formation (Alikani et al., 2000). Compaction was expected to occur on day 4 (Nikas et al., 1996); a delay of one day in compaction was considered an indication of slow development and thus sub-optimal. Compaction on day 4 was defined as complete, incomplete, regional, or fusion-like. Ideally, all cells within the embryo were included in the morula, either completely compacted or in the process of compaction. But, regional compaction also occurred with exclusion of one or more cells from the compacted / compacting mass (Figure 10A). The extent to which this particular morphology impacts the potential of the morula for further development probably depends on the extent of blastomere exclusion, which we did not assess. A less common form of compaction was the complete disappearance of intercellular spaces not by flattening of cells, but by what appeared as "fusion" of cells (Figure 10B).

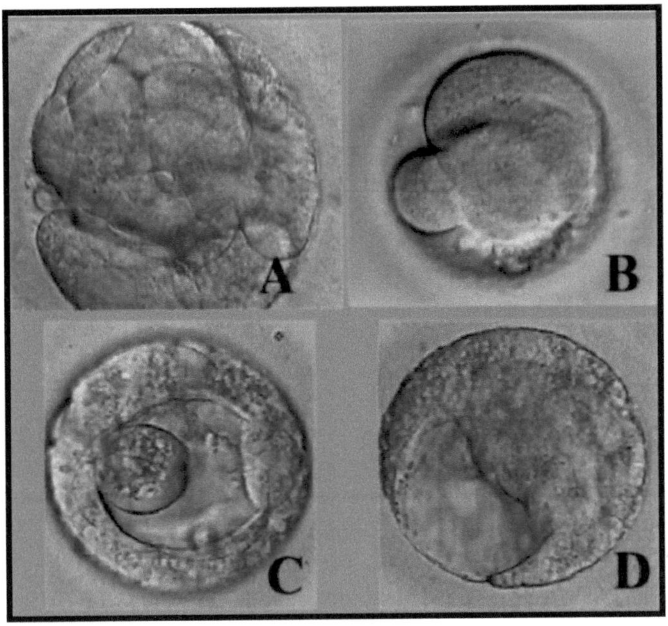

Figure 10: Abnormal / atypical forms of compaction and cavitation. (A) Regional compaction with exclusion of two cells from the compacted / compacting mass. (B) Fusion-like compaction and complete disappearance of intercellular spaces. (C) The persistence of large vacuoles, and (D) small cavity on day 5.

In our experience, this morphology is often exhibited by slow developing embryos and signals failure to reach the blastocyst stage; but this needs further assessment. We have classified regional, fusion-like, and delayed compaction (one full day delay) as sub-optimal forms of compaction.

It was not unusual to see the beginning of cavitation on day 4. However, a single defined cavity was never observed before day 5. A delay of one full day in cavitation was considered sub-optimal, but normal. The persistence of large vacuoles, small and multiple cavities on day 5 was considered abnormal (Dokras et al., 1991; 1993) (Figure 9C, D). A distinct ICM, organized as a compacted mass of numerous cells, was considered normal.

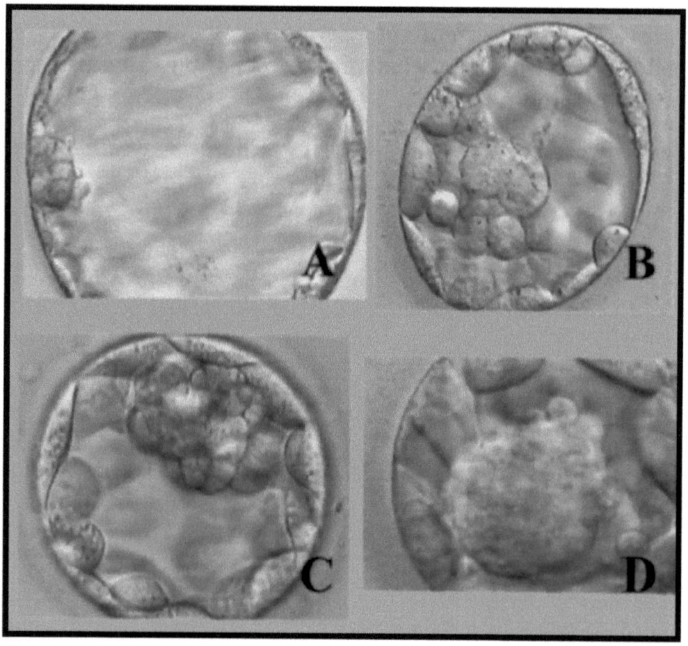

Figure 11: Human blastocysts with (A) small and highly compacted mass of ICM-cells, (B) loosely packed ICM-cells, (C) large ICM-cells, (D) few and large TE-Cells

If the inner cells formed a small mass (Figure 11A), or if the mass comprised loosely packed cells (Figure 11B and C), the formation was considered sub-optimal but normal. The absence of an ICM was abnormal. A cohesive layer of numerous tightly packed cells in the TE was normal; large TE-Cells were sub-optimal but normal (Figure 11D). Highly irregularly arranged ICM and/or TE-Cells were considered abnormal. A blastocyst was considered morphologically normal if it contained either a normal or sub-optimal ICM in combination with either a normal or sub-optimal TE on day 5. Representative embryos with dysmorphic ICM and TE are presented in Figure 12.

Figure 12: Human blastocysts with abnormal / atypical ICM and TE formation. Blastulation with few cells, disorganized and large TE and ICM-cells visible.

At our institute, selection of blastocysts for transfer has been based on morphological normality on day 5 of development. However, we have not so far excluded any patients from transfer if at least one of her embryos has reached the morula stage by day 5. (The late morula / early blastocyst depicted in figure 10 led to the birth of a female infant. A second comparable early blastocyst in another patient led to monozygotic twinning and two stillborn female infants). The only embryos that are absolutely excluded from transfer on day 5 are those with development arrest on day 3, multiple abnormal cavities, and a clear absence of an ICM. With this policy in place, clinical pregnancy (foetal heart detected at ultrasound) and implantation (foetal heart per embryo replaced) rates after day 5 transfer were 63.8% (88/138) and 45.4% (129/284), respectively.

6. Conclusions and future directions

Certain phenotypic features of human embryos in culture are correlated with viability and implanting potential. Embryos may be assessed pre and post genomic activation, at the zygote, cleavage, and blastocyst stages. Pronuclear and nucleolar morphology, cleavage rate, timing, and symmetry, fragmentation degree and pattern, zona pellucida thickness and variation, cytoplasmic integrity, cell-cell contact, inner cell mass and trophectoderm cell number, shape, and organisation are key markers of embryonic potential.

Incorporation of a diligent and detailed morphology evaluation protocol in the daily routine of the human IVF laboratory should and would contribute to improved clinical results. However, the future of embryo evaluation lies elsewhere. Diagnosis of aneuploidy in early embryos prior to intra-uterine replacement has already been shown to increase implantation (Gianaroli et al., 1997) and reduce embryonic loss after implantation (Munné et al., 1999). Eventually, application of novel technologies such as comparative genomic hybridization (CGH)(Wells and Delhanty, 2000) and gene expression analysis in single oocytes and embryos (Steuerwald et al., 1999) will push this frontier even further and may offer a panacea to the problem of embryo quality evaluation and embryo selection during therapeutic in vitro fertilization.

References

Alikani M, Cohen J. 1995a. Patterns of cell fragmentation in the human embryo. J Assist Reprod Gen 12:28S.

Alikani M, Palermo G, Bertoli M, Blake M, Cohen J. 1995b. Intracytoplasmic sperm injection in dysmorphic human oocytes. Zygote 3:283-288.

Alikani M, Cohen J, Tomkin G, Garrisi GJ, Mack C, Scott R. 1999. Human embryo fragmentation in vitro and its implications for pregnancy and implantation. Fertil Steril 71:836-842.

Alikani M, Calderone G, Tomkin G, Garrisi J, Kokot M, Cohen J. 2000. Cleavage anomalies in early embryos and survival after prolonged culture in vitro. Hum Reprod 15:2634-2643.

Alikani M. 2001. Cytoplasmic fragmentation in human embryos in vitro: implications and the relevance of fragment removal. In: Textbook of Assisted Reproductive Techniques, Laboratory and Clinical Perspectives. D Gardner, A Weissman, C Howles, Z Shoham, eds. Martin Dunitz: United Kingdom, p169-182.

Antczak M, Van Blerkom J. 1999. Temporal and spatial aspects of fragmentation in early human embryos: possible effects on developmental competence and association with the differential elimination of regulatory proteins from polarized domains. Hum Reprod 14:429-447.

Balakier H, Cadesky K. 1997. The frequency and developmental capability of human embryos containing multinucleated blastomeres. Hum Reprod 12:800-804.

Cao W, Brenner CA, Alikani M, Cohen J, Warner CM. 1999. Search for a human homologue of the mouse Ped gene. Mol Hum Reprod 5:541-547.

Cohen J, Simons RF, Edwards RG, Fehilly C, Fishel S. 1985. Pregnancies following the frozen storage of expanding human blastocysts. J in vitro Fert Embryo transfer 2:59-64.

Cohen J, Wiemer KE, Wright G. 1988. Prognostic value of morphological characteristics of cryopreserved embryos: a study using video-cinematography. Fertil Steril 49:827-834.

Cohen J, Inge KL, Suzman M, Wiker S, Wright G. 1989. Videocinematography of fresh and cryopreserved embryos: A retrospective analysis of embryonic morphology and implantation. Fertil Steril 51:820-827.

Cohen J, Malter H, Elsner C, Kort H, Massey J, Mayer MP. 1990. Immuno-suppression supports implantation of zona pellucida dissected human embryos. Fertil Steril 53:662-665.

Cohen J, Alikani M, Trowbridge J, Rosenwaks Z. 1992. Implantation enhancement by selective assisted hatching using zona drilling of embryos with poor prognosis. Hum Reprod 7:685-691.

Cohen J. Assisted hatching of human embryos. 1993. J Assist Reprod Genet 8:179-190.

Conaghan J, Hardy K, Leese H, Winston RL, Handyside AH. 1998. Culture of human preimplantation embryos to the blastocyst stage: a comparison of 3 media. Int J Dev Biol 42:885-893.

Desai NN, Goldstein J, Rowland DY, Goldfarb JM. 2000. Morphological evaluation of human embryos and derivation of an embryo quality scoring system specific for day 3 embryos: a preliminary study. Hum Reprod 15:2190-2196.

Dokras A, Sargent IL, Ross C, Gardner RL, Barlow DH. 1991. The human blastocyst: morphology and human chorionic gonadotrophin secretion in vitro. Hum Reprod 6:1143-1151.

Dokras A, Sargent IL, Barlow DH. 1993. Human blastocyst grading: an indicator of developmental potential? Hum Reprod 8:2119-2127.

Edwards RG, Steptoe PC, Purdy JM. 1970. Fertilization and cleavage in vitro of preovulatory human oocytes. Nature 227:1307-1309.

Edwards RG, Fishel SB, Cohen J, Fehilly CB, Purdy JM, Slater JM, Steptoe PC, Webster JM. 1984. Factors influencing the success of in vitro fertilization for alleviating human infertility. J in vitro Fertil Embryo transfer 1:3-23.

Edwards RG, Beard HK. 1997. Oocyte polarity and cell determination in early mammalian embryos. Mol Hum Reprod 3:863-906.

Fleming, TP, Johnson MH. 1988. From egg to epithelium. Ann Rev Cell Biol; 4:459-485.

Gardner DK, Vella P, Lane M, Wagley L, Schlenker T, Schoolcraft WB. 1998a. Culture and transfer of human blastocysts increases implantation rates and reduces the need for multiple embryo transfers. Fertil Steril 69:84-88.

Gardner DK, Schoolcraft WB, Wagley L, Schlenker T, Stevens J, Hesla J. 1998b. A prospective randomized trial of blastocyst culture and transfer in in vitro fertilization. Hum Reprod 13:3434-3440.

Gardner DK, Schoolcraft WB. 1999. In vitro culture of human blastocysts. In: Towards Reproductive Certainty, Fertility and Genetics Beyond 1999. Jansen R, Mortimer D, eds. New York: The Parthenon Publishing Group, p 378-388.

Gardner DK., Lane M., Stevens J, Schlenker T, Schoolcraft WB. 2000. Blastocyst score affects implantation and pregnancy outcome: towards a single blastocyst transfer. Fertil Steril 73:1155-1158.

Garello C, Baker H, Rai J, Montgomery S, Wilson P, Kennedy CR, Hartshorne GM. 1999. Pronuclear orientation, polar body placement, and embryo quality after intracytoplasmic sperm injection and in vitro fertilization: further evidence for polarity in human oocytes? Hum Reprod 14:2588-2595.

Garside WT, De Mola JR, Bucci JA, Tureck R. 1997. Sequential analysis of zona pellucida thickness during in vitro culture of human zygotes: Correlation with embryo quality, age, and implantation. Mol Reprod Dev 47:99-104.

Gerris J, De Neubourg D, Mangelschots K, Van Royen E, Van de Meersche M, Valkenburg M. 1999. Prevention of twin pregnancy after in vitro fertilization or intracytoplasmic sperm injection based on strict embryo criteria: a prospective randomized clinical trial. Hum Reprod 14:2581-2587.

Gianaroli L, Magli M, Ferraretti AP, Fiorentino A, Garrissi J, Munne S. 1997. Preimplantation genetic diagnosis increases the implantation rate in human in vitro fertilization by avoiding the transfer of chromosomally abnormal embryos. Fertil Steril 68:1128-1131.

Giorgetti C, Terriou P, Auquier P, Hans E, Spach JL, Salzmann J, Roulier R. 1995. Embryo score to predict implantation after in vitro fertilization: based on 957 single embryo transfers. Hum Reprod 10:2427-2431.

Goyanes VJ, Ron-Corzo A, Costas E, Maneiro E. 1990. Morphometric categorization of the human oocyte and early embryo. Hum Reprod 5:613-618.

Hardarson T, Hanson C, Sjogren, Lundin K. 2001. Human embryos with unevenly sized blastomeres have lower pregnancy and implantation rates: indications for aneuploidy and multinucleation. Hum Reprod 16:313-318.

Hardy K, Winston RML, Handyside AH. 1993. Binucleate blastomeres in preimplantation human embryos in vitro: failure of cytokinesis during early cleavage. J Reprod Fertil 98:549-558.

Harper JC, Robinson F, Duffy S, Griffin DK, Handyside AH, Delhanty JD, Winston RM. 1994. Detection of fertilization in embryos with accelerated cleavage by fluorescence in situ hybridization (FISH). Hum Reprod 9:1733-1737.

Jackson KV, Ginsberg ES, Hornstein MD, Mitchell RS, Clarke RN. 1998. Multinucleation in normally fertilized embryos is associated with an accelerated ovulation induction response and lower implantation and pregnancy rates in in vitro fertilization-embryo transfer cycles. Fertil Steril; 70:60-66.

Johnson MH, Maro B, Takeichi M. 1986. The role of cell adhesion in the synchronization and orientation of polarization in 8-cell mouse blastomeres. J Embryol Exp Morph 93:239-255.

Jurisicova A, Casper RF, McLusky NJ, Mills GB, Librach CL. 1996a. HLA-G expression during preimplantation human embryo development. Proc Natl. Acad. Sci. USA 93:161-165.

Jurisicova A, Varmuza S, Caspar RF. 1996b. Programmed cell death and human embryo fragmentation. Mol Hum Reprod 2:93-98.

Kligman I, Benadiva C, Alikani M, Munné S. 1996. The presence of multinucleated blastomeres in human embryos correlates with chromosomal abnormalities. Hum. Reprod 11:1492-1498.

Laverge H, De Sutter P, Verschraegen-Spae MR, De Paepe A, Dhont M. 1997. Triple colour fluorescent in situ hybridization for chromosomes X, Y, and 1 on spare human embryos. Hum Reprod 12:809-814.

Levron J, Munné S, Willadsen S, Cohen J. 1995. Male and female genomes associated in a single pronucleus in human zygotes. Biol Reprod 52:653-657.

Magli MC, Gianaroli L, Munné S, Ferraretti AP. 1998. Incidence of chromosomal abnormalities from a morphologically normal cohort of embryos in poor prognosis patients. J Assist Reprod Gen 15:297-301.

Márquez C, Sandalinas M, Bahçe M, et al. 2000. Chromosome abnormalities in 1255 cleavage stage human embryos. Reprod Bio Med online 1:17-26.

Munné S, Cohen J. 1993. Unsuitability of multinucleated human blastomeres for preimplantation genetic diagnosis. Hum Reprod 8:1120-1125.

Munné S, Alikani M, Grifo J, Cohen J. 1994a. Monospermic polyploidy and atypical embryo morphology. Hum Reprod 91:506-510.

Munné S, Grifo J, Cohen J, Weier HUG. 1994b. Chromosome abnormalities in arrested human preimplantation embryos: a multiple probe fluorescence in situ hybridization (FISH) study. Am J Hum Genet 55:150-159.

Munné S, Magli C, Cohen J, Morton P, Sadowy S, Gianaroli L, Tucker M, Marquez C, Sable D, Ferraretti AP, Massey JB, Scott R. 1999. Positive outcome after preimplantation diagnosis of aneuploidy in human embryos. Hum Reprod 14:2191-2199.

Munné S, Sandalinas M, Cohen J. 2001. Chromosome abnormalities in human embryos. In: Textbook of Assisted Reproductive Techniques, Laboratory and Clinical Perspectives. D Gardner, A Weissman, C Howles, Z Shoham, eds. Martin Dunitz: United Kingdom, p 297-318.

Neganova IE, Sekirina GG, Eichenlaub-Ritter U. 2000. Surface-expressed E-cadherin, and mitochondrial and microtubule distribution in rescue of mouse embryos from 2-cell block by aggregation. Mol Hum Reprod 6:454-464.

Nikas G, Ao A, Winston R, Handyside AH. 1996. Compaction and surface polarity in the human embryo in vitro. Biol Reprod 55:32-37.

Payne D, Flaherty SP, Barry MF, Mathews CD. 1997. Preliminary observations on polar body extrusion and pronuclear formation in human oocytes using time-lapse video cinematography. Hum Reprod 12:532-541.

Pellestor F, Dufour MC, Arnal F, Humeau C. 1994. Direct assessment of the rate of chromosomal abnormalities in grade IV human embryos produced by in vitro fertilization procedure. Hum Reprod 9:293-302.

Puissant F, Van Rysselberge M, Barlow P, Deweze J, Leroy F. 1987. Embryo scoring as a prognostic tool in IVF treatment. Hum Reprod 2:705-708.

Sadowy S, Tomkin G, Munné S, Ferrara-Congedo T, Cohen J. 1998. Impaired development of zygotes with uneven pronuclear size. Zygote 6:137-141.

Sadowy S, Tomkin G, Escudero T, Cohen J. Detailed pronuclear and nucleolar assessment of human zygotes and its possible role in predicting embryo development and implantation. In preparation.

Sathananthan AH, Bongso A, Ng SC, Ho J, Mok H, Ratnam S. 1990. Ultrastructure of preimplantation human embryos co-cultured with human ampullary cells. Hum Reprod 5:309-318.

Sathananthan H, Ng SC, Bongso A, Trounson A, Rtnam S. 1993. Visual Atlas of early human development for assisted reproductive technology. Singapore: National University of Singapore.

Sakkas D, Shoukir Y, Chardonennes D, Bianchi PG, Campana A. 1998. Early cleavage of human embryos to the 2 cell stage after intracytoplasmic sperm injection as an indicator of embryo viability. Hum Reprod 13:182-187.

Sakkas D. 2001. Evaluation of embryo quality: a strategy for sequential analysis of embryo development with the aim of single embryo transfer. Textbook of Assisted Reproductive Techniques, Laboratory and Clinical Perspectives. D Gardner, A Weissman, C Howles, Z Shoham, eds. Martin Dunitz: United Kingdom, p 223-232.

Scott LA, Smith S. 1998. The successful use of pronuclear embryo transfers the day following oocyte retrieval. Hum Reprod 13:1003-1013.

Scott L, Alvaro R, Leondires M, Miller B. 2000. The morphology of human pronuclear embryos is positively related to blastocyst development and implantation. Hum Reprod 15:2394-2403.

Schmiady H, Kentenich H. 1993. Cytological studies of human zygotes exhibiting developmental arrest. Hum Reprod 8:744-751.

Shoukir Y, Campana A, Farley T, Sakkas D. 1997. Early cleavage of in vitro fertilized human embryos to the 2-cell stage: a novel indicator of embryo quality and viability. Hum Reprod 12:1531-1536.

Staessen C, Janssenswillen C, Devroey P, Van Steirteghem AC. 1993a. Cytogenetic and morphological observations of single pronucleated human oocytes after in vitro fertilization. Hum Reprod 8:221-223.

Staessen C, Janssenwillen C, Van Den Abbeel E, Devroey P, Van Steirteghem AC. 1993b. Avoidance of triplet pregnancies by elective transfer of two good quality embryos. Hum Reprod 8:1650-1653.

Staessen C, Van Steirteghem A. 1998. The genetic constitution of multinuclear blastomeres and their derivative daughter blastomeres. Hum Reprod 13:1625-1631.

Steuerwald N, Cohen J, Herrara R, Brenner CA. 1999. Analysis of gene expression in single oocytes and embryos by real-time rapid cycle fluorescence monitored RT-PCR. Mol Hum Reprod 5:1034-1039.

Sultan KM, Munné S, Palermo GD, Alikani M, Cohen J. 1995. Chromosomal status of uni-pronuclear human zygotes following in vitro fertilization and intracytoplasmic injection. Hum Reprod 10:132-136.

Tesarik J, Kopecny V, Plachot M, Mandelbaum J. 1987. Ultrastructural and autoradiographic observations on multinucleated blastomeres of human cleaving embryos obtained by in vitro fertilization. Hum Reprod 2:127-136.

Tesarik J, Kopecny V, Plachot M, Mandelbaum J. 1988. Early morphological signs of genome expression in human preimplantation development as revealed by quantitative electron microscopy. Dev Biol 128:15-20.

Tesarik J, Mendoza C. 1996. Spermatid injection into human oocytes. I. Laboratory techniques and special features of zygote development. Hum Reprod 4:772-779.

Tesarik J, Greco E. 1999. The probability of abnormal preimplantation development can be predicted by a single static observation on pronuclear stage morphology. Hum Reprod 14:1318-1323.

Tesarik J, Junca AM, Hazout A, Aubriot FX, Nathan C, Cohen-Bacrie P, Dumont-Hassan M. 2000. Embryos with high implantation potential after intracytoplasmic sperm injection can be recognized by a simple, non-invasive examination of pronuclear morphology. Hum Reprod 15:1396-1399.

Tomkin G, Cohen J. 2001. Data management and interpretation- computerized database for an ART clinic: hardware and software requirements and solutions. Textbook of Assisted Reproductive Techniques, Laboratory and Clinical Perspectives. D Gardner, A Weissman, C Howles, Z Shoham, eds. Martin Dunitz: United Kingdom, p 367-380.

Van Blerkom J. 1989. Developmental failure in human reproduction associated with preovulatory oogenesis and preimplantation embryogenesis. In: Ultrastructure of Human Gametogenesis and Early Embryogenesis. Van Blekom J, Motta P., eds. Boston, MA: Kluwer Academic Publishing, p 125-180.

Van Blerkom J. 1991. Extrinsic and intrinsic influences on human oocyte and early embryonic developmental potential. In: Elements of Mammalian Fertilization. Wassarman PM, ed. Boca Roton FL: CRC Press, p 81-109.

Van Blerkom J. 1993. Development of human embryos to the hatched blastocyst tage in the presence or absence of a monolayer of Vero cells. Hum Reprod 8:1525-1539.

Van Blerkom J, Antczak M, Schrader R. 1997. The developmental potential of the human oocyte is related to the dissolved oxygen content of follicular fluid: association with vascular endothelial growth factor levels and perifollicular blood flow characteristics. Hum Reprod 12:1047-1055.

Van Royen E, Mangelschots K, De Neubourg D, Valkenburg M, Van de Meerssche M, Ryckaert G, Eestermans W, Gerris J. 1999. Characterization of a top quality embryo, a step towards single-embryo transfer. Hum Reprod. 14:2345-2349.

Van Royen E, Mangelschots K, De Neubourg D, Laureys I, Ryckaert G, Gerris J. 2001. Calculating the implantation potential of day 3 embryos in women younger than 38 years of age: a new model. Hum Reprod 16:326-332.

Warner CM, Exley GE, McElhinny AS, Tang C. 1998a. Genetic regulation of preimplantation mouse embryo survival. J Exp Zool 282:272-279.

Warner CM, Cao W, Exley GE, McElhinny AS, Alikani M, Cohen J, Scott RT, Brenner CA. 1998b. Genetic regulation of egg and embryo survival. Hum Reprod 13 (supplement):178-190.

Wassarman PM, Mortillo S. 1991. Structure of the mouse egg extracellular coat, the zona pellucida. Int Rev Cytol 130:85-110.

Wells D, Delhanty JD. 2000. Comprehensive chromosomal analysis of human preimplantation embryos using whole genome amplification and single cell comparative genomic hybridization. Mol Hum Reprod 6:1055-1062.

Winston NJ, Braude PR, Pickering SJ, George MA, Cant A, Currie J, Johnson MH. 1991. The incidence of abnormal morphology and nucleocytoplasmic ratios in 2-, 3-, and 5-day human pre-embryos. Hum Reprod 6:17-24.

Wright G, Wiker S, Elsner C, Kort H, Massey J, Mitchell D, Toledo A, Cohen J. 1990. Observations on the morphology of pronuclei and nucleoli in human zygotes and implications for cryopreservation. Hum Reprod 5:109-115.

Yanagimachi R. 1988. Mammalian fertilization. In: The Physiology of Reproduction. Knobil E, Neill J, eds. New York: Raven Press, p 135-185.

Ziebe S, Petersen K, Lindenberg S, Andersen AG, Gabrielsen A, Andersen AN. 1997. Embryo morphology or cleavage stage: how to select the best embryos for transfer after in vitro fertilization. Hum Reprod 12:1545-1549.

Chapter 2

MORPHOLOGICAL EVALUATION OF EMBRYOS IN DOMESTIC SPECIES

Sybrand Merton
Holland Genetics ET-IVP Laboratory,
Broekstraat 5b, 7217 SL Harfsen, The Netherlands

1. Introduction

The need to judge the quality of domestic animal embryos with regard to the establishment of pregnancies following embryo transfer (ET) is obvious. The use of embryo production techniques in a commercial set-up for e.g. bovine reproduction can only be successful with acceptable pregnancy rates. Therefore, the selection of 'good' embryos has since long been of great interest.

The simplest but also the most subjective way to judge an embryo prior to transfer is to evaluate its morphology. Consequently, this approach is the most common and oldest method.

The first embryo transfers in cattle were performed with in vivo produced embryos. Following superovulation and artificial insemination, bovine embryos are flushed out of the uterus at day 7 of the oestrous cycle. These embryos are predominantly at the morula stage, whereas only a small percentage is represented by blastocysts. In 1983, Lindner and Wright proposed a useful and practical system for scoring day 7 bovine embryos. This system combines a number of developmental and morphological criteria to classify embryos as excellent (I), good (II), fair (III) or poor (IV) and has been used in our laboratory for a number of years. A more simplified version of the Lindner and Wright scoring system has been published by the

International Embryo Transfer Society (IETS). It combines embryos of grades I and II into the IETS grade 1, whereas IETS grade 2 embryos are of fair and IETS grade 3 are of poor quality (IETS Manual, 1998). More importantly, the International Embryo Transfer Society scoring system has become the standard reference for embryos that are exported/imported internationally.

The introduction of in vitro production (IVP) techniques increased the heterogeneity of the embryo population. The morphological appearance of day 7 IVP embryos is not nearly so homogeneous as that of day 7 embryos which are flushed out of the uterus. Day 7 IVP embryos exhibit a wider and more evenly distributed range of developmental stages. This difference in the distribution of developmental stages between in vitro and in vivo embryos on day 7 is at least partially due to the traditional differences in the way embryo age is determined. In vivo embryo recoveries are most frequently performed seven days after oestrus (day 0) is observed in the donor. Ovulation and subsequently the first contact between oocyte and spermatozoa occur approximately one day after the start of oestrus (day 1). In contrast, the age of IVP embryos is determined from the time of first contact between oocyte and spermatozoa during IVF (day 0). Consequently, IVP embryos are approximately 1 day older on any given day than corresponding in vivo embryos. Thus, day 7 IVP embryos approximate day 8 instead of day 7 in vivo-derived cattle embryos as far as their age is concerned. This is also reflected in the developmental stages of the respective embryos (Figure 1).

	Day of first appearance									
Day	0	1	2	3	4	5	6	7	8	9
In vivo	Oestrus	Fert					M	EB	XB	HB
In vitro	Fert					M	EB	XB	HB	

Figure 1: Comparison of the timing of fertilization (Fert) and embryo development in vivo versus in vitro (M = morula, E B= early blastocyst, XB = expanded blastocyst, HB = hatched blastocyst) (Adapted from Hasler et al., 1995).

As already mentioned, morphological evaluation of embryos is quite subjective. However, criteria described in the IETS manual help to maintain a uniform classification system, which is used on an international basis.

This chapter will describe embryo morphology of domestic animals but will mainly be focused on cattle. However, criteria to assess the quality of bovine embryos can in principle be used for embryos of other domestic species.

2. Developmental processes in embryos

After a successful contact between the oocyte and spermatozoa, the then called zygote will undergo a number of cleavages during the first 5 days of development. At 5 days post fertilization, the embryo starts to compact. Compaction is a developmental process which is one of the morphological parameters important in the assessment of embryo quality.

2.1. Compaction

It is important to take embryo age, developmental stage and morphological appearance into consideration when evaluating embryos. As already mentioned, in vivo embryos are normally recovered at the 7th day of the oestrous cycle. By determining the embryo stage of development in relation with embryonic age, the possible retardation of individual embryos can often be judged. The process of embryo compaction starts at day 5-6 post fertilization in the bovine with the formation of tight junctions between blastomeres (Van Soom et al., 1997). For both sheep and pig, this process starts at 3 days post fertilization (Watson et al., 1993). The formation of tight junctions, along with shrinkage of blastomeres, results in a rounder and more compact embryo with a larger perivitelline space. Morphologically, this means that it becomes more difficult to see individual cells and the embryo resembles a tight clump of cells. Under some IVP conditions, this compaction process is less obvious (Van Soom et al., 1992, Hasler et al., 1995, Holm et al., 1998, Merton et al., 1998, Figure 2). However, when in vitro fertilized zygotes are cultured in host oviducts (e.g. sheep), compaction is normal (Galli and Lazzari, 1996).

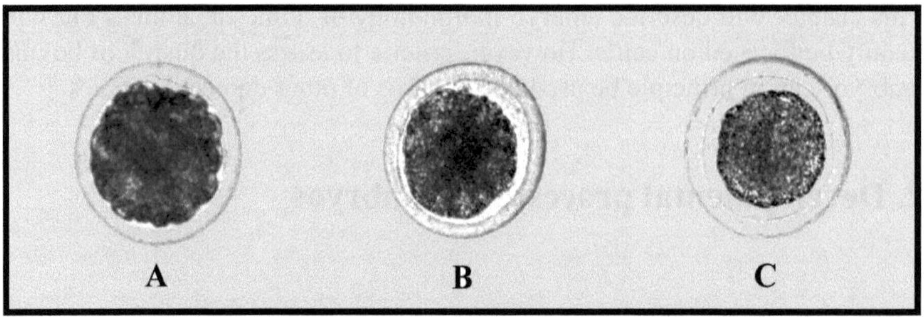

Figure 2: *Compaction in morulae produced in different IVP systems compared to in vivo flushed morulae: in vitro co-culture FCS (A); in vitro SOFaaBSA (B); in vivo-derived (C).*

During the process of compaction, the embryo is called a morula or early morula, whereas following compaction it is known as a tight morula. This process of compaction normally starts around the 32-cell stage in the bovine both in vivo and in vitro (Van Soom, 1996). However compaction seems to be independent from the total number of cells. Especially during in vitro culture, compacted 8-16 cell embryos are sometimes observed. The cell biological clock controlling compaction in vitro is poorly understood.

2.2. Blastulation

As a result of the formation of tight junctions, the embryo forms inner and outer cell types. This is a prerequisite for the process of blastocoel formation, which takes place at day 6 - 7 post fertilization in the cow. In the sheep and pig, blastocysts appear at 4-5 days and 3-5 days post fertilization, respectively (Watson et al., 1993). Due to the active transport of sodium passing through the outer cell layer, an ion gradient is formed, leading to the osmotic accumulation of fluid (Gardiner and Menino, 1993, Brison and Leese, 1994, Watson, 1992). This results in the formation of the blastocoel cavity just beneath the outer layer of cells.

At the early blastocyst stage, one can distinguish two types of blastomeres: differentiated trophoblast cells, lining the blastocoel, and undifferentiated inner cells. Blastocoel formation normally starts on one side of the embryo and the resulting developmental stage is known as an early blastocyst. Interestingly, in some in vitro cultured cattle embryos, several blastocoel cavities develop simultaneously. It is clear that the morphological evaluation of such an embryo becomes more difficult (Figure 3). The blastocoel occupies less than 50% of the total embryo in early blastocysts. When the process of blastocoel formation continues, the inner cells/blastocoel ratio changes further. When the blastocoel occupies more than 50% of the embryo, the embryo is classified as a blastocyst or mid-blastocyst.

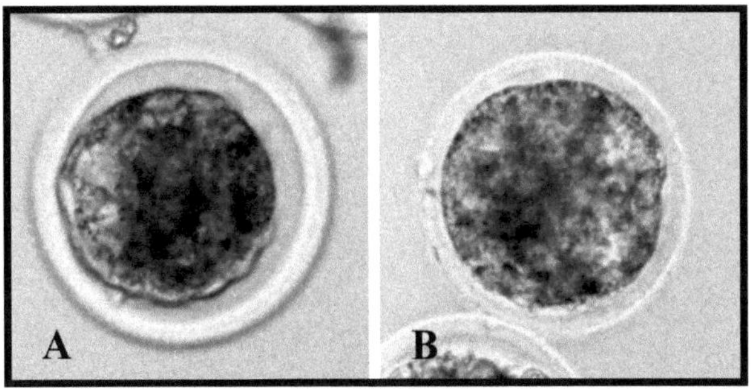

Figure 3: Early blastocysts. Blastocoel formation at one side of the embryo (A) or at several places simultaneously (B).

2.3. Blastocoel expansion

As the osmotic pressure inside the blastocoel cavity rises, the embryo starts to expand and the overall diameter of the embryo increases. At the same time, the zona pellucida stretches and becomes thinner. At this stage, the embryo is classified as an expanded blastocyst. One of the most striking characteristics of an expanded blastocyst is the inner cell mass (ICM). It is a tightly compacted, spherical clump of cells attached to the wall of the trophoblastic vesicle. However, both in vitro and in vivo, depending on the orientation of the embryo, this ICM sometimes may resemble a disc or spoke (Figure 4).

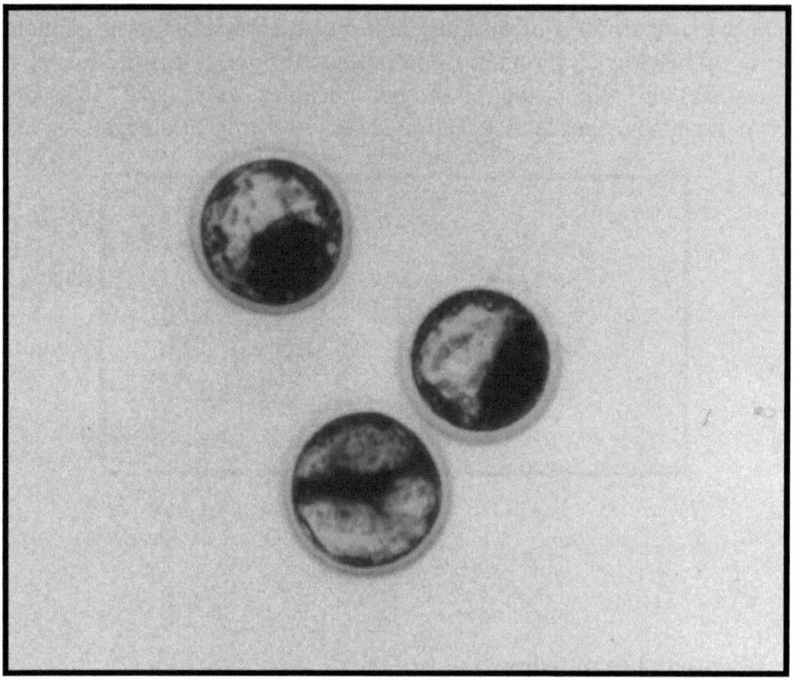

Figure 4: Different shapes of ICM in bovine embryos (IVP, grade 1)

The number of ICM-cells may differ between in vivo-derived and embryos produced in vitro in different culture systems (Iwasaki et al., 1990; Du et al., 1996; Van Soom, 1996). During the expanded stage, which persists for one or two days, the trophoblast may sometimes collapse. This has been clearly demonstrated by in vitro cinematographic studies (Massip and Mulnard, 1980; Massip et al., 1982). An imperfect cell division (formation of tight junctions) in the trophoblast may be hypothesized as one of the reasons for such a collapse.

2.4. Hatching

Ultimately, at day 9-11 post fertilization, the zona pellucida ruptures and the embryo hatches (Figure 5). This process is probably caused by increased pressure of the embryo within the zona, eventually combined with cycles of expansion and collapse and/or an enzymatic process initiated by the trophoblast.

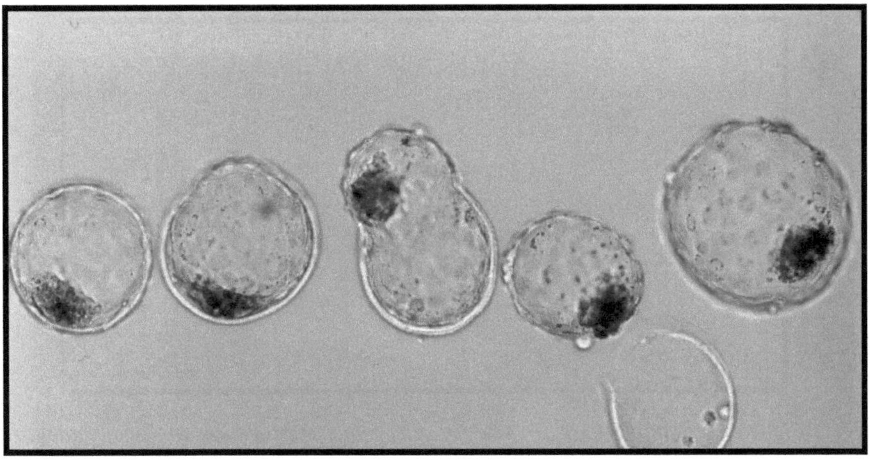

Figure 5: The process of hatching in vitro.

Probably dependent on the elasticity of the zona, in vitro, the size of the zona rupture may vary between a small pore and a large hole (Figure 6). In case of a small pore, this hatching process is often not completed in vitro! Whether the process of hatching is similar in vivo is questionable. Massip et al. (1982) already showed in vitro that some blastocysts do expand and hatch without collapsing. Both in mouse and hamster it has been shown that in vivo cycles of expansion and collapse probably do not occur (Checiu and Checiu, 1996; Gonzales and Bavister, 1995). Also the process of zona solubilisation may differ between in vivo and in vitro (Gonzales and Bavister, 1995).

Preliminary results of a study in our laboratory showed no effect of the size of the zona rupture (small versus large hole) on pregnancy rate. These results suggest either that the uterus environment positively affects the hatching process of in vitro produced bovine embryos or that elongation and implantation of in vitro produced bovine embryos is not hampered by an incomplete hatching process.

When the process of hatching is completed in vitro, the embryo is often pear shaped or collapsed. After a while the embryo increases in size and will regain its spherical form. At this stage, there is a large increase in the number of trophoblast cells and the embryo starts elongating. This phenomenon is not reproducible in vitro, although in vivo incubation of IVP blastocysts in the sheep uterus can be performed to assess the elongation capacity of experimental embryos (Rexroad and Powell, 1999; Talbot et al., 2000).

Figure 6: Differences in size of the zona rupture in in vitro hatching blastocysts.

3. Classification of embryos by stage of development

Based on the processes described above, international codes for successive developmental stages have been established by the IETS in order to uniformize and simplify the administration in international trading. Accurate staging of embryo development is important in both trading and transfer of embryos. In international trading, only embryos with an intact zona pellucida (stage 3 – 7) are used. In case of embryo transfer, when embryos have to be transferred, more advanced embryos are preferentially put into recipients that are more advanced in their cycle. Also with in vitro produced embryos more advanced stages are preferentially frozen.

In Table 1 and Figure 7, the coding for embryonic stages in cattle embryos is shown as used in our laboratory. This coding differs from the IETS code in the last two stages. Whereas IETS describes stage 8 and 9 as hatched or expanded hatched, we discriminate a hatching or hatched blastocyst by stage 8 and 9 respectively.

Table 1: Coding for embryonic stages in cattle embryos.

Code	Stage	Morphological description
3	Early morula	A clump of 16-32 non-compact larger blastomeres
4	Morula	A clump of 32-64 compact smaller blastomeres
5	Early blastocyst	Blastocoel < 50 %
6	Blastocyst	Blastocoel > 50 % ICM and trophoblast discernible
7	Expanded blastocyst	Overall diameter of embryo increases, zona pellucida becomes thinner
8	Hatching blastocyst	Rupture of zona pellucida, partially hatched embryo
9	Hatched blastocyst	Fully hatched embryo further increasing in size, empty zona pellucida with some extruded cells left

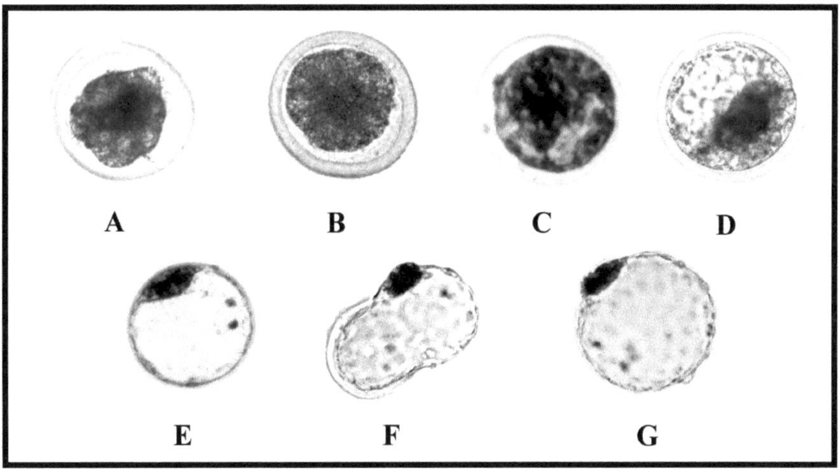

Figure 7: Different stages of in vitro bovine embryos, all classified grade 1; early morula (A), morula (B), early blastocyst (C), blastocyst (D), expanded blastocyst (E), hatching blastocyst (F) and hatched blastocyst (H).

4. Classification of embryos by quality grade

The assessment of the embryo is based on a number of morphological parameters, which might interfere with embryo quality. Classification of embryos into different morphological grades is performed by means of a stereo microscope. Different aspects of the morphological appearance are evaluated to give the embryo its final grade:

- Presence or absence of cellular fragments;

  Due to the prominent perivitelline space, the number of extruded cells or the amount of cellular debris can easily be assessed in bovine morulae and early blastocysts (Figure 8). In some in vitro production systems however, such as serum containing coculture systems, morulae and early blastocysts are less compacted, which makes this assessment more difficult (Holm et al., 1998; Merton et al., 1998). This problem can be overcome by putting the embryos in a sucrose solution. Because of the hypertonicity of the medium, viable cells shrink and cellular fragments appear more clearly in the perivitelline space (Van Soom et al., 1996). For accurate assessment, it is important to roll over embryos under the microscope in order to visualize them in a three-dimensional perspective. In later developmental stages, extruded cells are forced against the zona pellucida and are therefore often difficult to distinguish. With appropriate magnification and careful observation however, extruded cells can often be detected, although the quantification of debris in (expanded) blastocysts remains very difficult.

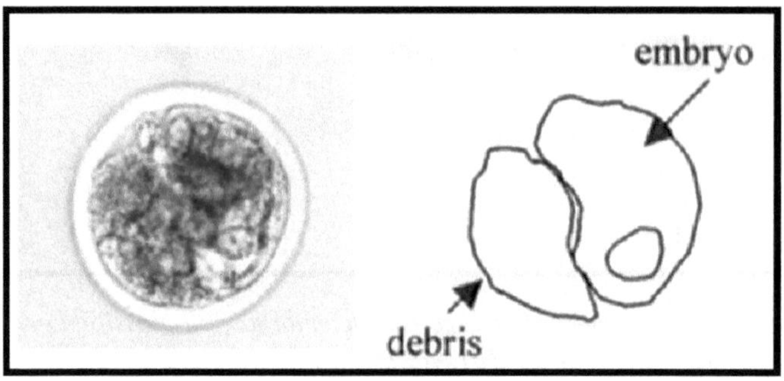

Figure 8: In vitro early blastocyst grade 2; approximately 45 % cellular debris.

- The degree of compaction;

  The degree of compaction is important in assessing both morulae and ICM of blastocysts. Compaction can be assessed by the roundness and/or the sharpness of the contours of the outer periphery of the morula. The sharpness of the inner border of the ICM can be visualized by rolling the embryos with a pipette. At least in vitro, all three types of ICM, i.e., spherical, disc and spoke-shaped, can have any degree of compactness (Lazzari, personal communication, Merton, unpublished data).

  Besides the degree of compactness, also the size of the ICM (normal, small, absent), is taken into account in the evaluation of embryo quality. In vitro, the number of cells allocated to the ICM varies enormously, sometimes resulting in embryos which resemble 'empty' trophoblastic vesicle (Van Soom, 1996).

- Colour and texture of the blastomeres

  The 'colour' of an embryo or ICM, which is partially influenced by the amount of transmitted light under the microscope, ranges from light to dark and is dependent on a number of factors:

  1) The uniform distribution of lipid vesicles and cytoplasmic organelles,
  2) The amount and size of the lipid vesicles in the cytoplasm,
  3) The number of pycnotic cells,
  4) The total number of blastomeres in the viable embryonic mass,
  5) The source of the embryo (in vivo versus in vitro),
  6) The nutrition of the donor cows (for in vivo embryos) and the culture conditions of the embryo (for IVP embryos).

  Especially these last two statements deserve more explanation. It is already known for a long time that IVP embryos have a darker texture than in vivo embryos (Van Soom and de Kruif, 1992). It has been suggested that the degree of darkness of the embryo and its granulation depends on the source of serum which has been used for embryo culture (Thompson, 1997). These granules are particularly visible at the blastocyst stage in the trophectoderm cells (Figure 9). Serum free embryo culture conditions produce much lighter embryos; although some batches of BSA can also lead to dark embryos (Thompson, 1997).

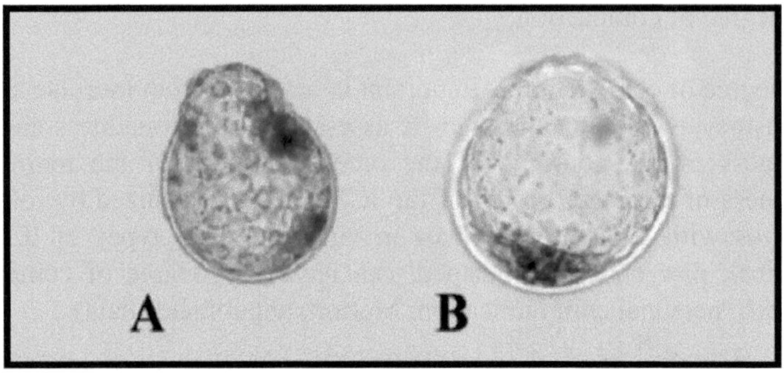

Figure 9: Granules in trophectoderm cells. Embryos produced in vitro in co-cultureFCS (A) and SOFaaBSA (B).

Also some types of oil overlay which are being used for embryo culture can affect the colour of the embryos (Van Soom et al., 2001). For in vivo produced cattle embryos, it has been shown that there is a significant correlation between embryo colour and donor blood cholesterol (Hill and Kuehner, 1998). Moreover, cattle embryos from high-yielding dairy cattle frequently exhibit a darker texture of the blastomeres than embryos from beef cattle (Geldhof, personal communication).

- Uneven cleavage divisions

  Asynchronous cell division may occur at any time after the first cell cycle. Thus, three, five, six, and seven cell stage embryos are not uncommon. As a consequence, especially at the morula stage, cell size may affect the symmetry and spherical appearance of the embryo and may be related to embryo quality.

Evaluation of embryo quality grade is important to predict pregnancy rates after transfer and to select embryos for cryopreservation. In the latest IETS manual (third edition, 1998), the codes for embryo quality are described as follows:

Code 1: Excellent or good

Symmetrical and spherical embryo mass with individual blastomeres (cells) that are uniform in size, colour, and density. This embryo is consistent with its expected stage of development. Irregularities should be relatively minor, and at least 85% of the cellular material should be an intact, viable embryonic mass. This judgement should be based on the percentage of embryonic cells represented by the extruded material in the perivitelline space. The zona pellucida should be smooth and have no concave or flat surfaces that might cause the embryo to adhere to a petri dish or a straw.

Code 2: Fair

Moderate irregularities in overall shape of the embryonic mass or in size, colour and density of individual cells. At least 50% of the cellular material should be an intact, viable embryonic mass.

Code 3: Poor

Major irregularities in shape of the embryonic mass or in size, colour and density of individual cells. At least 25% of the cellular material should be an intact, viable embryonic mass.

Code 4: Dead or degenerating

Degenerating embryos, oocytes or 1-cell embryos: non-viable.

5. Relationship between embryo morphology evaluation and pregnancy rates

Embryo classification is only useful when morphological differences are correlated with differences in pregnancy rates. Due to a lack of a significant difference in pregnancy rate of quality I and II embryos in the study by Lindner and Wright (1983), both types of embryos are now classified as IETS class 1 embryos. In Table 2, the relationship between embryo stage/quality and pregnancy is shown. It should be taken into account, however, that such a relation is dependent on a number of factors and factors other than the embryo are important sources of variation in embryo survival or loss (McMillan, 1998).

Factors influencing pregnancy rates are the person evaluating the embryo (Farin et al., 1999), the method of embryo production (Hasler, 1998), the method of embryo cryopreservation (Chupin et al., 1984, Wurth et al., 1994), the embryo biopsy method and last but not least recipient quality (McMillan, 1998) or an interaction of these factors. Although less obvious in vivo, it is clear that in vitro, within a given embryo production system, the most advanced embryos are of higher quality based on cell number (Jiang et al., 1992) and pregnancy rate (Hasler et al., 1998). In a study of Merton et al. (1998), it has also been shown that embryo quality affected birth weight of IVP calves.

Table 2: The effect of bovine fresh IVP-derived embryo stage, age and grade (IETS) on pregnancy rate. (Data adapted from Hasler, 1998).

	% pregnant (number of transfers)			
	Day 7		Day 8	
Embryo stage	Grade 1	Grade 2	Grade 1	Grade 2
Morula	45 (335)	31 (83)	0 (1)	0 (2)
Early blastocyst	53 (1114)[a]	36 (259)[b]	39 (31)	27 (30)
Blastocyst	56 (1305)[a]	46 (296)[b]	49 (96)	31 (45)
Expanded blastocyst	63 (1069)	57 (54)	47 (188)[a]	33 (24)[d]
Hatched	57 (91)	--	45 (45)	--
Total	56 (3914)[a]	41(692)[c]	48(361)[a]	30 (101)[b]

a,b,c,d: Values in rows differ significantly; a vs. b: $p < 0.01$, a vs. c: $p < 0.001$,a vs. d, b vs. c: $p < 0.05$ Chi-square analysis.

Although the morphological evaluation for in vivo and in vitro derived embryos is based on the same parameters, pregnancy rates differ between both type of embryos (Table 3) (Farin and Farin, 1995: Galli and Lazzari 1996, Hasler 1998, Holland Genetics, unpublished data). This difference is mainly due to a lower viability of in vitro produced bovine embryos, which was clearly demonstrated in a study of McMillan (1998).

Table 3: Pregnancy rate following transfer of fresh and frozen bovine in vivo and in vitro derived embryos. (Data adapted from Hasler, 1998)

		Pregnancy rate	
Embryo	No. Transfers	Grade 1	Grade 2
In vitro fresh	4606	56[a]	41[a]
In vitro frozen	67	42[b]	30[b]
In vivo fresh	599	76[c]	67[c]
In vivo frozen	517	64[d]	

[a,b,c,d]: Values within a column differ significantly: $p < 0.05$, Chi-square analysis.

6. Factors affecting embryo evaluation

Embryo evaluation based solely on morphological parameters is a subjective method. Although 'rules' has been made to come to an internationally reliable grading system, variation will always be present. Main factors are evaluator, embryo source, microscope and magnification.

6.1. Evaluator

To reduce variation between evaluators, the use of a simple embryo grading system is beneficial (Farin et al., 1995). It is very difficult for different embryologists to agree in assigning specific grades to bovine embryos recovered from uterine flushings, as has been shown repeatedly throughout literature (Shea, 1981; Callesen et al., 1995; Farin et al., 1995; Farin et al., 1999). Interestingly, excellent agreement among evaluators was found for differentiating degenerate embryos from non-degenerate embryos (Farin et al., 1999). Variation among evaluators involves mainly the quality grading of embryos, whereas embryo stage is less frequently a point of disagreement (Callesen et al., 1995; Farin et al., 1995; Van Soom et al., 1996; van Wagtendonk-de Leeuw, 1996). To establish and continue a standardized evaluation level within an ET/IVP laboratory, a frequent group wise discussion of a number of embryos is therefore recommended.

6.2. Embryo source

As already mentioned, in vitro produced embryos are more difficult to evaluate than their in vivo produced counterparts. This has been shown in several studies, where agreement among evaluators was always lower for in vitro produced bovine embryos (Farin et al., 1995; Farin et al., 1999). One major aspect involved here, is the increased heterogeneity of the in vitro embryo population both in stage and quality grade (Figure 10). Agreement among evaluators for IVP embryos can however be improved by incubating the embryos during evaluation in a 0.3 M sucrose solution (Van Soom et al., 1996).

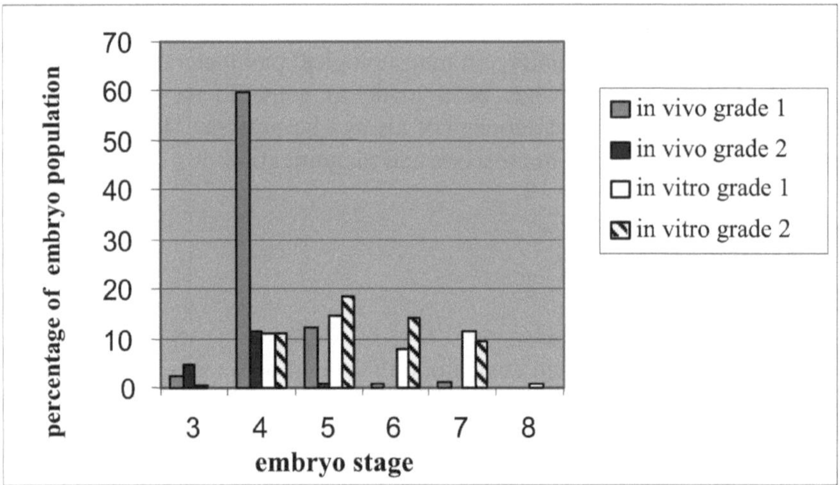

Figure 10: Distribution of embryo stage and quality within the population of in vivo and in vitro derived embryos (Data Holland Genetics).

6.3. Microscope

In addition to the evaluator, the microscope also influences the evaluation of an embryo. In most cases a stereo microscope will be used for embryo evaluation. However, different models of microscope have different, specific views. Technicians often become most comfortable with their own 'personal' microscope. In the case of a large-scale laboratory, where more than one microscope will be used for quality assessment, one model of microscope is recommended to enhance standardisation.

6.4. Magnification

In the IETS-manual it is stated that quality assessment should be performed at a magnification of at least 50x to ensure good discrimination. Each magnification has its own advantages but will therefore give different impressions of an embryo (Figure 11). A fixed magnification is therefore also recommended for standardized embryo quality assessment.

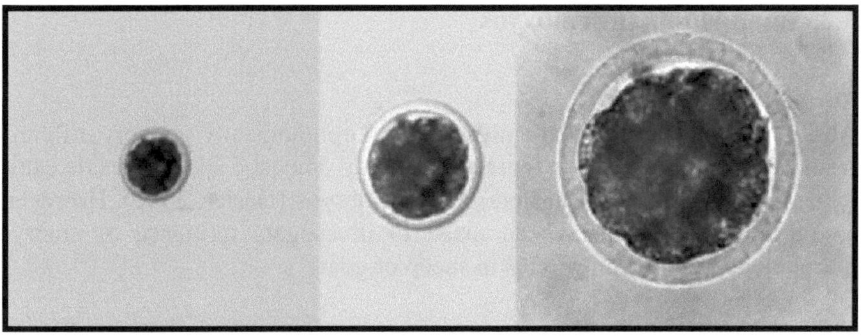

Figure 11: In vitro early morula at 3 different microscopic magnifications (about X70, X120, X250).

7. Embryo morphology in other domestic species

7.1. Pigs

In contrast to the cow, the pig is a polytocous animal. Therefore, in non-surgical transfer procedures, 10 to 40 embryos can be transferred per recipient. Embryo quality assessment of in vivo produced porcine embryos is mainly focused on embryo stage in relation to age and retarded embryos are normally discarded. Better pregnancy rates were obtained when "large" blastocysts (> 159 μm) were transferred than with "smaller" blastocysts (< 159 μm) (Hazeleger and Kemp, 1999). Development of in vitro embryos is slower than their in vivo counterparts. On day 5 post fertilization, cell number of in vivo blastocysts average 80 cells. In vitro blastocysts reach a cell number of 60 only on day 7. At this stage an apparent relation exists between blastocyst diameter and cell number, as well as nuclei quality with larger blastocysts being of better quality (Kidson, unpublished data).

7.2. Small ruminant embryos

At first sight sheep and goat embryos are morphologically not very different from cattle embryos. Sheep represent the only domestic species beside cattle with a large number of transfers of IVP embryos (Hasler, 1998). However, so far no attempts have been made to investigate influence of embryo morphology on pregnancy rates in sheep or goats.

7.3. Equine embryos

Morphology assessment has been performed in equine in vivo embryos and has been related with pregnancy rates. Grade 1 (excellent) embryos produced 78 % pregnant mares at day 50, whereas the pregnancy rate is reduced to 33 % when grade 4 (poor) embryos are transferred (Squires et al., 1999).

7.4. Evaluation of mammalian embryos in general

For all types of embryos, a standard 'route' can be followed in order to come to a final classification (Figure 12). First, the stage of the embryo should be determined. From this point on, an embryo is judged by the criteria already described.

In case of morulae or early blastocysts, first the percentage of extruded cells is determined and an initial score is given. Subsequently the quality of the cellular mass is assessed which will lead to the final score.

For blastocysts and more advanced stages, first the quality of the ICM is assessed. The initial score given here can never become higher independent of the quality of the trophoblast, which is subsequently judged.

8. Conclusion

Due to a lack of alternative non-invasive methods, embryo quality estimation based on morphological parameters is still the most commonly used method. Advantages of this method are the limited time needed per embryo and the ease of learning it. A disadvantage however, especially with in vitro produced embryos, is that embryo viability, based on pregnancy rate, still varies within groups of embryos with the same grade. Therefore, a combination of embryo morphology and other non-invasive methods e.g. timing of development or quick metabolic tests would very likely result in a better discrimination between viable and non-viable embryos.

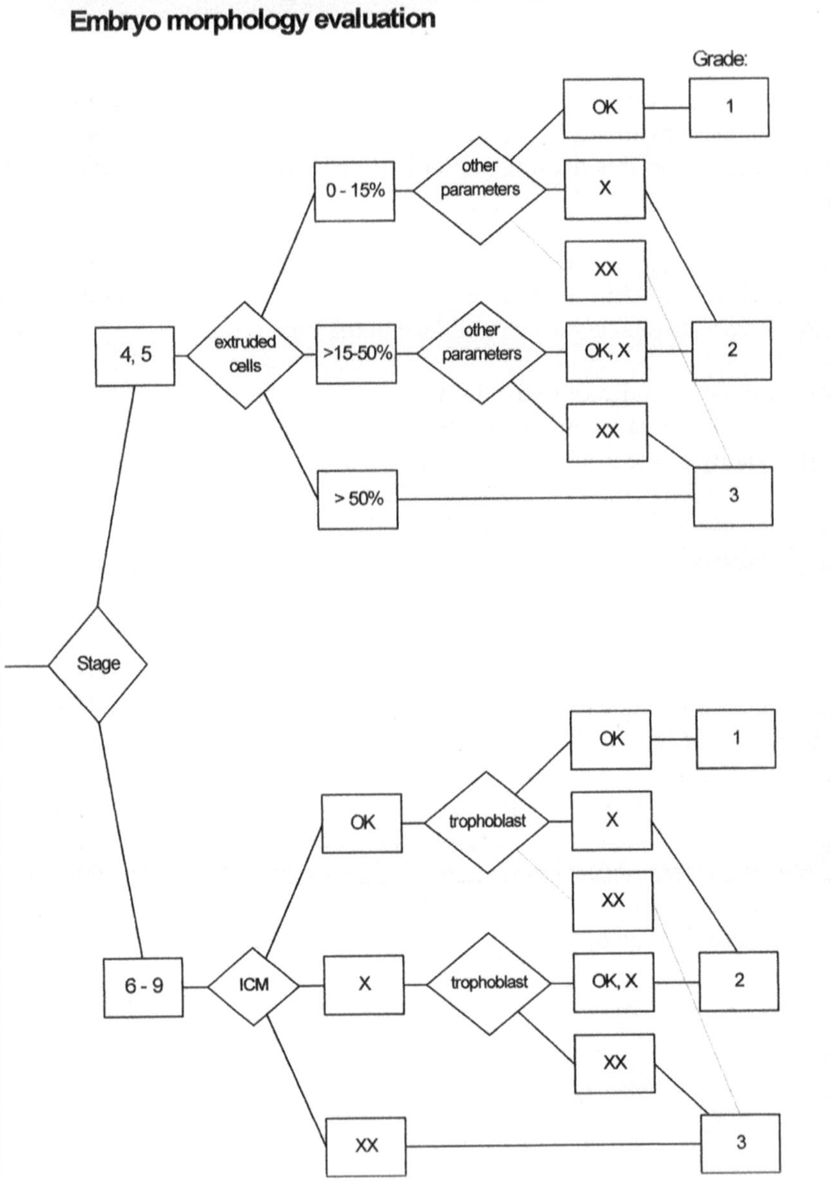

Figure 12: Schematic presentation of the assessment of embryo quality.
4 – 9: Embryo stage
X: moderate irregularities, XX: major irregularities

References

Brison DR, Leese HJ. 1994. The role of exogenous energy substrates in blastocoel fluid accumulation in the rat. Zygote 2:69-77.

Callesen H, Lovendahl P, Bak A, Greve T. 1995. Factors affecting the developmental stage of embryos recovered on day 7 from superovulated dairy cattle. J Anim Sci 73:1539-1543.

Checiu I, Checiu M. 1996. There are no in vivo pulsations of mouse blastocysts. Rom J Morphol Embryol 42:147-154.

Chupin D, Florin B, Procureur R. 1984. Comparison of two methods for one-step in-straw thawing and direct transfer of cattle blastocysts. Theriogenology 21:455-459.

Du F, Looney CR, Yang X. 1996. Evaluation of bovine embryos produced in vitro vs. in vivo by differential staining of inner cell mass and trophectoderm cells. Theriogenology 45:211 abstr.

Farin PW, Britt JH, Shaw DW, Slenning BD. 1995. Agreement among evaluators of bovine embryos produced in vivo or in vitro. Theriogenology 44:339-349.

Farin PW, Farin CE. 1995. Transfer of bovine embryos produced in vivo and in vitro: survival and fetal development. Biol Reprod 52:676-682.

Farin PW, Crosier AE, Farin CE. 1999. Estimates of pregnancy outcomes based on selection of bovine embryos produced in vivo or in vitro. Theriogenology 52:659-670.

Galli C, Lazzari G. 1996. Practical aspects of IVM/IVF in cattle. Anim Reprod Sci 42:371-379.

Gardiner CS, Menino AR Jr. 1991. Development of Na/K ATPase activity and blastocoel formation. Proceedings of the symposium on Preimplantation Embryo Development 1991:200-211.

Gonzales DS, Bavister BD. 1995. Zona pellucida escape by hamster blastocysts in vitro is delayed and morphologically different compared with zona escape in vivo. Biol Reprod 52:470-480.

Hasler JF, Henderson WN, Hurtgen PJ, Jin ZQ, Mc.Cauley AD, Mower SA, Neely B, Shuey LS, Stokes JE, Trimmer SA. 1995. Production, freezing and transfer of bovine IVF-embryo's and subsequent calving results. Theriogenology 43:141-152.

Hasler JF. 1998. The current status of oocyte recovery, in vitro embryo production and embryo transfer in domestic animals with emphasis on the bovine. J Anim Sci 76 (suppl. 3):52-74.

Hazeleger W, Kemp B. 1999. State of the art in pig embryo transfer. Theriogenology 51:81-90.

Hill BR, Kuehner LF. 1998. Bovine donor blood cholesterol, embryo colour and embryo freezability. Theriogenology 49:168.

Holm P, Callesen H. 1998. In vivo versus in vitro produced bovine ova: similarities and differences relevant for practical application. 14th reunion A.E.T.E. Venice, 11-12 September 1998.

Iwasaki S, Yoshiba N, Ushijima H, Watanabe S, Nakahara TJ. 1990. Morphology and proportion of inner cell mass of bovine blastocysts fertilized in vivo and in vitro. J Reprod Fertil 90:279-284.

Jiang HS, Wang WL, Lu KH, Gordon I, Polge C. 1992. Examination of cell numbers of blastocysts derived from IVM, IVF and IVC of bovine follicular oocytes. Theriogenology 37:229.

Lindner GM, Wright RW. 1983. Bovine embryo morphology and evaluation. Theriogenology 20:407-416.

Manual of the International Embryo Transfer Society, 3rd edition, 1998.

Massip A, Mulnard J. 1980. Time-lapse cinematographic analysis of hatching of normal and frozen-thawed cow blastocysts. J Reprod Fertil 58:475-478.

Massip A, Mulnard J, Vanderzwalmen P, Hanzen C, Ectors F. 1982. The behaviour of cow blastocyst in vitro: cinematographic and morphometric analysis. J Anat 134:399-405.

McMillan WH. 1998. Statistical models predicting survival to term in cattle after embryo transfer. Theriogenology 50:1053-1070.

Merton JS, van Wagtendonk-de Leeuw AM, den Daas JHG. 1998. Factors affecting birth weigth of IVP calves. Theriogenology 49:293.

Merton JS, Mullaart E. 1999. Comparison of the effect of two bovine in vitro culture systems, TCM199/FCS/BRL-co-culture and SOFaaBSA, on embryo production and pregnancy rate. Theriogenology 51:248.

Rexroad CE, Powell AM. 1999. The ovine uterus as a host for in vitro produced bovine embryos. Theriogenology 52:351-364.

Shea BF. 1981. Evaluating the bovine embryo. Theriogenology 15:31-42.

Squires EL, McCue PM, Vanderwall D. 1999. The current status of equine embryo transfer. Theriogenology 51:91-104.

Talbot NC, Powell A, Garrett E, Edwards JL, Rexroad C. 2000. Ultrastructural and karyotypic examination of in vitro produced bovine embryos developed in the sheep uterus. Tissue Cell 32:9-27.

Thompson JG. 1997. Comparison between in vivo derived and in vitro produced pre-elongation embryos from domestic ruminants. Reprod Fertil Dev 9:341-354.

Van Soom A, de Kruif A. 1992. A comparison between in vivo and in vitro produced bovine embryos: Scientific and practical implications. Proc 12th Int Con Anim Reprod AI 3:1365-1367.

Van Soom A. 1996. Bovine embryo quality: timing of development, morphological differentiation and cell allocation. PhD thesis, Ghent University.

Van Soom A, Boerjan M, Ysebaert M-T, de Kruif A. 1996a. Cell allocation to the iner cell mass and the trophectoderm in bovine embryos cultured in two different media. Mol Reprod Dev 45:171-182.

Van Soom A, Ysebaert M-T, Vanhoucke-Demedts A., Van De Velde A, Merton S, Delval A, Van Langendonckt A, Donnay I, Vanroose G, Bols PEJ, De Kruif A. 1996b. Sucrose induced shrinkage of in vitro produced bovine morulae: effect on viability, morphology and ease of evaluation. Theriogenology 46:1131-1147.

Van Soom A, Boerjan ML, Bols PEJ, Vanroose G, Lein A, Coryn M, de Kruif A. 1997. Timing of compaction and inner cell allocation in bovine embryos produced in vivo after superovulation. Biol Reprod 57:1041-1049.
Van Soom A, Mahmoudzadeh AR, Christophe A, Ysebaert MT, de Kruif A. 2001. Silicone oil used in microdrop culture can affect bovine embryonic development and freezability. Reprod Dom Anim 36:169-176.
Van Wagtendonk-de Leeuw AM. 1996. Evaluation of uniformity among persons in embryo grading from video recordings. Theriogenology 45:230.
Watson AJ. 1991. Activation of the embryonic genome: comparison between mouse and bovine development. Proceedings of the symposium on Preimplantation Embryo Development 1991:115-130.
Watson AJ. 1992. The cell biology of blastocyst development. Mol Reprod Dev 33:492-504.
Wurth YA, Reinders JMC, Rall WF, Kruip TAM. 1994. Developmental potential of in vitro produced bovine embryos following cryopreservation and single embryo transfer. Theriogenology 42:1275-1284.

Chapter 3

METABOLIC MARKERS OF EMBRYO VIABILITY

Isabelle Donnay
Université Catholique de Louvain, Unité des Sciences Vétérinaires, Place Croix du Sud 3, B-1348 Louvain-la-Neuve, Belgium

1. Introduction

Assisted reproductive techniques (ART) have spread rapidly both in humans and farm animals. Whatever the species, a major challenge has been how to select the best embryo for transfer. Indeed, in human medicine, the objective is to obtain the highest pregnancy rate with the lowest risk of multiple pregnancies (Conaghan et al., 1993; Geber and Sampaio, 1999). In farm animals, especially in cattle, where single transfer is preferred, pregnancy rates are variable, particularly after the transfer of in vitro matured, fertilized and cultured (ie. in vitro produced - IVP) embryos: their quality is lower and more heterogeneous than that of embryos recovered after multiple ovulation and embryo transfer. IVP embryo viability is also highly reduced after cryopreservation. Embryo manipulation by means of biopsy, nuclear transfer or freezing enhances the risk of damage, which has to be evaluated before transferring the embryo. Selection of viable embryos before transfer will thus improve the overall efficiency of ART procedures by decreasing the number of recipient females required in transfer programmes and increasing the commercial value of embryos selected. Moreover, if embryo quality could be estimated by the presence of a "viability marker", this could also be used as a tool for improving culture systems and techniques of embryo manipulation without having to transfer the embryos. We can thus distinguish between markers that can be used to select embryos prospectively for transfer and those that can be used for a more rapid and precise evaluation of the impact of a procedure (eg. culture system, biopsy, nuclear transfer, cryopreservation…) on embryo viability.

The ideal marker to select embryos prospectively for transfer should fulfil the following criteria: it should be completely non-invasive, of high sensitivity (capable of evaluating single embryos); able to distinguish clearly between viable and non-viable embryos; be simple and rapid without the need for cryopreservation and applicable in all conditions. Such a marker has not yet been reported. The non-invasive method traditionally used for selecting embryos is based upon morphological characteristics of the embryo (as described by Merton in Chapter 2). Kinetics of cleavage can also be used (McKiernan and Bavister; Grisart et al., 1994; Shoukir et al., 1997; Ziebe et al., 1997).

In this review, I will present a general overview of the parameters related to mammalian embryo metabolism that can be used to evaluate embryo quality non-invasively. Parameters related to energy metabolism are commonly used since the need to form ATP is a prerequisite for the survival of any cell. However, other embryonic metabolic activities can be measured non-invasively. Among them are the uptake and release of amino acids, and the production of autocrine or paracrine factors involved in embryo development, in the recognition of pregnancy or in implantation.

2. General characteristics of embryo metabolism

2.1. Maternal and zygotic control of metabolism

The metabolic activity of preimplantation embryos can be divided into two major phases: (1) the "maternal" period, controlled by mRNA and protein accumulated in the oocyte, which is characterized by a low metabolic activity; (2) a dramatic increase in embryonic metabolism occurring in the period following activation of the embryonic genome, particularly during blastocyst expansion (Leese, 1995). Measuring metabolism during the period under maternal control is thus not really a measure for embryo quality but rather a way of evaluating the quality of the oocyte and the fertilization step (Gardner and Lane, 1997; Mermillod et al., 1999; Gardner and Leese, 2000; Van Blerkom et al., 2000; Spindler et al., 2000). When metabolism is evaluated after the major onset of the embryonic genome, the first selection is already done: embryos that did not cleave, or became arrested, can be discarded. Metabolism now reflects the quality of the embryo itself, notably its capacity to transcribe and translate the zygotic genome (Lequarré et al., 1997), to differentiate into two cell lineages showing different metabolic activities (Hewitson and Leese, 1993), to regulate blastocyst expansion (Donnay and Leese, 1999) and achieve the appropriate number of cells (Kaidi et al., 2001).

2.2. Characteristics of embryonic energy metabolism

The activity of the glycolytic pathway can be evaluated by the uptake, use and transformation of glucose. The end product of glycolysis is pyruvate which can be transformed either to acetyl CoA or to lactate. One molecule of glucose gives rise to two molecules of pyruvate (Figure1). The equilibrium between pyruvate and lactate depends mainly on the redox status of the cell. Glycolytic activity is calculated as the ratio between lactate production and glucose uptake by the cells. However, after phosphorylation, glucose can also enter the pentose phosphate pathway (PPP), the main cellular source of purine nucleotides and NADPH. The proportion of the PPP versus glycolysis depends on the status of the cell (mitotic and transcriptional activity (purine synthesis) and level of oxidation (NADPH turnover)). Although glycolysis has a low efficiency in terms of ATP production (2 molecules of ATP from 1 molecule of glucose) in comparison with oxidative metabolism (36 molecules of ATP from 1 molecule of glucose), it can function in the absence of oxygen or when mitochondria are not active.

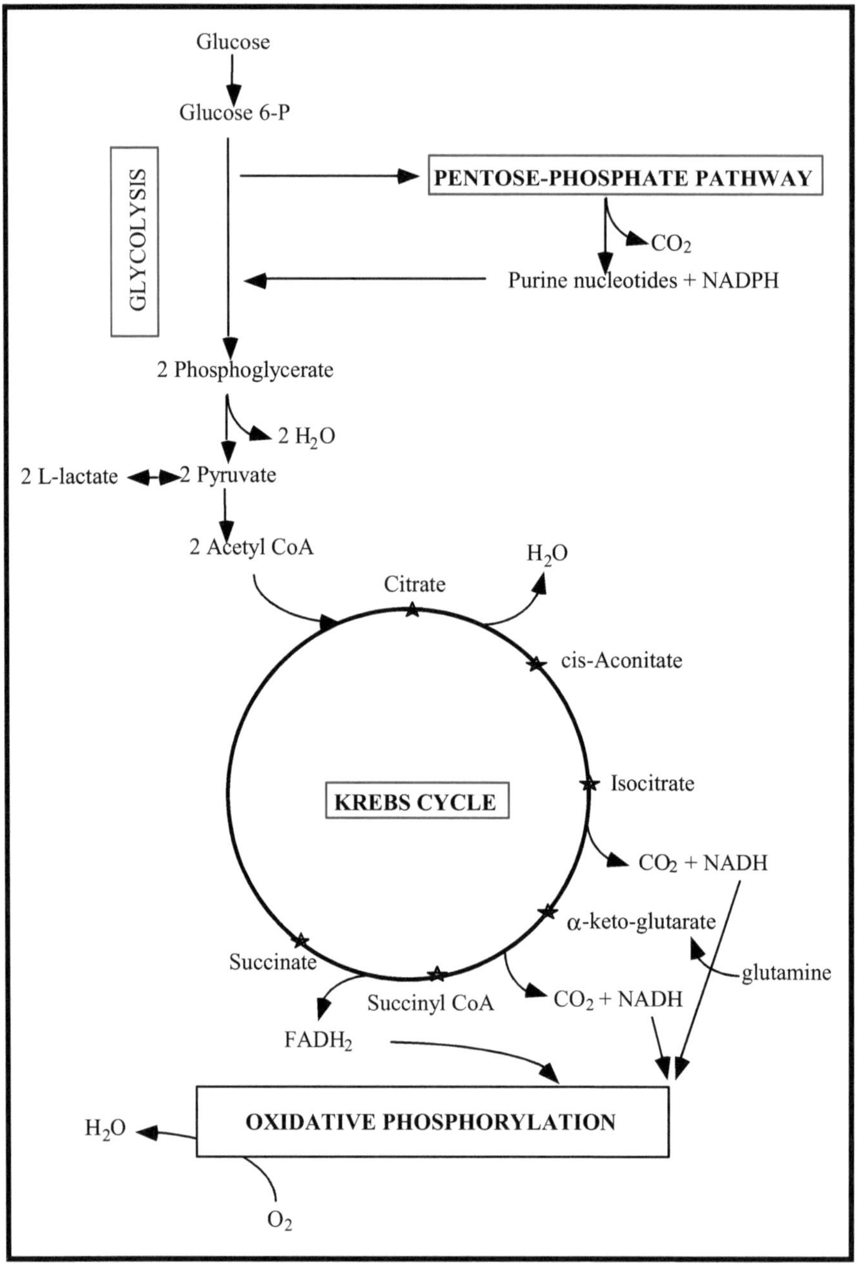

Figure 1: Pathways of energy metabolism

Little glucose is used by early embryos before the morula or blastocyst stage. Cleavage-stage embryos show limited glycolytic activity in the studied species (rabbit: Brinster, 1968; pig: Flood and Wiebold, 1988; human: Wales et al., 1987; bovine: Rieger et al., 1992; hamster: Barnett, 1993). It is also well known that serum glucose concentrations (5.5mM) can be toxic for the first steps of embryo development in vitro. In contrast, glycolysis represents an important pathway for blastocyst metabolism in several species. This is the case in the rat embryo and also in ruminants (Brison and Leese, 1991; Thompson et al., 2000) and could be related to the relative anoxic conditions encountered by the blastocyst in the uterine cavity before implantation (Fischer and Bavister, 1993). The inner cell mass (ICM) cells of the blastocyst seem more dependent on glycolysis than the trophectoderm (TE) cells, at least in the mouse embryo (Hewitson and Leese, 1993). Trophectoderm cells regulate the availability of energy substrates for ICM metabolism. The relative activity of the PPP is high during the early stages but is substantially lower in the blastocyst (for a review, see Barnett and Bavister, 1996).

The tricarboxyllic acid cycle and oxidative phosphorylation pathway, both located in mitochondria, represent the final pathways in the oxidation of all energy substrates: carbohydrates, amino acids and fatty acids. Fuel molecules can enter the cycle as acetyl CoA (derived from pyruvate, fatty acids and several amino acids) as well as α–ketoglutarate (from glutamate in turn derived, for example, from glutamine), succinyl CoA (for nonpolar amino acids), fumarate or oxaloacetate (Stryer, 1995). The Krebs cycle produces electron carriers (NADH and $FADH_2$) that finally donate their electrons to oxygen during the oxidative phosphorylation, generating ATP from ADP. Although little is known about the physiological substrates that enter the embryonic Krebs cycle, exogenous pyruvate and glutamine seem the most important in vitro. Glutamine is known to be used as an energy source by rapidly dividing cells in vitro (Zielke et al., 1984). If the pyruvate present in the culture medium is a major energy substrate, little pyruvate resulting from the glycolysis seems to enter the Krebs cycle, even at the blastocyst stage, when both glycolysis and oxidative pathways are active. This may be explained by the transformation of endogenous pyruvate to lactate in order to control the redox status of the cell (Butcher et al., 1998). Exogenous pyruvate may also be transformed into lactate for the same reason. Embryos may use other potent substrates of the Krebs cycle including lactate, acetate, fatty acids and amino acids other than glutamine. Some studies showed that such compounds are taken up by embryos but their use as fuel molecules seems minimal (for a review, see Barnett and Bavister, 1996; Partridge and Leese, 1996; Gomez, 1997). The first embryonic cleavages are completely dependent on oxidative metabolism. At that stage, low concentrations of inhibitors of oxidative phosphorylation block embryonic development (Kane and Buckley, 1977; Thomson, 1967; Brison and Leese, 1991). Oxidative metabolism is dramatically increased with blastocyst expansion in cattle (Donnay and Leese, 1999) and mouse embryos (Houghton et al., 1996). This increase could be related to the rising number and/or activity of trophectoderm cells through which blastocoel expansion is mediated. However, blastocysts seem less sensitive to the partial blockage of oxidative phosphorylations and could compensate by an increased glycolytic activity (rat: Brison and Leese 1991; bovine: Donnay and Leese, 1999; Thompson et al., 2000). It has even been shown that partial inhibition or uncoupling of oxidative phosphorylation at the blastocyst stage could have a beneficial effect on bovine and porcine blastocyst development and quality (Thompson et al., 2000; Machaty et al., 2000).

2.3. Variation and adaptation of embryonic metabolism

Embryo metabolism is characterized by a high degree of adaptability (Leese, 1995). However, adaptation does not necessarily mean debilitation or impairment. The main difficulty in the use of metabolic parameters for assessing embryo viability is to distinguish between modifications that are, or are not, related to impaired developmental potential. Variations related to the following factors have to be added to those linked with embryonic stage.

2.3.1. Variations with sex

Glucose uptake by female mouse embryos tends to be higher than for male embryos, although glycolytic rates are not different (Gardner and Leese, 1987; Lane and Gardner, 1996). This could be related to the incomplete X-chromosome inactivation in these early stages. In female cleavage stage embryos both X-chromosomes are active and therefore twofold activity of the X-linked enzyme glucose 6-phosphate dehydrogenase (G6PDH) involved in the regulation of the PPP is present (Chapman, 1978; Gardner and Leese, 1987). X-chromosome inactivation seems to start at the blastocyst stage in the trophectoderm cells (Chapman, 1978). It was demonstrated that the activity of the PPP and the expression of G6PDH is higher in female than male bovine embryos at the blastocyst stage (Tiffin et al., 1991; Gutierrez-

Adan et al., 2000). However, such metabolic differences are not sufficient to allow embryo sexing in practice.

2.3.2. Variations with embryonic origin: in vivo/in vitro

In general, the pattern of utilisation/production of glucose, pyruvate, and lactate is qualitatively similar between bovine embryos derived in vitro and in vivo (Khurana and Niemann, 2000a). However, IVP embryos in the bovine (Khurana and Niemann, 2000a), sheep (Thompson et al., 1991) and mouse (Gardner and Sakkas, 1993) tend to convert more glucose to lactate than those derived in vivo. Increased glycolysis, altered substrate oxidation patterns, decreased respiration and increased (aerobic) lactate production all occur in embryos produced in vitro (for a review, see Barnett and Bavister, 1996). However, glycolytic activity close to the in vivo situation is not necessarily associated with an increase in embryo viability (Leppens et al., 1996). One of the main differences between in vivo- and in vitro-derived ruminant embryos lies in the accumulation of lipids, mainly triglycerides, during embryo culture in the presence of serum (Ferguson and Leese, 1999). However, current techniques for evaluating embryonic lipid metabolism are invasive and could thus not be used for selecting embryos before transfer. Amino acid uptake is also about 30% lower in IVP bovine embryos cultured in serum-free conditions compared with in vivo-derived (Partridge and Leese, 1996).

Differences may exist in human chorionic gonadotropin (hCG) secretion between in vivo and in vitro derived human embryos: daily hCG production during the first week of implantation is 100-fold higher in vivo compared with the output at equivalent times in vitro (Lopata and Oliva, 1993a). IVF mouse embryos release less platelet activating factor (PAF) than those fertilized in situ. Moreover, the amount of PAF activity released by embryos cultured in vitro is reduced as development proceeds (reviewed by O'Neill, 1998).

2.3.3. Variations with culture conditions

Substrate concentrations in the culture medium influence embryo metabolism. For example, pyruvate uptake is proportional to pyruvate concentration (bovine: Eckert et al., 1998; human: Conaghan et al., 1993) and lactate oxidation is proportional to lactate concentration (mouse: Lane and Gardner, 2000). Pyruvate concentration also affects glucose uptake (Butler and Williams, 1990). Lactate and pyruvate each affect the metabolism of the other (Leese et al., 1986; Lane and Gardner, 2000). The addition of glucose to culture media decreases oxygen uptake by bovine morulae/blastocysts and increases lactate production without affecting blastocyst quality (Donnay et al., unpublished observations). The oxidative metabolism of in vitro produced bovine embryos is altered by the availability of exogenous protein sources in the culture medium. Replacing protein with PVA led to a significant increase in pyruvate uptake (Eckert et al., 1998). However, the production of CO_2 from pyruvate through the Krebs cycle is decreased (Lee et al., 1998), as is oxygen uptake (Eckert et al., 1998). The rate of glycolysis is also lower in bovine embryos cultured in protein-free medium and closer to that reported for in vivo embryos (Krisher et al., 1999).

The composition of the culture medium also influences the secretion of factors such as Pregnancy-specific ß1-glycoprotein (SP-1) (human: Juriscova et al., 1999); Platelet activating factor (PAF) (mouse: Wells and O'Neill, 1992) and Interferon tau (IFN-τ) (bovine: Stojkovic et al., 1995).

3. Non-invasive techniques of evaluation

3.1. General aspects

Several techniques are available for the evaluation of embryo metabolism. We can distinguish two major categories:

1. Techniques based on the evaluation of changes in the composition of the culture medium; specifically the uptake or release of compounds such as energy substrates, oxygen, hormones, growth factors and cytokines.

2. Techniques based on the use of radiolabelled or fluorescent substrates or dyes; for example: the production of ^3H_2O or $^{14}CO_2$ from the catabolism of radiolabelled glucose, pyruvate, glutamine and acetate.

Only the first category can be considered as completely non-invasive: the embryos are usually cultured in the same medium as for normal culture, without any additive. The main change in their environment is the reduction of the number of embryos cultured together. This is known to have an impact on the development and quality of the resulting embryos, at least at some embryonic stages (Gardner et al., 1994; Donnay et al., 1997). However, it has been shown not to interfere with human embryo viability (Conaghan et al., 1993). The main disadvantages are the delicacy of the measurement technique and the lack of automatisation and the inability to measure specific metabolic pathways.

The evaluation of metabolic pathways is important, as substrates can have several functions. For example, pyruvate is considered as a major energy substrate for most mammalian adult as well as embryonic cells. However, pyruvate can play other important roles. For example, it can react with hydrogen peroxide and protect the embryo from oxidative stress. This has been demonstrated in the mouse (O'Fallon and Wright, 1995) and bovine (Morales et al., 1999). Pyruvate can also be converted into alanine, which may serve as a means of removing ammonia from the early embryo (Fridhandler, 1968; Wales and Whittingham, 1970; Butcher et al., 1998; Donnay et al., 1999). Another example of nutrients having pleiotropic effects are amino acids which benefit embryo development by several mechanisms: they can act as precursors for protein synthesis, energy substrates, osmolytes, antioxidants and help maintain intracellular pH (for a review, see Barnett and Bavister, 1996).

The second category of techniques are often more sensitive and allow for the evaluation of some metabolic pathways. However, the addition of labelled substrates that can be toxic to cells or the exposure of embryos to radioactivity or UV light that could impair further embryonic development (Goto et al., 1993) limit the application of such techniques before transfer, especially in the human.

3.2. Evaluation Of Changes In The Composition Of The Culture Medium

The principle is simple: single embryos are incubated in small volumes of culture medium and changes in the concentration of compounds are evaluated. Control drops not containing embryos are incubated in parallel. The main technical problems encountered with these techniques are:

- The adjustment of culture volume and the duration of incubation. In order to evaluate precisely changes in the concentration of a given compound, differences due to embryo metabolism must ideally represent around 10% of the initial concentration (for the accuracy of measurement) but the decrease (or increase) in the compound concentration cannot be too great, or it may influence embryo metabolism itself. To fulfil those conditions, it is often necessary to decrease the volume of the culture droplets. Another possibility is to decrease the initial concentration of the compound in the medium during the incubation period but care must be taken not to impair embryo metabolism (Gardner and Leese, 2000). At the end of the incubation period, embryos are recovered. The culture medium can then be analysed immediately or frozen until assay. It is also possible to take serial samples of medium during the incubation period. In this case, the volume of the sample has to be negligible in comparison with the volume of the culture droplet. To evaluate pyruvate or glucose uptake and lactate release by single bovine blastocysts, an incubation period of 12h in 1-µl droplets or of 24h in 2-µl drops is required.

- The adaptation of techniques of measurement to small volumes of medium. Well established biochemical enzymatic reactions are used to evaluate substrate concentrations. This is the case, for example, for glucose and lactate evaluations. Automated analytical systems may be used, but these are not adapted to small volumes, and most techniques require miniaturisation to allow for the analysis of very small volumes. This means that the procedures often have to be performed manually by well-trained individuals, which reduces the number of samples which can be analysed in a single run, and increases the cost. Moreover, special equipment is often required (calibrated nanopipettes, photometer and photomultiplier attachments to a fluorescence microscope, etc.).

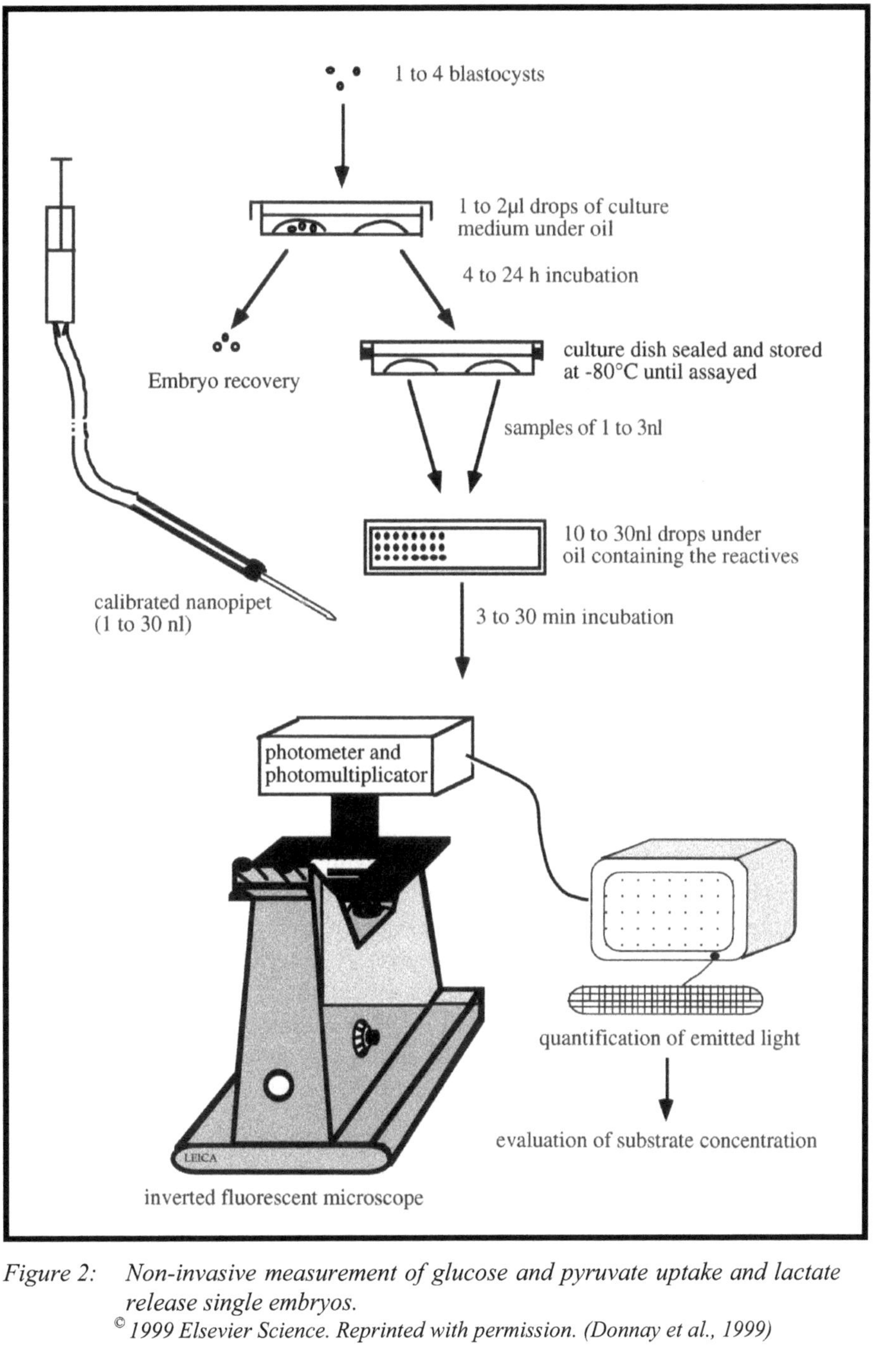

Figure 2: *Non-invasive measurement of glucose and pyruvate uptake and lactate release single embryos.*

3.2.1. Measurement of uptake/release of glucose, lactate, pyruvate

The assays for measuring glucose, lactate and pyruvate concentrations are based on the generation or consumption of the reduced pyridine nucleotides NADH and NADPH in coupled enzymatic reactions. These nucleotides fluoresce when excited with light at 340 nm, whereas the oxidized forms, NAD^+ and $NADP^+$ do not. Fluorescence is measured directly in small droplets (10 to 30 nl) of reagents using a photometer and photomultiplier attached to a fluorescence inverted microscope. One to 3 nl samples of spent and control culture medium are taken using homemade capillary pipettes previously calibrated (Gardner and Leese, 1993 - Figure 2).

Using larger volumes, Renard et al. (1980) showed that day 10 in vivo-derived bovine blastocysts with a glucose uptake superior to 5 µg/h gave rise to more pregnancies after transfer than those with a glucose uptake below 5 µg/h. Gardner and Leese (1987) performed a similar study on individual mouse blastocysts: pregnancy to term was correlated with a higher glucose uptake in culture. However, parthenogenetic embryos also showed a significantly higher glucose uptake at the blastocyst stage compared to fertilized embryos (Hardy and Handyside, 1996).

Using glucose conversion to lactate to quantify glycolysis, it was possible to identify prospectively from a population of morphologically identical IVP mouse blastocysts, high quality embryos which would develop after transfer from those embryos which had little, if any, developmental potential (Lane and Gardner, 1995; 1996). Embryos with low levels of glycolysis corresponding to levels observed in in vivo grown embryos, were shown to have high viability (80% foetal development). On the contrary, embryos showing a high lactate production gave an inferior foetal development (10%). High glycolytic activity could reflect the level of stress resulting from embryo culture (Leese et al., 1998). However, in another study, glycolytic activity comparable to the in vivo situation was not associated with an enhancement of viability after transfer of mouse blastocysts (Leppens et al., 1996).

Pyruvate uptake by fertilized or activated human oocytes was significantly higher than by unfertilized oocytes (Devreker et al., 2000): fertilization and activation increased pyruvate uptake by the human embryo by approximately 30%. This was previously shown in the mouse embryo (Gardner and Leese, 1996 ; Hardy and Handyside, 1996). Pyruvate uptake was higher in human embryos developing to the blastocyst stage than those arresting at cleavage stages (Hardy et al., 1989). However, in contrast with these results pyruvate uptake was significantly lower for cleavage stage embryos from women who became pregnant after embryo transfer, compared with those who did not (Conaghan et al., 1993). High pyruvate uptake could be due to uncoupling of mitochondria, making the inner mitochondrial membrane permeable to protons. The small differences within groups of embryos from individual patients make embryo selection on the basis of pyruvate uptake alone impossible. Moreover, a similar study on embryos derived from natural cycles (one embryo per transfer) failed to demonstrate a significant difference in pyruvate uptake post fertilization in embryos which did or did not succeed in establishing a pregnancy after transfer (Turner et al., 1994). However, the variation of pyruvate uptake in the non-pregnant patients was significantly greater than in the pregnant group. Morphologically good embryos were thus more likely to implant if they demonstrated an intermediate pyruvate uptake. In another study, analysis of oocyte and early embryo pyruvate and glucose uptake for 48 h before transfer did not improve the selection of viable human embryos (Devreker et al., 2000). Analysis of nutrient uptake during a longer period or after the onset of the embryonic genome (for example up to the blastocyst stage) may prove to be a more successful approach (Devreker et al., 2000).

3.2.2. Measurement of amino acids uptake / release

The spent culture droplets can also be used to evaluate amino acid uptake / release following analysis by reverse-phase HPLC (Lamb and Leese, 1994; Partridge and Leese, 1996; Donnay and Leese, 1999). Samples of spent culture medium are diluted and amino groups reacted with o-phthaldialdehyde to yield highly fluorescent products. Amino acids are then separated and quantified using reverse-phase HPLC on an ODS-18 column connected to a fluorescence detector. Although in vitro culture can influence amino acid uptake (Partridge et al., 1996), no data are available on a possible correlation between amino acid uptake / release and embryo viability.

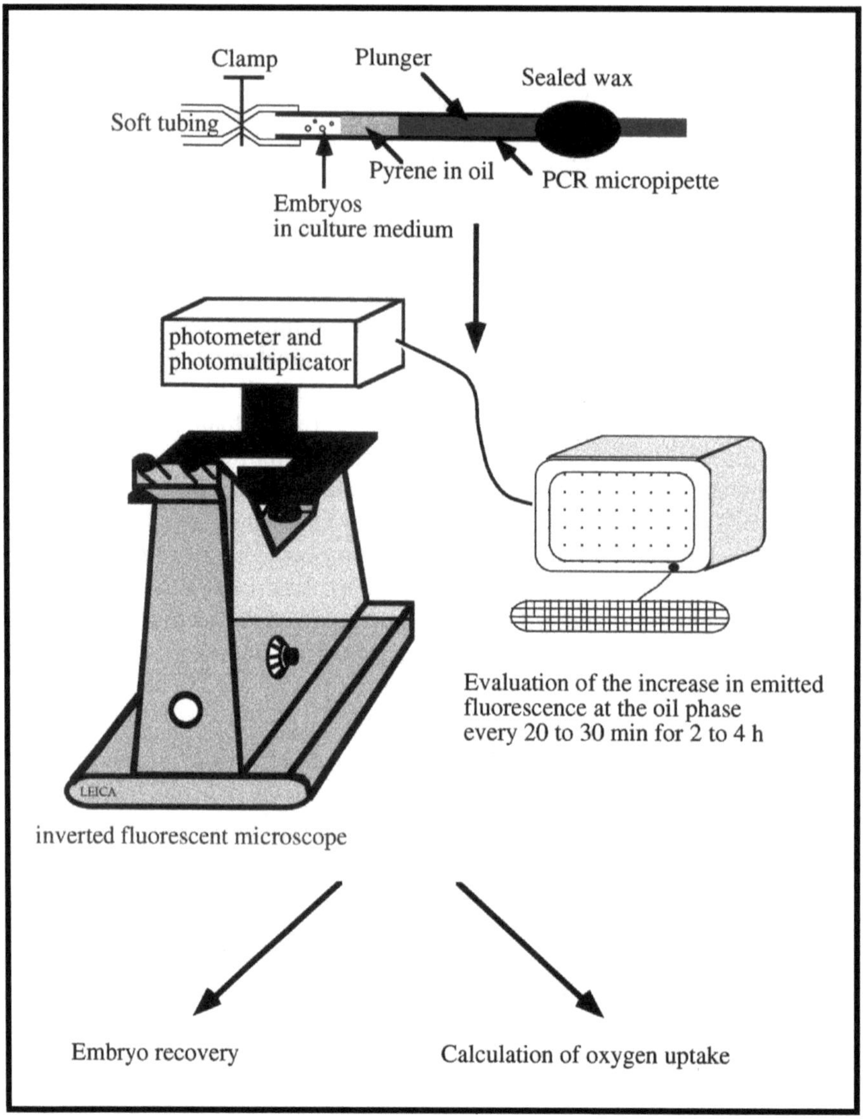

Figure 3: Non-invasive measurement of oxygen uptake by embryos
© 1999 Elsevier Science. Reprinted with permission (Donnay et al., 1999)

3.2.3. Measurement of oxygen uptake

Several techniques can be used for measuring oxygen uptake, including microelectrodes (Overström et al., 1992, 1996, 1998), microspectrophotometry (Magnusson et al., 1977, 1986) or microfluorescence (Houghton et al., 1996). Magnusson et al. (1977, 1986) used a sensitive micro-spectrophotometric technique allowing the evaluation of single oocytes or blastocysts incubated in a small (12.3 nl) gas-tight cylindrical cuvette containing oxyhaemoglobin. Seshagiri and Bavister (1991) used micro-electrodes inserted into glass vials containing 50 to 100 embryos. Overström et al. (1992, 1996, 1998) developed a multichannel apparatus using a high-resolution gas-sensing microoxygen electrode allowing for the evaluation of oxygen uptake simultaneously by 10 individual embryos in 1-µl chambers. Embryo respirometry was conducted over a period of 20 to 40 min. Using this technique, the only published study relating viability after transfer and oxygen uptake was performed on a small scale on in vivo bovine blastocysts (Overström et al., 1992). Slightly higher pregnancy rates (36% vs 27%) were observed when oxygen consumption was higher than 2 nl/embryo/h.

Another technique involves pyrene, an oil-soluble fluorescent compound, the fluorescence of which is quenched in proportion to the oxygen concentration (Houghton et al., 1996). It has been applied to groups of at least 2 embryos incubated in a 2-µl volume of Hepes-buffered culture medium in a 5-µl PCR micropipette in contact with 1µl paraffin oil containing the pyrene (Figure 3). The pipette is sealed at both ends and the fluorescence is measured at the oil-phase every 20 to 30 minutes for 2 to 4 h. The increase in fluorescence is proportional to the oxygen uptake by the embryos. This technique has been shown not to interfere with bovine blastocyst expansion and hatching (Donnay and Leese, 1999) and could be adapted for the evaluation of single embryos (J.G. Thompson, personal communication). However, embryos have to be loaded into the pipettes and are exposed to variations in temperature during the measurements. The embryos are not directly exposed to UV light, as the readings are taken in the oil phase, but reduced oxygen concentration cannot be used during the procedure and the incubation medium has to be Hepes-buffered to maintain [pH]i in the absence of CO_2 in the gas phase.

This technique was used to evaluate oxidative metabolism in mouse (Houghton et al., 1996), bovine (Thompson et al., 1996) and human embryos (Butcher et al., 1998). It was successfully combined with the evaluation of glucose, pyruvate, lactate and amino acids uptake / release on the same embryos and during the same incubation period (Butcher et al., 1998; Donnay and Leese, 1999). It has not yet been applied to correlate oxygen uptake with viability after transfer. It was however recently used to compare the metabolism of in vitro-produced bovine embryos before and after cryopreservation using conventional freezing method or vitrification (Kaidi et al., 2001). In this study, oxygen uptake, as well as the uptake of glucose and pyruvate, reflected embryonic cell number, and particularly the number of trophectoderm cells.

Recently, a new sensitive technique using self-referencing electrodes allowed non-invasive measurements of oxygen uptake by individual embryos by detecting differences between the composition of the medium immediately surrounding cells relative to that of the bulk medium (Portefield et al., 1998; Trimarchi et al., 2000). Several embryos can be evaluated at the same time and only 60 min were required to complete the measurements from a set of embryos. Viability of mouse embryos exposed to the conditions of measurements was not affected (Trimarchi et al., 2000). Using this technique it was demonstrated that healthy control mouse one-cell embryos and blastocysts consumed more oxygen than did those undergoing cell death (Trimarchi et al., 2000).

Measuring oxygen uptake allows for the direct evaluation of oxidative metabolism from both endogenous and exogenous substrates. Another advantage is that the results can be obtained rapidly. However, oxygen uptake can be affected by uncoupling of mitochondria and does not reflect only oxidative phosphorylation (OXPHOS). Moreover, the proportion of oxygen consumption by OXPHOS-dependent and independent processes could change during preimplantation development; in the mouse and rabbit blastocysts, roughly 70% of the oxygen taken up is used through OXPHOS (Benos and Balaban, 1980; Trimarchi et al., 2000). The remaining 30% could be used by H_2O_2-producing oxygenases in the rabbit blastocyst and may play a role during implantation (Manes and Lai, 1995). However, at the early cleavage stages only 20 to 30% of oxygen seems to be used through OXPHOS (Trimarchi et al., 2000).

3.2.4. Release of enzymes into the culture medium

The measurement of L-lactate-dehydrogenase (LDH) release into the culture medium is routinely used to evaluate the proportion of degenerating or dead somatic cells in vitro. The test is very sensitive due to the use of bioluminescent substrates (Brolin, 1977) and can be completed in 5 to 6 min after the initial incubation (Johnson et al., 1991). Due to the presence of LDH in serum, incubation has to be performed in the absence of serum supplementation. Renard et al. (1982) and Johnson et al. (1991) evaluated LDH release by in vivo cattle embryos. Embryos that failed to develop in vitro released more enzymes into the surrounding culture medium. The assay resulted in a complete separation of live and dead embryos, with no overlap between the two populations. However, embryos at a very early stage of development at the time of evaluation or which appear severely degenerated, released little LDH. This test has thus to be combined with morphological evaluation. Although the results were promising, this approach has not yet been used to select embryos prospectively for transfer (Gardner and Leese, 2000).

3.2.5. Release of hormones, cytokines or growth factors

Several factors are produced by the embryos which may act as autocrine or paracrine factors in embryo development, in recognition of pregnancy or in implantation (for a review, see Kane et al., 1997). Amongst these are hormones, growth factors and cytokines.

Human chorionic gonadotropin (hCG). Embryonic secretion of hCG is critical to maintain corpus luteum function and subsequent increases in oestrogen and progesterone in primates. It could also be involved in paracrine interactions with the endometrium (Levin et al., 1990). In human embryos, hCG expression is detected from the 2-cell stage onwards (Juriscova et al., 1999). Secretion by blastocysts starts around day 7 and peaks around day 10 (Dokras et al., 1991; Woodward et al., 1993). Only a subset of blastocysts seems to be able to produce detectable concentrations of hCG in vitro (around 50%, Dimitriadou et al., 1992; Woodward et al., 1994). hCG secretion is a time-dependent phenomenon and is independent of hatching (Woodward et al., 1993). Favourable culture conditions may increase hCG secretion (Turner and Lenton, 1996) but no correlation has been found between the rate of hCG transcription and cell numbers or cell death rates at the blastocyst stage (Juriscova et al., 1999). Moreover, output of hCG is not associated with cleavage rates or morphological appearance of the embryo (Woodward et al., 1994). In one study, cumulative hCG secretion by blastocysts derived from parthenogenic and tripronucleate embryos was significantly lower than controls; an observation which may reflect impaired function of the trophectoderm (Dokras et al., 1991). However, this finding was not confirmed in other studies (Lopata and Oliva, 1993 a and b).

Interferon tau (IFN-τ). IFN-τ is a major secretory protein produced by the trophectoderm in domestic ruminants and plays an important role as first signal for pregnancy recognition via its antiluteolytic effect (for a review, see Bazer et al., 1994). Although maximal IFN-τ cellular concentrations in the bovine are detected around day 16 of pregnancy (Farin et al., 1990), IFN-τ mRNA is present at day 8 in non-hatched blastocysts, which also release significant amounts of the protein into the culture medium (Hernandez-Ledezma et al., 1992a, 1993). The onset of IFN-τ secretion is coincident with the expansion of the blastocoel cavity and may be a reflection of the differentiation state of the trophectoderm, the cell layer to which IFN-τ expression is restricted. IFN- τ production can be detected by immunoassay or antiviral activity. There seems to be considerable variation in IFN-τ production between blastocysts. This might be related to the fact that if the assay is highly sensitive the accuracy may be low. Hatched day 8 blastocysts secrete significantly more IFN-τ than non-hatched embryos (Stojkovic et al., 1995). Moreover, high-quality hatched blastocysts secrete more IFN- τ than hatched blastocysts of poor quality (Hernandez-Ledezma et al., 1993). However, in another study no correlation was found between the amount of IFN-τ produced and the number of cells in the blastocyst or the quality of the embryo (Kubisch et al., 1998). IFN-τ production may be lower in blastocysts that have cleaved only once during the initial 44 h post-insemination suggesting a relationship between kinetics of development and protein synthesis or secretion. It has to be kept in mind that the kinetics of early cleavage in the bovine are related to the developmental potential of an embryo, at least to the blastocyst stage (Grisart et al., 1994). However blastocysts formed at day 7 or 8 produce less IFN-τ than those which form later, on day 9 or 10, although they were more viable after transfer. This could be explained if the production of IFN-τ, as for hCG, is initiated at a specific age, independent of developmental stage (Kubisch et al., 1998). On the other hand, group-cultured blastocysts secrete significantly more IFN- τ than those cultured individually (Larson and Kubisch, 1999). This might be related to the quality of the blastocyst, as it is well known that culturing bovine embryos in groups increases developmental rates and blastocyst cell number (Donnay et al., 1997).

Plasminogen activators (PAs). Plasminogen activators are serine proteases which convert plasminogen into the active serine enzyme plasmin. PAs participate in several reproductive functions including ovulation, implantation, tissue remodelling and cell migration during embryogenesis. PAs are produced during oocyte maturation in vitro in several species (reviewed by Park et al., 1999). Embryonic production of PAs has been detected in the rat, mouse, pig, cattle and sheep (Leidholm and Astedt, 1975; Strickland et al., 1976; Fazleabas et al., 1983; Menino and William, 1987; Menino et al., 1989; Zhang et al., 1996). In mouse embryos, PAs production is associated with trophoblastic penetration of the endometrium and with parietal endodermal cell migration. In swine, PAs production occurs when the embryo is undergoing tissue remodelling and cellular proliferation. Bovine blastocysts release PAs during blastocoel expansion and hatching and PAs production is positively correlated with the overall diameter and cell number attained by bovine embryos developing in vitro (Kaaekuahiwi and Menino, 1990). In ovine embryos, PAs production increases during the morula to blastocyst transition and remains elevated through blastocoel expansion and hatching (Menino et al., 1989). PAs production is also correlated with cell number of trophectoderm outgrowths in the sheep (Bartlett and Menino, 1995). Production of PAs by ruminant blastocysts could play a role in hatching, by increasing zona pellucida solubility, and in cell migration. In these species, PAs are secreted by the endoderm into the blastocoel, and by the trophectoderm cells into the external environment (Berg and Menino, 1992). A similar pattern was observed in mouse embryos (Marotti et al., 1982). The technique used for evaluation of PAs is based on the lysis of casein. It is sensitive enough to measure the production of PAs by single embryos after 24h incubation in 25-μl culture drops (Kaaekuahiwi and Menino, 1990) but, to the best of our knowledge, has not been applied to evaluate blastocyst quality prospectively.

<u>Prostaglandins</u>. Prostanoids production by embryos occurs in a number of species and may play a role in the regulation of embryonic development, maternal immunological recognition of pregnancy and the maintenance of a suitable hormonal environment. In humans, the production of PGE_2 and $PGF_{2\alpha}$ has been reported (Shutt and Lopata, 1981; Holmes et al., 1990) but thromboxane was the only prostanoid consistently found in media conditioned by ova or embryos, although its concentration was not correlated with pregnancy outcome (van der Weiden et al., 1994). However, the embryonic origin of the prostanoid was in doubt as maternal serum was added to the culture media. In bovine embryos, PGE_2 is secreted during the first 24h of maturation and 6h after fertilization, whereas $PGF_{2\alpha}$ is undetectable. There is a dramatic enhancement in PGE_2 and $PGF_{2\alpha}$ secretion 48h after fertilization. At 72h post-fertilization, PGE concentrations were much lower and $PGF_{2\alpha}$ undetectable (Gurevich and Shemesh, 1994).

<u>Interleukins</u>. A high level of interleukins production (IL1 and IL6) occurs during IVF in human embryos. This level is then reduced (Zolti et al., 1991; Austgulen et al., 1991). In a study by Sheth et al. (1991), all cycles with an IL1 level > 60 pg/ml resulted in a pregnancy. However, no correlation was found in another study (Austgulen et al., 1991) between IL1 or IL6 concentrations and morphological score or cell number. In the studies of Sheth et al. (1991) and Austgulen et al. (1991), measurements were

performed during IVF or during the first cleavage and the presence of cumulus cells and spermatozoids could have influenced cytokine production.

Immunosuppressive activity. Clark et al. (1989) found a significant correlation between the percentage of immunosuppressive activity in culture media and implantation after human embryo transfer. The data indicated that no successful pregnancy occurred in the absence of immunosuppressive activity. The combination of measurement of interleukin-1-α and immunosuppressive factors in embryo culture fluid from the beginning of IVF was shown to predict 100% of pregnancies following IVF. However, the analysis was performed on pooled culture media of all embryos from an individual patient (Sheth et al., 1991). Moreover, detection of immunosuppressive activity took 6 days to perform and had poor repeatability. Amongst the potent immunosuppressive factors secreted by embryos are Platelet Activating Factor (PAF) and Pregnancy Specific ß-1 glycoprotein SP1 (SP-1).

Platelet Activating Factor (PAF). Preimplantation embryos of mice (O'Neill, 1985), humans (Collier et al., 1988, 1990) and sheep (Battye et al., 1991) secrete PAF which can be detected in culture media. PAF affects uterine prostaglandin secretion (Battye et al., 1992). The secretion of PAF by human embryos showed a positive correlation with their morphological quality, kinetics of cleavage and pregnancy potential following embryo transfer (O'Neill and Saunders, 1984; O'Neill et al., 1987; Collier et al., 1990; Punjabi et al., 1990, Ammit and O'Neill, 1991). However, PAF production cannot alone predict pregnancy as 13% of patients who became pregnant received embryos that did not produce PAF in vitro (Collier et al., 1990). Moreover, a number of embryos secreted substantial amounts of PAF in vitro but did not give rise to pregnancy following transfer. PAF detection can be performed by bioassay or radioimmunoassay after extraction of the spent culture medium (Ammit and O'Neill, 1991) but seems to yield variable results (Ammit and O'Neill, 1991; Ammit et al., 1992).

Pregnancy-specific ß1 glycoprotein (SP-1). Secretion of SP-1 by the human embryo is thought to prevent rejection of the foetus by the maternal immune system. SP-1 can be detected in human embryo culture medium from day 1 following IVF, which is earlier than for hCG (Saith et al., 1996). There is a trend towards higher daily mean cumulative SP-1 concentrations with increasing morphological grade but no significant difference was observed with poor quality blastocysts. However, vacuolated morulae secreted less SP-1 than blastocysts of the same age (Saith et al., 1996). No relationship was found with the in vitro developmental potential of human embryos

(Dimitriadou et al., 1992); a puzzling observation since the higher percentage of blastulation observed in KSOM was associated with a higher secretion of SP-1 (Juriscova et al., 1999).

Tumor Necrosis Factor alpha (TNF-α). TNF-α is produced by the embryos of several mammalian species (Zolti et al., 1991; Witkin et al., 1991; Gurevich and Shemesh, 1994). In the bovine, TNF-α is secreted during oocyte maturation and 6h after fertilization. There is a dramatic enhancement in TNF-α secretion 48h after in vitro fertilization and high TNF-α activity continues to be detected after 72h (Gurevich and Shemesh, 1994). In humans, TNF-α was only detected in the culture medium at the 6 to 8-cell stage (Zolti et al., 1991).

Colony Stimulating Factor-1 (CSF-1). CSF-1 may promote embryonic development (Pampfer et al., 1991; Bhatnagar et al., 1995). Its secretion by human embryos in culture increases gradually from the oocyte stage onwards, reaching a peak at the 6 to 8-cell stage (Zolti et al., 1991).

Transforming Growth Factor-ß (TGF-ß). TGF-ß is produced throughout the culture period by human embryos. However, no correlation was found between TGF-ß concentrations and morphological score or cell number (Austgulen et al., 1991).

Other growth factors expressed in preimplantation embryos of several species, include: EGF, TGF-alpha, IGF-I, IGF-II, FGF-4, bFGF, PDGF, TGF-beta, LIF and growth hormone (for a review, see Kane et al., 1997). Most of those factors could act as autocrine or paracrine factors. However their evaluation in embryo culture media as markers of embryo viability has yet to be performed. Other molecules are known to be produced by preimplantation embryos at specific stage of development and might be used in the future to assess developmental potential. Amongst them are retinoic acid (Parrow et al., 1998) and nitric oxide (Gouge et al., 1998).

3.3. Metabolism of labelled substrates

3.3.1. Evaluation of metabolic pathways using radiolabelled substrates

The technique first described by O'Fallon and Wright (1986) was adapted by Rieger to study single bovine embryo metabolism (Rieger and Guay, 1988; Rieger et al., 1992). It allows for the evaluation of metabolic pathways of several substrates such as glucose, glutamine, or pyruvate. Embryos are incubated in the presence of one or two labelled substrates (one labelled with ^{14}C, one labelled with ^{3}H) in a drop of culture medium hanging under the cap of a conical tube containing bicarbonate. Radioactive CO_2 and H_20 produced during the metabolism of the labelled substrates are trapped in the bicarbonate solution. After a 3h incubation period, the embryo is recovered and the radioactivity in the bicarbonate solution is counted after addition of NaOH to convert the dissolved CO_2 and bicarbonate into carbonate (Figure 4). Using this technique, it was demonstrated that the metabolism of glucose is markedly higher in morphologically normal than in degenerate bovine embryos (Rieger, 1984) but it remained to be demonstrated if it can improve the selection made on the basis of morphology alone.

The main advantages of the technique are its high sensitivity and the ability to evaluate the metabolic pathways of chosen substrates. Viable normal offspring were obtained after transfer of embryos after evaluation (O'Fallon and Wright, 1986). However, this technique uses radioactive substrates that could impair further embryonic development or quality and could thus not be applied in the human. Moreover, the technique does not allow for evaluating the use of endogenous stores such as glycogen, protein and lipids.

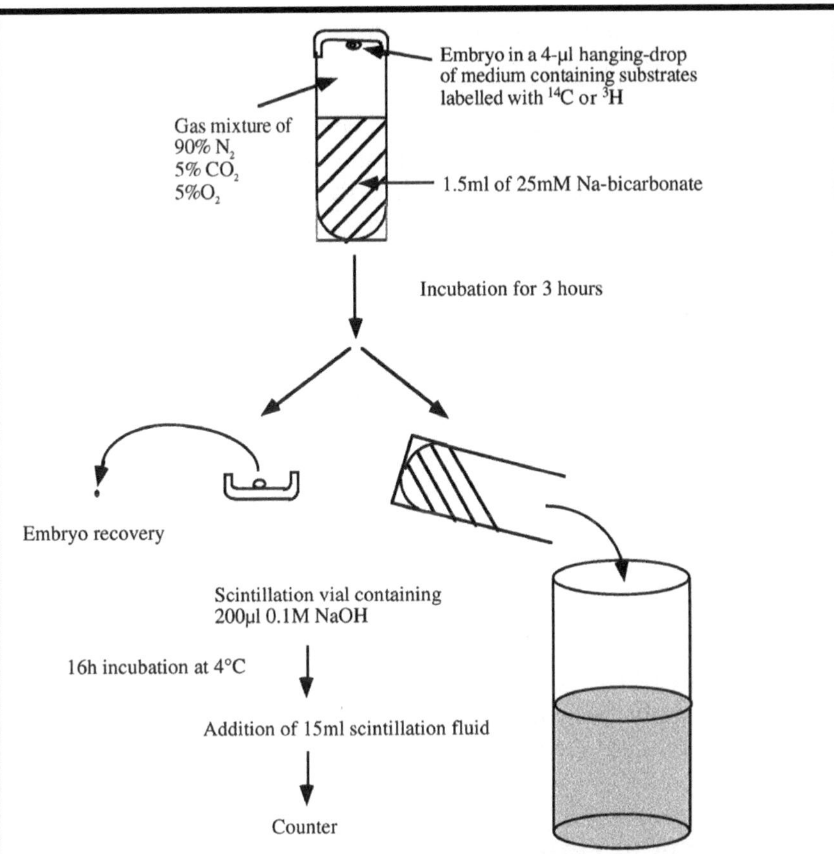

<u>Examples of labelled substrates</u>

D-[1-^{14}C] glucose	→	Production of $^{14}CO_2$ via the pentose-phosphate pathway or the Krebs cycle
D-[6-^{14}C] glucose	→	Production of $^{14}CO_2$ via the Krebs cycle
D-[5-^{3}H] glucose	→	Production of $^{3}H_2O$ via the glycolysis
L-[3,4-^{3}H] glutamine	→	Production of $^{3}H_2O$ via the Krebs cycle
[2-^{14}C] pyruvate	→	Production of $^{14}CO_2$ via the Krebs cycle

Figure 4: Evaluation of metabolic pathways using radiolabelled substrates.

3.3.2. Metabolism of fluorescent substrates

The fluorogenic compound diacetylfluoresceine (FDA) can enter cells where endogenous esterases transform FDA to the highly fluorescent compound fluorescein. If cell membranes are intact, fluorescence is restricted to the cytoplasm and can be detected under a fluorescence microscope after UV excitation. The intensity of the fluorescence depends both on esterase activity and the presence of an intact cell membrane. Schilling and Doepke (1978) and Schilling et al. (1979) examined the possible use of FDA to assess embryo viability in the rabbit and cow. More detailed studies were performed by Mohr and Trounson in the mouse (1980). The technique was also applied to human embryos (Mohr et al., 1983). The measurement of the intensity of fluorescence exhibited by fluorescein and its derivatives may also be used to evaluate intracellular pH (Butler and Biggers, 1989).

The main disadvantage is that this technique measures only basic cell function and not developmental potential. Moreover, embryos are exposed to UV light, which is known to be detrimental for their further development (Goto et al., 1993). Using a related compound, biscarboxyethyl-carboxy-fluorescein (BCECF), it was found that fragmented mouse embryos fluoresced strongly, although they were not viable (Butler, Biggers and Lechene, unpublished, cited by Butler and Bigger, 1989).

4. Evaluation of embryo viability after embryo manipulation

4.1. After cryopreservation

Several studies have used metabolic parameters to evaluate the damage induced by cryopreservation. As freeze/thaw embryonic damage is often associated with plasma membrane disruption, measurement of LDH in the medium in which the embryos were frozen might indicate the degree to which the cells were lysed during freezing and thawing (Johnson et al., 1991). Rieger et al. (1993) showed that bovine blastocysts that subsequently hatched after thawing had a greater glycolytic activity than those that did not hatch. Using glucose uptake and lactate production by embryos after thawing, it is possible to identify blastocysts capable of re-expanding their blastocoel (Gardner et al., 1996; Eckert et al., 1998). The same is the case for CO_2 production from glucose oxidation (Khurana and Niemann, 2000 b). However, nutrient uptake or utilisation could not be used prospectively to identify embryos which would survive freezing and thawing (Gardner et al., 1996; Khurana and Niemann, 2000 b). Similarly, glucose uptake was significantly higher for mouse embryos that continued development after freezing/thawing to the blastocyst stage than for those that did not. A bimodal distribution can be plotted which suggests that a preliminary decision-making rule for subsequent development can be formulated based on embryo's ability to take up glucose in vitro (Williams and Butler, 1993).

4.2. After biopsy

It has been shown in the marmoset that partial removal of trophectoderm cells affects chorionic gonadotrophin secretion and subsequent implantation (Summers et al., 1988). Such a drastic approach will of course affect all parameters related to the number of trophectoderm cells. It remains to be determined if a threshold can be set for metabolic parameters to distinguish between embryos that will re-expand and implant after biopsy and embryos that will not. The biopsies themselves, performed for Preimplantation Genetic Diagnostics (PGD) in humans or for Marker Assisted Selection (MAS) in farm animals, may be used for evaluating the developmental potential of an embryo. Development of isolated biopsied blastomeres is related to the in vitro developmental potential of the remaining embryo (Geber and Sampaio, 1999). However, this technique necessitates the cryopreservation of the embryo before transfer. It will probably be easier to evaluate embryo metabolism or the expression of developmentally related genes in the biopsied cells, assuming that these are representative of the whole embryo. For example, improved preimplantation development was observed in mouse embryos in which early expression of IGF-I was detected on a blastomere, removed by biopsy (Kowalik et al., 1999).

4.3. After nuclear transfer (NT)

Few data are available on the metabolism of NT embryos evaluated non-invasively. It has been shown that embryos derived in vivo or by conventional in vitro methods secrete significantly more interferon tau than embryos derived following nuclear transfer and embryo splitting (Stojkovic et al., 1999).

5. Conclusions

Although several techniques are available for evaluating embryo metabolism in vitro, few data are available on their use as markers of embryo viability. This is quite surprising as the potential use of such metabolic markers was reviewed by Whittingham as early as 1978.

Several explanations can be proposed:

Related to the techniques of evaluation:

- The delicacy and cost of the experimental techniques necessitate special equipment and well-trained people.

- The lack of sensitivity of some techniques renders them inapplicable to single embryos.

- Few techniques can be considered as wholly non-invasive. Moreover, the degree of invasiveness that can be tolerated may change depending on the specific application. For example, the introduction of a fluorescent substance into an embryo and its subsequent exposure to UV light may be acceptable for use with domestic animals but unacceptable for use with human embryos (Butler and Biggers, 1989).

Related to the organisation in an IVF lab:

- Metabolism evaluation has to be co-ordinated with the transfer of fresh embryos since cryopreservation is known to affect embryo survival, especially for IVP embryos.

- Most tests are too time-consuming for routine use in a busy clinical programme (Devreker and Englert, 2000).

Related to the embryo:

Variations of embryo metabolism with stage and culture conditions mean that metabolic parameters have to be established for each system. Indeed, the challenge is to detect metabolic changes which relate to alterations in embryo development.

Related to the results obtained with the first pilot studies:

Even though some metabolic parameters differed significantly between viable and non-viable embryos, the main problem remains the absence of a clear cut-off for a given parameter between developmentally competent embryos and embryos which will arrest. Thus a threshold has to be defined that will either eliminate some viable embryos or retain some non-viable embryos, depending on the application. On the other hand, most of the metabolic parameters described in this review concern the activity of the trophectoderm cells and are correlated with the number of these cells (Conaghan et al., 1993; Saith et al., 1996; Krisher et al., 1999; Khurana and Niemann, 2000; Kaidi et al., 2001). This introduces a double bias as (i) a correct ratio between ICM and TE-Cells is necessary for establishing a viable pregnancy; (ii) cell number does not correlate with cell functions and the number of cells of a blastocyst (or the number of TE-Cells) does not reflect the viability of the embryo after transfer.

In the future, embryo-derived factors that have yet to be evaluated could be tested as markers of embryo health. However, it is unlikely that measurement of a single "key" cellular parameter will predict the survival of an embryo unequivocally (Overström, 1998; Donnay et al., 1999). In the future, a combination of several metabolic factors, or of metabolic factors with other parameters such as morphology and the kinetics of development could be used (Figure 5 - Turner et al., 1994). This combination should allow different cellular activity / process to be measured within the same embryo. The development of sensitive techniques allowing for the simultaneous, rapid, and if possible automated, evaluation of several parameters during embryo culture is likely to be a prerequisite for the routine application of such tests in clinical IVF.

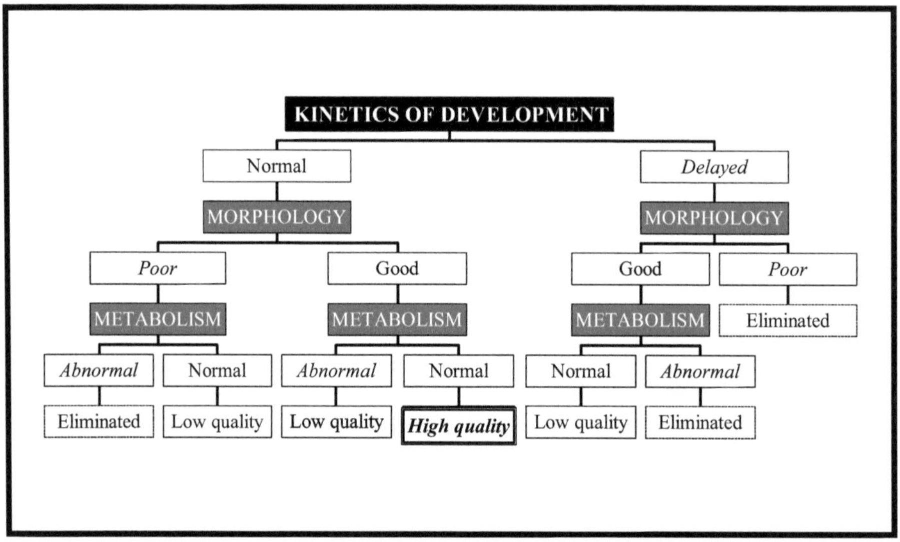

Figure 5: How to combine 3 different markers to select the best embryos for transfer: schematic representation.

Another approach could be to use methods that measure the ability of an embryo to respond to, or to survive, an external stimulus (Butler and Biggers, 1989); for example, the recovery of intracellular pH (Butler and Biggers,1989), the resistance to oxidative stress (Feugang et al., submitted) or the resistance to cryopreservation (Greve et al., 1992). Even a test that renders a proportion of embryos non-viable may be acceptable, at least in domestic livestock, if the selection achieved results in improved pregnancy rates (Butler and Biggers, 1989).

Acknowledgements

We are grateful to Prof. H. Leese for his critical review of the manuscript. The author is Chercheur qualifié du "Fonds national de la Recherche scientifique" and receives financial support from the "Direction générale de la Communauté française de Belgique", Action de recherche concertée n° 96/01-196 and from the European Commission, grant EU/QKL-CT-1999-00104.

References

Ammit AJ, O'Neill C. 1991. Comparison of a radioimmunoassay and bioassay for embryo-derived platelet-activating factor. Hum Reprod 6:872-878.

Ammit AJ, Wells XE, O'Neill C. 1992. Structural heterogeneity of platelet activating factor produced by murine preimplantation embryos. Hum Reprod 7:865-870.

Austgulen R, Arntzen KJ, Vatten LJ, Kahn J, Sunde A. 1995. Detection of cytokines interleukin-1, interleukin-6, transforming growth factor-ß and soluble tumour necrosis factor receptors in embryo culture fluids during in vitro fertilization. Hum Reprod 10:171-176.

Barnett DK. 1993. Sensitivity of hamster 2-cell embryos to glucose and phosphate is restricted to a brief period immediately before cleavage. Biol Reprod 48 (suppl) 1:176.

Barnett DK, Bavister BD. 1996. What is the relationship between the metabolism of preimplantation embryos and their developmental competence. Mol Reprod Dev 43:105-133.

Bartlett SE, Menino Jr AR.1995. Evaluation of extracellular matrices and the plasminogen activator system in sheep inner cell mass and trophectodermal outgrowth in vitro. Biol Reprod 52:1436-1445.

Battye KM, Ammit AJ, Evans G, O'Neill C. 1991. Production of platelet activating factor in vitro by preimplantation sheep embryos. J Reprod Fertil 93:507-514.

Battye KM, O'Neill CO, Evans G. 1992. Evidence that platelet-activating factor suppresses uterine oxytocin-induced 13, 14-dihydro-15-keto-prostaglandin F2 alpha release and phosphatidylinositol hydrolysis in the ewe. Biol Reprod 47:213-219.

Bazer FW, Ott Tl, Spencer TE .1994. Pregnancy recognition in ruminants, pigs and horses:signals from the trophoblast. Theriogenology 41:79-94.

Benos DJ, Balaban RS.1980. Energy requirements of the developing mammalian blastocyst for active ion transport. Biol Reprod 23:941-947.

Berg DA, Menino AR. 1992. Bovine embryos produce an urokinase-type plasminogen activator. Mol Reprod Dev 31:14-19.

Bhatnagar P, Papaioannou VE, Biggers JD. 1995. CSF-1, mouse preimplantation development in vitro. Development, 121:1333-1339.

Brinster RL. 1968. Carbon dioxide production from glucose by the preimplantation rabbit embryo. Exp Cell Res 51:330-334.

Brison DR, Leese HJ. 1991. Energy metabolism in late preimplantation rat embryos. J Reprod Fertil 93:245-251.

Brison DR, Leese HJ. 1994. The role of exogenous energy substrates in blastocoel fluid accumulation in the rat. Zygote 2:69-77.

Brolin SE .1977. Attempts to simplify the analytical performance in micro assay of metabolites with bacterial luciferase. Bioelectrochem Bioenerg 4:647-659.

Butcher L, Coates A, Martin KL, Rutherford AJ, Leese HJ. 1998. Metabolism of pyruvate by the early human embryo. Biol Reprod 58:1054-1056.

Butler JE, Biggers JD.1989. Assessing the viability of preimplantation embryos in vitro. Theriogenology 31:115-126.
Butler JE, Williams JE. 1990. Preliminary characterization of pyruvate uptake by one-cell ovine embryos. Theriogenology 33:423-432.
Clark DA, Lee S, Fishell S, Mahadevan M, Goodall H, Moye M, Schechter O, Stedronska-Clark J, Daya S, Underwood J, Craft I,, Mowbray J. 1989. Immunosuppressive activity in human in vitro fertilization IVF culture supernatants and prediction of the outcome of embryo transfer multicenter trial. J In Vitro Fert Embryo Transfer 6:51-57.
Collier M, O'Neill C, Ammit AJ, Saunders DM. 1988. Biochemical and pharmacological characterisation of human embryo-derived platelet activating factor. Hum Reprod 3:993-998.
Collier M, O'Neill C, Ammit AJ, Saunders DM. 1990. Measurement of human embryo-derived platelet-activating factor (PAF) using a quantitative bioassay of platelet aggregation. Hum Reprod 5:323-328.
Conaghan J, Hardy K, Handyside AH, Winston RML, Leese HJ. 1993. Selection criteria for human embryo transfer: a comparison of pyruvate uptake and morphology. J Assist Reprod Gen 10:21-30.
Devreker F, Hardy K, Van den Bergh M, Winston J, Biramane J, Englert Y. 2000. Noninvasive assessment of glucose and pyruvate uptake by human embryos after intracytoplasmic sperm injection and during the formation of pronuclei. Fertil Steril 73:947-954.
Devreker F, Englert Y. 2000. In vitro development and metabolism of the human embryo up to the blastocyst stage. Eur J Obst Gyn Reprod Biol 92:51-56.
Dimitriadou F, Phocas I, Mantzavinos T. 1992. Discordant secretion of pregnancy specific ß-1 glycoprotein and human chorionic gonadotropin by human pre-embryos cultured in vitro. Fertil Steril 57:631-636.
Dokras A, Sargent IL, Ross C, Gardner RL, Barlow DH. 1991. The human blastocyst: morphology and human chorionic gonadotrophin secretion in vitro. Hum Reprod 6:1143-1151.
Donnay I, Van Langendonckt A, Auquier P, Grisart B, Vansteenbrugge A, Massip A, Dessy F. 1997. Effects of co-culture and embryo number on the in vitro development of bovine embryo. Theriogenology 47:1549-1561.
Donnay I, Leese HJ. 1999. Embryo metabolism during the expansion of the bovine blastocyst. Mol Reprod Dev 53:171-178.
Donnay I, Partridge RJ, Leese HJ.1999. Can embryo metabolism be used for selecting bovine embryos before transfer? Reprod Nutr Dev 39:523-533.
Donnay I. Feugang JM, Bernard S, Marchandise J, Pampfer S, Dessy F. 2001. Impact of glucose on the development and metabolism and quality of in vitro produced bovine embryos from the morula to the blastocyst stage. Submitted.
Eckert J, Pugh PA, Thompson JG, Niemann H, Tervit HR. 1998. Exogenous protein affects developmental competence and metabolic activity of bovine pre-implantation embryos in vitro. Reprod Fertil Dev 10:327-332.
Farin CE, Imakawa K, Hansen TR, McDonnell JJ, Murphy CN, Farin PW, Roberts RM. 1990. Expression of trophoblastic interferon genes in sheep and cattle. Biol Reprod 43:210-218.

Fazleabas AT, Geisert RD, Bazer FW, Roberts RM. 1983. Relationship between release of plasminogen activator and estrogen by blastocysts and secretion of plasmin inhibitor by uterine endometrium in the pregnant pig. Biol Reprod 29:225-238.

Ferguson EM, Leese HJ. 1999. Triglyceride content of bovine oocytes and early embryos. J Reprod Fertil 116:373-378.

Feugang JM, Van Langendonckt A, Rees JF, Pampfer S, Dessy F, Donnay I. Delayed degeneration of bovine blastocysts induced by endogenous or exogenous oxidative stress: prevention by Trolox or ß-mercaptoethanol. Submitted

Fischer B, Bavister BD. 1993. Oxygen tension in the oviduct and uterus of rhesus-monkeys, hamsters and rabbits. J Reprod Fertil 99:673-679.

Flood MR, Wiebold JL. 1988. Glucose metabolism of preimplantation pig embryos. J Reprod Fertil 84:7-12.

Fridhandler L. 1968. Intermediary metabolic pathways in preimplantation rabbit blastocysts. Fertil Steril 19:424-434.

Gardner DK, Leese HJ. 1987. Assessment of embryo viability prior to transfer by the non-invasive measurement of glucose uptake. J. Exp Zool 242:103-105.

Gardner DK, Leese HJ. 1993. Assessment of embryo metabolism and viability. In: Handbook of in vitro fertilization, A. Trounson, DK Gardner editors, CRC Press, London, pp.195-211.

Gardner DK, Sakkas D.1993. Mouse embryo cleavage, metabolism and viability: role of medium composition. Hum Reprod 8:288-295.

Gardner DK, Leese HJ.1996. Non-invasive measurement of nutrient uptake by single cultured pre-implantation mouse embryos. Hum Reprod 1:25-27.

Gardner DK, Lane M. 1997. Culture and selection of viable blastocysts: a feasible proposition for human IVF? Human Reprod Update 3:367-382.

Gardner DK, Leese HJ. 2000. Assessment of embryo metabolism and viability. In: Handbook of in vitro fertilization, A. Trounson, DK Gardner edd, CRC Press, London, pp. 347-371.

Gardner DK, Land M, Spitzer A, Batt PA.1994. Enhanced rates of cleavage, development for sheep zygotes cultured to the blastocyst stage in vitro in the absence of serum and somatic cells: amino-acids, vitamins, culturing embryos in groups stimulate development. Biol Reprod 50:390-400.

Gardner DK, Pawelczynski M, Trounson AO. 1996. Nutrient uptake and utilization can be used to select viable day 7 bovine blastocysts after cryopreservation. Mol Reprod Dev 44:472-475.

Geber S, Sampaio M. 1999. Blastomere development after embryo biopsy: a new model to predict embryo development and to select for transfer. Human Reprod 14:782-786.

Gomez E. 1997. Acetate, ß-D-hydroxybutyrate as energy substrates during early bovine embryo development in vitro. Theriogenology 48:63-74.

Goto Y, Noda Y, Mori T, Nakano M. 1993. Increased generation of reactive oxygen species in embryo cultured in vitro. Free Radical Biol Med 15:69-75.

Gouge RC, Marshburn P, Gordon BE, Nunley W, Huet-Hudson, YM. 1998. Nitric oxide as regulator of embryonic development. Biol Reprod 58:875-879.

Greve T., Loskutoff NM, Buckrell BC, Christian CR, Leibo SP, Betteridge KJ. 1992. Morphology and freezing tolerance of in vitro derived bovine embryos after culture in vivo or in vitro. In: Flechon JE ed. 5ème colloque Franco-Tchécoslovaque sur la reproduction des animaux domestiques, Jouy-en-Josas, France p36.

Grisart B, Massip A, Dessy F. 1994. Cinematographic analysis of bovine embryo development in serum-free oviduct-conditioned medium. J Reprod Fertil 101:257-264.

Gurevich M, Shemesh M.1994. Induction of cyclooxygenase and prostaglandin E2 production by the bovine pre-embryo. Reprod Fert Dev 6:687-691.

Gutierrez-Adan A, Oter M, Martinez-Madrid B, Pintando B, de la Fuente J. 2000. Differential expression of two genes located on the X chromosome between male and female in vitro-produced bovine embryos at the blastocyst stage. Mol Reprod Dev 55:146-151.

Hardy K, Hooper MAK, Handyside AH, Rutherford AJ, Winston RML, Leese HJ. 1989. Non-invasive measurement of glucose and pyruvate uptake by individual human oocytes and preimplantation embryos. Hum Reprod 4:188-191.

Hardy K, Handyside AH. 1996. Metabolism and cell allocation during parthenogenetic preimplantation mouse development. Mol Reprod Dev 43:313-322.

Hernandez-Ledezma JJ, Sikes JD, Murphy CN, Watson AJ, Schultz GA, Roberts RM. 1992. Expression of bovine trophoblast interferon in conceptuses derived by in vitro techniques. Biol Reprod 47:374-380.

Hernandez-Ledezma JJ, Mathialagan N, Villanueva C, Sikes JD, Roberts RM.1993. Expression of bovine trophoblast interferon by in vitro-derived blastocysts is correlated with their morphological quality and stage of development. Mol Reprod Dev 36:1-6.

Hewitson LC, Leese HJ. 1993. Energy metabolism of the trophectoderm and inner cell mass of the mouse blastocyst. J Exp Zool 267:337-343.

Holmes PV, Sjogren A, Hamberger L. 1990. Prostaglandin-E2 released by pre-implantation human conceptuses. J Reprod Immunol 17:79-86.

Houghton FD, Thompson JG, Kennedy CJ, Leese HJ. 1996. Oxygen consumption and energy metabolism of the early mouse embryo. Mol Reprod Dev 44:476-485.

Johnson SK, Jordan JE, Dean RG, Page RD. 1991. The quantification of bovine embryo viability using a bioluminescent assay for lactate dehydrogenase. Theriogenology 35:425-433.

Jurisicova A, Antenos M, Kapasi K, Meriano J, Casper RF. 1999. Variability in the expression of trophectodermal markers ß-human chorionic gonadotrophin, human leukocyte antigen-G and pregnancy specific ß-1 glycoprotein by the human blastocyst. Hum Reprod 14:1852-1858.

Kaaekuahiwi MA, Menino AR. 1990. Relationship between plasminogen activator production and bovine embryo development in vitro. J Anim Sci 68:2009-2014.

Kaidi S, Bernard S, Lambert P, Massip A, Dessy F, Donnay I. 2001. Effect of conventional controlled-rate freezing and vitrification on morphology, metabolism of bovine blastocysts produced in vitro. Biol Reprod accepted for publication.

Kane MT, Buckley NJ. 1977. The effects of inhibitors of energy metabolism on the growth of one-cell rabbit ova to blastocysts in vitro. J Reprod Fertil 49:261-266.

Kane MT, Morgan PM, Coonan C. 1997. Peptide growth factors and preimplantation development. Hum Reprod 3:137-157.

Khurana NK, Niemann H. 2000a. Energy metabolism in preimplantation bovine embryos derived in vitro or in vivo. Biol Reprod 62:847-856.

Khurana NK, Niemann H. 2000b. Effects of cryopreservation on glucose metabolism and survival of bovine morulae and blastocysts derived in vitro or in vivo. Theriogenology 54:313-326.

Kowalik A, Liu HC, He ZY, Mele C, Barmat L, Rosenwaks Z. 1999. Expression of the insulin-like growth factor-1 gene and its receptor in preimplantation mouse embryos; is it a marker of embryo viability? Mol Hum Reprod 5:861-865.

Krisher RL, Land M, Bavister BD. 1999. Developmental competence and metabolism of bovine embryos cultured in semi-defined and defined culture media. Biol Reprod 60:1345-1352.

Kubisch HM, Larson MA, Roberts RM. 1998. Relationship between age of blastocyst formation and interferon-tau secretion by in vitro-derived bovine embryos. Mol Reprod Dev 49:254-260.

Lamb VK, Leese HJ. 1994. Uptake of a mixture of amino acids by mouse blastocysts. J Reprod Fertil 10:169-175.

Lane M, Gardner DK. 1995. Glycolytic activity can be used to select viable mouse embryos prior to transfer. Biol Reprod, suppl 52:137 abstr 321.

Lane M, Gardner DK. 1996. Selection of viable mouse embryos prior to transfer using metabolic criterion. Hum Reprod 11:1975-1978.

Lane M, Gardner DK. 2000. Lactate regulates pyruvate uptake and metabolism in the preimplantation mouse embryo. Biol Reprod 62:16-22.

Larson MA, Kubisch HM. 1999. The effects of group size on development and interferon-tau secretion by in vitro fertilized and cultured bovine blastocysts. Hum Reprod 14:2075-2079.

Lee ES, Pugh PA, Allen N, Fukui Y, Thompson JG. 1998. [5-3H]-glucose, [1-14C]-pyruvate utilization by fresh, frozen-thawed day 7 IVP bovine blastocysts cultured in PVA- or BSA-supplemented medium. Theriogenology 49:233.

Leese HJ.1995. Metabolic control during preimplantation mammalian development. Hum Reprod Update 1:63-72.

Leese HJ, Hooper MAK, Edwards RG, Ashwood-Smith MJ. 1986. Uptake of pyruvate by early human embryos determined by a non-invasive technique. Hum Reprod 1:181-182.

Leese HJ, Donnay I, Thompson J. 1998. Human assisted conception: a cautionary tale. Lessons from domestic animals. Hum Reprod 13 (suppl 4):184-202.

Leidholm P, Astedt B. 1975. Fibrinolytic activity of the rat ovum and appearance during tubal passage, disappearance at implantation. Int J Fert 20:24-26.

Lequarré AS, Grisart B, Moreau B, Schuurbiers N, Massip A, Dessy F. 1997. Glucose metabolism during bovine preimplantation development: analysis of gene expression in single oocytes, embryos. Mol Reprod Dev 48:216-226.

Leppens G., Gardner DK, Sakkas D. 1996. Co-culture of 1-cell outbred mouse embryos on bovine kidney epithelial cells: effect on development, glycolytic activity and inner cell mass: trophectoderm ratios and viability. Hum Reprod 11:598-603.

Levin JH, Tonetta SA, Lobo RA. 1990. Human chorionic gonadotrophin hCG enhances progestin stimulation of prolactin PRL production by human endometrial stromal cells in culture: evidence for trophoblast-endometrial paracrine interaction. 38th Annual Meeting of the pacific Coast Fertility Society, Prog. Suppl., Abstr. 7.

Lopata A, Oliva K. 1993a. Chorionic gonadotrophin secretion by human blastocysts. Hum Reprod 8:932-938.

Lopata A, Oliva K. 1993b. Regulation of chorionic gonadotropin secretion by cultured human blastocysts. In Bavister BD ed, Preimplantation Embryo Development. Springer Verlag, New York, pp 276-295.

Machaty Z, Abeydeera LR, Thompson JG, Day BN, Prather RS. 2000. Inhibition of oxidative phosphorylation and its effect on porcine embryonic development. Theriogenology 53:277.

Magnusson C, Hillensjö T, Tsafriri A, Hultborn R, Ahren K. 1977. Oxygen consumption of maturing rat oocytes. Biol Reprod 17:9-15.

Magnusson C, Hillensjo T, Hamberger L, Nilsson L. 1986. Oxygen consumption by human oocytes and blastocysts grown in vitro. Hum Reprod 1:183-184.

Manes C, Lai NC. 1995. Non-mitochondrial oxygen utilization by rabbit blastocysts and surface production of superoxide radicals. J Reprod Fertil 104:69.

Marotti KR, Belin D, Strickland S. 1982. The production of distinct forms of plasminogen activator by mouse embryonic cells. Dev Biol 90:154-159.

McKiernan SH, Bavister BD. 1994. Timing of development is a critical parameter for predicting successful embryogenesis. Hum Reprod 9:2123-2129.

Menino AR Jr, Williams JS. 1987. Activation of plasminogen by the early bovine embryo. Biol Reprod 36:1289-1295.

Menino AR, Dyk AR, Gardiner CS, Grobner MA, Kaaekuahiwi MA, Williams JS. 1989. The effects of plasminogen on in vitro ovine embryo development. Biol Reprod 41:899-905.

Mermillod P, Oussaid B, Cognie Y. 1999. Aspects of follicular, oocyte maturation that affect the developmental potential of embryos. J Reprod Fertil (Suppl) 54:449-460.

Mohr LR, Trounson AO. 1980. The use of fluorescein acetate to assess embryo viability in the mouse. J Reprod Fertil 58:189-196.

Mohr LR, Trounson AO, Leeton JF, Wood C. 1983. Evaluation of normal and abnormal human embryo development during procedures in vitro. In: Beier HM, Lindner HR ed Fertilization of the Human Egg Embryo In Vitro. Springer-Verlag, Heidelberg, pp.211-221.

Morales H, Tilquin P, Reese J.F, Massip A, Dessy F, Van Langendonckt A. 1999. Pyruvate prevents peroxide-induced injury of *in vitro* preimplantation bovine embryo. Mol Reprod Dev 52:149-157.

O'Fallon JV, Wright RW JR. 1986. Quantitative determination of the pentose phosphate pathway in preimplantation mouse embryos. Biol Reprod 34:58-64.

O'Fallon JV, Wright RW. 1995. Pyruvate revisited: a non-metabolic role for pyruvate in preimplantation embryo development. Theriogenology 43:288.

O'Neill C. 1985. Partial characterisation of embryo-derived platelet-activating factor in mice. J Reprod Fertil 88:105-112.

O'Neill C. 1998. Autocrine mediators are required to act on the embryo by the 2-cell stage to promote normal development and survival of mouse preimplantation embryos in vitro. Biol Reprod 58:1303-1309.

O'Neill C, Saunders DM. 1984. Assessment of embryo quality. Lancet ii:1035.

O'Neill C, Gidley-Baird AA, Pke IL, Saunders DM. 1987. Use of a bioassay for embryo-derived platelet-activating factor as a means of assessing quality and pregnancy potential of human embryos. Fertil Steril 47:969-975.

Overström EW, Duby RT, Dobrinsky J, Roche JF, Boland MP. 1992. Viability and oxidative metabolism of the bovine blastocyst. Theriogenology 37:269.

Overström EW. 1996. In vitro assessment of embryo viability. Theriogenology 45:3-16.

Overström EW. 1998. Embryo respiration as an objective measure of embryo quality. Proceedings of the Satellite Workshop on Embryo development in vitro. Current challenges, future concepts. International Embryo Transfer Society, Boston, Jan 1998, pp.22-23.

Pampfer S, Arceci RJ, Pollard JW. 1991. Role of colony stimulating factor-1 CSF-1 and other lympho-hematopoietic growth factors in mouse preimplantation development. BioEssays 13:535-540.

Park KW, Choi SH, Song XX, Funahashi G, Niwa K. 1999. Production of plasminogen activators PAs in bovine cumulus-oocytes complexes during maturation in vitro: effect of epidermal growth factor on production of PAs in oocytes and cumulus cells. Biol Reprod 61:298-304.

Parrow V, Horton C, Maden M, Laurid S, Notarianni E. 1998. Retinoids are endogenous to the porcine blastocyst and secreted by trophectoderm cells at functionally-active levels. Int J Dev Biol 42:629-632.

Partridge RJ, Leese HJ. 1996. Consumption of amino acids by bovine preimplantation embryos. Reprod Fertil Dev 8:945-950.

Punjabi U, Vereecken A, Delbeke L, Angle M, Gielis M, Gerris J, Johnston J, Buytaert P. 1990. Embryo-derived platelet activating factor, a marker of embryo quality and viability following ovarian stimulation for in vitro fertilization. J in vitro Fertilization Embryo Transfer 7:321-326.

Renard JP, Menezo Y, Heyman Y. 1982. Alternative tests to assess viability of bovine embryos. Theriogenology 17:106.

Renard JP, Philippon A, Menezo Y. 1980. In vitro uptake of glucose by bovine blastocysts. J Reprod Fertil 58:161-164.

Rieger D. 1984. The measurement of metabolic activity as an approach to evaluating viability and diagnosing sex in early embryos. Theriogenology 21:138-149.

Rieger D, Guay P. 1988. Measurement of the metabolism of energy substrates in individual bovine blastocysts. J Reprod Fertil 83:585-591.

Rieger D, Loskutoff NM, Betteridge KJ. 1992. Developmentally related changes in the uptake and metabolism of glucose, glutamine and pyruvate by cattle embryos produced in vitro. Reprod Fertil Dev 4:547-557.

Rieger D, Pollard JW, Leibo SP. 1993. The effect of cryopreservation on the metabolic activity of in vitro produced cattle blastocyst. Cryobiology 30:631.

Saith RR, Bersinger NA, Barlow DH, Sargent IL. 1996. The role of pregnancy-specific ß-1 glycoprotein SP1 in assessing human blastocyst quality in vitro. Hum Reprod 11:1038-1042.

Schilling E, Doepke HH. 1978. A rapid diagnostic test for the viability of early cattle and rabbit embryos using diacetylfluorescein. Naturwissenschaften 65:685.

Schilling E, Smidt D, Sacher G, Petac D, El Kaschab S. 1979. Diagnosis of the viability of early bovine embryos by fluorescence microscopy. Ann Biol Anim Biochem Biophys 19:1625-1629.

Seshagiri PB, Bavister BD. 1991. Glucose, phosphate inhibit respiration, oxidative metabolism in cultured hamster embryos: evidence for the "Crabtree effect". Mol Reprod Dev 30:105-111.

Sheth KV, Roca GL, Al-Sedairy ST, Parhar RS, Hamilton CJCM, Jabbar FA. 1991. Prediction of successful embryo implantation by measuring interleukin-1-alpha and immunosuppressive factors in preimplantation embryo culture fluid. Fertil Steril 55:952-957.

Shoukir Y, Campana A, Farley T, Sakkas D. 1997. Early cleavage of in vitro fertilized human embryos to the 2-cell stage: a novel indicator of embryo quality and viability. Hum Reprod 12:1531-1536.

Shutt DA, Lopata A. 1981. The secretion of hormones during the culture of human preimplantation embryos with corona cells. Fertil Steril 35:413-416.

Stojkovic M, Wolf E, Büttner M, Berg U, Charpigny G, Schmitt A, Brem G. 1995. Secretion of biologically active interferon tau by in vitro-derived bovine trophoblastic tissue. Biol Reprod 53:1500-1507.

Stojkovic M, Büttner M, Zakhartchenko V, Riedl J, Reichenbach HD, Wenigerkind H, Brem G, Wolf E. 1999. Secretion of interferon-tau by bovine embryos in long-term culture:comparison of in vivo derived and in vivo produced, nuclear transfer and demi-embryos. Anim Reprod Sci 55:151-162.

Strickland S, Reich E, Sherman MI. 1976. Plasminogen activator in early embryogenesis: enzyme production by trophoblast and parietal endoderm. Cell 9:231-240.

Stryer L. 1995. Biochemistry, 4th edition, WH Freeman, Company, New York.

Summers PL, Campbell JM, Miller MW. 1988. Normal in vivo development of marmoset monkey embryos after trophectoderm biopsy. Hum Reprod 3:389-393.

Thompson JG, Simpson AC, Pugh PA, Wright RW Jr, Tervit HR. 1991. Glucose utilization by sheep embryos derived in vivo or in vitro. Reprod Fert Dev 3:571-576.

Thompson JG, Partridge RJ, Houghton FD, Cox CI, Leese HJ. 1996. Oxygen uptake and carbohydrate metabolism by in vitro derived bovine embryos. J Reprod Fertil 106:299-306.

Thompson JG, McNaughton C, Gasparrini B, McGowan LT, Tervit HR. 2000. Effect of inhibitors, uncouplers of oxidative phosphorylation during compaction and blastulation of bovine embryos cultured in vitro. J Reprod Fertil 118:47-55.

Thomson JL. 1967. Effect of inhibitors of carbohydrate metabolism on the development of preimplantation mouse embryos. Exp Cell Res 41:411-427.

Tiffin GJ, Rieger D, Betteridge KJ, Yadav BR, King WA. 1991. Glucose and glutamine metabolism in pre-attachment cattle embryos in relation to sex and stage of development. J Reprod Fertil 93:125-132.

Trimarchi JR, Liu L, Porterfield DM, Smith PJS, Keefe DL. 2000. Oxidative phosphorylation-dependent and -independent oxygen consumption by individual preimplantation mouse embryos. Biol Reprod 62:1866-1874.

Turner K, Martin KL, Woodward BJ, Lenton EA, Leese HJ. 1994. Comparison of pyruvate uptake by embryos derived from conception and non-conception natural cycles. Human Reprod 9:2362-2366.

Turner K, Lenton EA. 1996. The influence of Vero cell culture on human embryo development and chorionic gonadotropin production in vitro. Hum. Reprod 11:1966-1974.

Van Blerkom J, Davis P, Alexander S. 2000. Differential mitochondrial distribution in human pronuclear embryos leads to disproportionate inheritance between blastomeres: relationship to microtubular organization, ATP content and competence. Human Reprod 15:2621-2633.

van der Weiden RMF, Helmerhorst FM, Noort WA, Keirse JNC, Naaktgeboren N. 1994. Prostanoid levels in in vitro fertilization culture medium are not related to the likelihood of implantation. Fertil, Steril 62:1217-1220.

Wales RG, Whittingham DG. 1970. Metabolism of specifically labelled pyruvate by mouse embryos during culture from the two-cell stage to the blastocyst. Aust J Biol Sci 23:877-887.

Wales RG, Whittingham DG, Hardy K, Craft IL. 1987. Metabolism of glucose by human embryos. J Reprod Fertil 79:289-297.

Wells XE, O'Neill C. 1992. Biosynthesis of platelet-activating factor by two-cell mouse embryos. J Reprod Fert 96:61-71.

Whittingham DG .1978. Viability assays for mammalian ova. Cryobiology, 15:245-248.

Williams JE, Butler JE. 1993. Glucose uptake of viable and non-viable mouse embryos following freezing. Biol Reprod, 48 (suppl) 1:96 abstract no. 149.

Witkin SS, Liu HC, Davis OK, Rosenwaks Z. 1991. Tumor necrosis factor is present in maternal sera and embryo culture fluids during in vitro fertilization. J Reprod Immunol 19:85-93.

Woodward BJ, Lenton EA, Turner K. 1993. Human chorionic gonadotropin:embryonic secretion is a time dependent phenomenon. Hum Reprod 8:1463-1468.

Woodward BJ, Lenton EA, Turner K, Grace WF. 1994. Embryonic human chorionic gonadotropin secretion and hatching: poor correlation with cleavage rate and morphological assessment during preimplantation development in vitro. Hum Reprod 9:1909-1914.

Zhang X., Shu MA, Harvey MB, Schultz GA. 1996. Regulation of urokinase plasminogen activator production in implanting mouse embryo: effect of embryo interaction with extracellular matrix. Biol Reprod 54:1052-1058.

Ziebe S, Petersen K, Lindenburg S, Andersen AG, Gabrielsen Andersen AN. 1997. Embryo morphology or cleavage stage: how to select the best embryos for transfer after in vitro fertilization. Hum Reprod 12:1545-1549.

Zielke HR, Zielke CL, Ozand PT. 1984. Glutamine: a major energy source for cultured mammalian cells. Fed Proc 43:121-125.

Zolti M, Ben-Rafael Z, Meirom R, Shemesh M, Bider D, Mashiach S, Apte RN. 1991. Cytokine involvement in oocytes and early embryos. Fertil Steril 56:265-272.

Chapter 4

EMBRYO QUALITY AND FREEZING TOLERANCE: CRYOPRESERVATION OF HUMAN EMBRYOS

Josiane Van der Elst
Infertility Centre, Ghent University Hospital,
De Pintelaan 185, 9000 Ghent, Belgium

1. Introduction

Freezing of mammalian embryos has become part of cryobiology around 30 years ago when the embryologist-cryobiologist team of Whittingham, Leibo and Mazur (1972) published for the first time the successful freezing of the mouse embryo leading to the birth of living young. Cryopreservation of the human embryo has entirely been based on experience gained with cryopreservation of mouse embryos. It was in 1983 that the first human pregnancy after transfer of a frozen-thawed eight-cell embryo was announced (Trounson and Mohr, 1983). At present, freezing of human embryos has become a firmly established adjunct standard routine in clinical programmes of assisted reproduction.

Despite the world-wide implementation of human embryo freezing in assisted reproduction, it is still harder to obtain pregnancies with frozen-thawed embryos than with fresh embryos. The establishment of an efficient clinical cryopreservation programme remains a challenge. The success rate of human embryo cryopreservation can be expressed in terms of the potential of the frozen embryo to survive after thawing, to implant and to give rise to a successful pregnancy leading to the birth of a healthy child. Whether an embryo can tolerate freezing or in other words whether survival, implantation and pregnancy can be obtained after embryo cryopreservation will depend on aspects at the embryological, cryobiotechnical and maternal level. Selection of the embryos capable of survival and implantation, the choice of the appropriate technical freezing protocol, as well as uterine receptivity are all factors playing a role in the final pregnancy outcome.

In this chapter attention will not be focused on the cryobiotechnology of freezing nor on uterine factors that may play a role in implantation. With regard to cryobiotechnology of embryo freezing we do want to refer to publications dealing with basic cryobiology of mammalian embryos since procedures used for cryopreservation of embryos of laboratory animals really served as a model to cuisine the recipes of human embryo freezing (Mazur, 1970, Whittingham et al., 1972, Leibo et al., 1974; Leibo, 1977; Mazur 1977; Leibo and Mazur, 1978; Leibo, 1981; Mazur, 1984; Leibo, 1986; Mazur, 1988; Mazur 1990). Mouse and laboratory animal cryopreservation will therefore be dealt with further in this chapter only where relevant.

When freezing is being related to embryo quality, two approaches are possible: on one hand the embryo has to possess a basic quality to survive the freezing process and on the other hand, the ability of the embryo to survive the freezing process can be a reflection of its quality. In this chapter, attention will be entirely focused on the quality a human embryo has to possess to withstand freezing and thawing successfully, to respond positively to the stress of cryopreservation without loosing the potential for implantation. In other words, how good should an embryo be to have a chance to survive freezing and to implant? In chapter 5 (Massip and Leibo), attention will be focused upon the possible influences of environmental conditions upon embryo freezability, a problem which became obvious when researchers started to freeze in vitro produced bovine embryos (Massip et al., 1995).

When the best embryo has to be selected for cryopreservation, the central issue is to find a means of assessing and judging embryo quality at the moment of freezing. Which characteristics of an embryo determine its quality and can influence the way in which the embryo will tolerate freezing? Therefore we must describe as accurately as possible which characteristics of an embryo give information on embryo quality at the time of freezing and also take into account factors in the embryo's history that can influence quality. Therefore we will document the human embryo by the following features and relate them to freezing tolerance:

1. Embryo morphology,

2. Embryo cleavage capacity,
3. Oocyte quality: influence of follicular stimulation,
4. Culture conditions,
5. Micromanipulation of the embryo,
6. Embryo genetics.

2. Selection of embryos for cryopreservation: features related to freezing tolerance

2.1. Embryo morphology

The morphological evolution of an embryo can be followed entirely under the microscope. Embryo morphology is described by the stage of development and features specific of that stage (see also chapter 1). This means that an embryo will be defined in terms of its number of blastomeres and its specific morphological appearance. The rate of development for a given embryo is described in relation to an expected scheme: at day one following oocyte insemination, the human embryo is still unicellular, two- to four-cell embryos are expected on day two, five- to eight-cell embryos on day three, compacted morulae on day four and blastocysts on day five. The embryo is thus developing from a unicellular fertilized zygote over a cleaved multicellular composite stage to a differentiated stage where a fluid-filled cavity containing the embryonic pole is lined by a unicellular trophectoderm.

The morphologic transformations from the unicellular over a multicellular system to a mini-tissue model at the blastocyst stage poses different cryobiological requirements. Freezing involves the use of cryoprotective agents and dehydration of the embryo and hence movement of cryoprotectants into and water out of cells. From a cryobiological point of view, the unicellular system is the most simple to approach. As the embryo starts to divide, the heterogeneity of the system increases and several subunits compose one item, the cleaving embryo. As the embryo divides the volume per blastomere is decreasing, the surface-to-volume ratio of the individual blastomeres increases, blastomere interactions change, membrane characteristics vary and permeability to cryoprotectants and water changes. There is a larger membrane surface at risk with more advanced stages of cleavage and the membrane becomes more permeable with subsequent divisions. The membrane permeability of blastomeres to water and cryoprotectants is influenced by the stage and the quality grade of the embryo. Fundamental cryobiological differences between stages of embryo development may require developmental stage-specific embryo cryopreservation protocols (Pfaff et al., 1998). Practically this involves that different embryo developmental stages will require the use of different cryopreservation protocols. Procedures that yield high survival of one developmental stage are often less effective for other developmental stages. This is also true for other mammalian embryos such as pig (Dobrinsky, 1997) and cattle embryos (Mahmoudzadeh et al., 1995).

Successful cryopreservation is feasible for one- to eight-cell embryos and blastocysts using a variety of protocols. Experience in the mouse model has led to the design of protocols that have nearly been copied for the human embryo with fair success rates. Different embryonic stages can be frozen with success, given the appropriate protocol is being chosen. With the current knowledge, it is advisable to use a propanediol-based freezing protocol for one- to four-cell embryos, a dimethylsulphoxide-based protocol for four- to eight-cell embryos and a glycerol-based protocol for blastocysts. Ultrarapid freezing or vitrification of human embryos has until now not been more successful than slow freezing (Rall and Fahy, 1985; Trounson et al., 1987; Trounson and Sjöblom, 1988; Barg et al., 1990; Gordts et al., 1990; Feichtinger et al., 1991; Lai et al., 1996; Van den Abbeel et al., 1997b; Lane et al.,1999; Mauri et al., 1999; Saito et al., 2000; Yokota et al., 2000). The reports on nonequilibrium freezing of human embryos suggest that further consideration is worthwhile but their safety and reproducibility has to be demonstrated.

In conclusion, the stage of an embryo prescribes, dictates, orients which type of freezing protocol should be used. Thus, freezing tolerance of an embryo depends definitely on the stage. Given the correct protocol is used, good survival and implantation can be obtained with different stages. When choosing the appropriate freezing protocol for a specific stage, the outcome per embryo will depend on specific morphological features associated with that stage. Pronuclear configuration will be essential for one-cell embryos, the degree of fragmentation for multicellular embryos and the expansion of the blastocoelic cavity for blastocysts.

2.1.1. Morphology of unicellular embryos

The morphology of one-cell embryos is highly defined by the intracellular processes of pronucleus formation (pronuclear appearance) and preparation for the first cell division (pronucleus migration and dissolution). The protocol of choice for freezing pronucleate embryos is a slow freezing protocol with propanediol as the cryoprotective agent. Survival is generally expressed as morphological survival after thawing (integrity of one cell) and the capacity to divide in culture in the 24h following thawing.

Freezability of unicellular pronucleate embryos is highly dependent on the dynamics of these processes of pronucleus formation, migration and dissolution. For the freezing of pronucleate (PN) embryos, timing is critical. Freezing has to happen after completion of PN migration to the centre of the zygote but before the onset of syngamy (Trounson and Jones, 1993; Van Steirteghem and Van den Abbeel., 1993; Veeck, 1993). Wright et al. (1990) suggest that this moment be determined accurately by studying PN association and nucleolar alignment. Preferably zygotes showing strongly adhered pronuclei with aligned nucleoli on adjacent sides should be frozen as these signs indicate the end of migration. Balakier et al. (1993) documented these morphological observations by biochemical studies. Results suggest that the DNA-synthetic S-phase of the human zygote is initiated by approximately 9 to 10 hours after insemination and is completed approximately 3 to 5 hours later. The duration of the G2-phase and mitosis is in the range of 4 to 6 and 3 to 3.5 hours, respectively. Because zygotes may be particularly susceptible to damage during the S-phase of the cell cycle, these results indicate that the optimal freezing time for human zygotes may be 20 to 22 hours after insemination. At this point in time of embryo development, the majority of zygotes should have finished PN migration and S-phase and have entered G2-phase before PN dissolution and chromosome condensation. Cryodamage to the sperm aster during pronuclear movement may cause an arrest or delay of pronuclear migration and may cause a change in the timely dismantling of the nuclear envelopes and condensation of chromosomes at the prophase of syngamy. The timing of PN migration may differ from patient to patient, especially when other types of follicular recruitment are used, possibly yielding unripe oocytes at the time of collection (Wright et al., 1990).

Fugger et al. (1988) and Veeck et al. (1993) reported that pregnancy rates of cryopreserved PN zygotes are very similar to those of their fresh sibling. In PN freezing, the chance of random quality being distributed over fresh and frozen embryos is apparently higher than with freezing of multicellular embryos where blastomere morphology has become a substantial factor in selecting embryos for fresh transfer or freezing. Therefore it is not surprising that pregnancy rates are not significantly different for fresh and frozen-thawed PN zygotes.

2.1.2. Morphology of multicellular embryos

An embryo evolves from the unicellular to the multicellular stage through consecutive cell divisions. The regularity of the divisions will foretell the morphological quality of the resulting embryos. Ideal embryo development assumes timely scheduled, synchronous divisions leading to equal blastomeres. Deficiencies at each of these points will result in slowly dividing or (partially) arrested embryos, asynchronously dividing embryos with uneven blastomere numbers, unequal blastomere size and fragmentation. With the number of divisions increasing and with increasing cell number, the risk of deficiencies will increase. In other words, blastomere arrest, asynchronous division and fragmentation are likely to increase as division progresses over the days. Fragmentation is the most significant criterion used for describing the morphological status of a multicellular embryo and is handled as a quality index.

The protocols used for freezing multicellular embryos are determined according to the cell number leading to the use of either PROH or DMSO for two-cell and four cell embryos (day 2 embryos) and of DMSO for four-to-sixteen cell embryos (day 2 and 3 embryos) (Van Steirteghem et al.,1987).

There is agreement on the definition of post-thaw survival for multicellular embryos. A multicellular embryo is considered surviving when at least 50% of the initial number of blastomeres is conserved after thawing hereby following the definition of Mohr et al. (1985). Embryos keeping at least 50% of their initial number of blastomeres intact have higher chance for survival and subsequent implantation. This does not exclude that pregnancy can be obtained from embryos having less than 50% of blastomeres intact (Mandelbaum, 1990).

The morphological aspect of the multicellular embryo before freezing, in particular the degree of fragmentation, is one of the determining criteria in its potential for survival and implantation, briefly in its freezability. Survival after thawing, potential of further cleavage after thawing and implantation are strongly correlated to the degree of fragmentation before freezing. Over the years of IVF-routine, many reports have been generated substantiating this from different angles of working attitudes (Lassalle et al., 1985; Ashwood-Smith, 1986; Cohen et al., 1986; Testart et al., 1986; Testart et al.,1987a and b; Cohen et al., 1988a; Mandelbaum et al., 1988; Van den Abbeel et al., 1988; Hartshorne et al., 1990; Mandelbaum, 1990; Trounson., 1990; Van Steirteghem et al., 1992; Schalkoff et al., 1993; Queenan et al., 1995; Karlström et al., 1997).

Mandelbaum et al. (1988) and Mandelbaum (1990) confirm in both papers that pre-freezing embryo morphology in particular in terms of fragmentation, is a major parameter involved in the success of freezing of multicellular embryos. The chance that an embryo survives freezing is higher the lesser the degree of fragmentation before freezing. Survival rates were found to be 67% for morphologically normal embryos without fragments versus 49% for embryos with fragments. This percentage may still seem high but it has to be clarified that in these papers it was the strategy to freeze multicellular embryos only when they were intact or showed maximally up to 10% of fragments. This explains also that freezing is undertaken in only about 10-20% of IVF-cycles whereas for programmes freezing embryos with up to 20% of fragments the percentage of IVF-cycles with freezing mounts up to 50%. Of course, the other strategy followed to prevent freezing of fragmented embryos is to freeze one-cell pronucleate embryos where fragmentation is not playing a role yet in survival as discussed above.

Amazingly, in the early reports by the Australians who managed to establish the first pregnancy from a frozen-thawed embryo (Trounson and Mohr, 1983; Mohr et al., 1985*)* no correlation was found between embryo morphology and survival rate but, then, only very good quality embryos were frozen.

In a paper by the Brussels group (Van Steirteghem et al., 1987) a wide array of multicellular embryos showing various degrees of fragmentation up to 20 % were frozen. Survival was clearly related to the degree of fragmentation of embryos prior to freezing but also the interplay with the stage of freezing was elegantly documented. It was shown that beyond the two-cell stage embryos with fragmentation might profit more from the use of the DMSO protocol which is apparently better designed to cope with embryos beyond two-cell but also with increasingly fragmented embryos. After the two-cell stage, embryos with 0% up to 10% and up to 20% of fragmentation remained replaceable (50-100% of blastomeres intact after thawing) in 83, 48 and 52 % of cases when frozen with DMSO, compared with in 58, 48 and 14% when frozen with PROH, respectively. These observations were further exemplified in the sequel paper by Van den Abbeel et al. (1988) where at the two-cell stage DMSO and PROH protocols seem to be interchangeable but where for later cleavage stages and increasing fragmentation survival was better when DMSO was the cryoprotectant. In a prospective randomized study by Van der Elst et al. (1995) investigating the optimal protocol for a heterogeneous cohort of embryos as obtained in daily IVF routine after transfer of the morphologically best embryos which means that only excess embryos of lesser quality are used for freezing, DMSO showed an advantage over PROH.

In a study by Camus et al. (1989), the relation between pre-freezing embryo fragmentation and implantation potential after thawing was unequivocally documented by the outcome of single embryo transfers. An implantation rate of 21% was found for excellent pre-freezing embryos with regular blastomeres without any fragments, of 14.5% for pre-freezing embryos with up to 20% of fragments and none of the pre-freezing embryos with more than 20% of fragments resulted in a pregnancy.

In a multi-factorial analysis on the pregnancy outcome in a cryopreservation programme the idea of the sponsoring embryo is mentioned (Schalkoff et al., 1993). A sponsoring embryo has at least four cells on day 2 after insemination, relative evenness of each of the four blastomeres, good cell-to-cell contacts and minimal fragmentation. Patients who had at least one good quality embryo frozen called the sponsoring embryo conceived 35% of the time which is more than double compared to the group having no sponsoring embryo (15%). With this concept of sponsoring, it becomes clear that not quantity but quality is an important facet in considering which embryos should be frozen. Testart et al. (1986) and Mandelbaum et al. (1988) also found a high pregnancy rate with single embryo transfers of frozen good quality embryos. In the same spirit Van den Abbeel et al. (1997a) investigated pregnancy outcome after transfer of embryos which were fully intact after thawing versus partially damaged ones. The percentage of gestational sacs showing foetal heart beat was almost three times higher after transfer of fully intact embryos (11.5%) versus partially damaged ones (3.5%).

Cohen et al. (1988b) investigated whether, except for the degree of fragmentation, other morphological parameters may be predictive of implantation after thawing. Twelve morphological characteristics were monitored after thawing and shortly before replacement using video-cinematography, including intactness of blastomeres, membrane surface smoothness, visibility of organelles, cell-cell adherence, zona pellucida thickness. Variable zona pellucida thickness in zygotes and smooth surface of the blastomere's membrane in cleaved embryos were the only parameters with predictive value of implantation. Furthermore, a highly significant correlation was found between the implanting capacity of frozen-thawed embryos and the number of abnormal morphological features after thawing, two abnormal features being almost not reconcilable with implantation. Thawed zygotes which had gone on to cleave in vitro and thawed-cleaved embryos have an excellent chance of implantation if they have near-normal morphological features (Cohen et al., 1988b). Frozen-cleaved embryos only implanted when the blastomeres were intact and of comparable size, when cell organelles were visible, smooth blastomere membranes were present and cell-cell binding was apparent. Any form of disorganisation of frozen-cleaved embryos following thawing can be considered a poor prognostic factor for implantation. So probably if before thawing these parameters are not present, the embryo is pre-compromised and maybe ought not to be frozen.

One other morphological criterion to judge suitability of embryos for intrauterine transfer in human IVF or ICSI is the presence of multinucleated blastomeres (MNB). MNB usually are arrested cells, therefore embryos with one or more MNB are expected to be developmentally incompetent (Hardy et al., 1993). The presence of multinucleated embryos is considered a negative prognostic factor for implantation of embryos. Although pregnancy can occur after transfer of embryos with MNB, the transfer of multinucleated embryos is discouraged unless insufficient or no embryos with only mononucleated blastomeres are available (Balakier and Cadesky, 1997; Pelinck et al., 1998). Because of the already lowered implantation potential of embryos with multinucleated blastomeres, it is therefore strongly advisable to adopt the strategy of not cryopreserving embryos with multinucleated blastomeres.

2.1.3. Blastocyst morphology

The blastocyst is vastly different in morphology from earlier pre-implantation embryos. Once past the eight-cell stage, cells of the embryo begin to share surfaces and the formerly well distinct separate cells start to adhere, leading to cell boundaries becoming vague or non-visible. The embryo turns into a compacting mass in which divisions are ongoing. A landmark indicating the transition of the compacted embryo to the blastocyst stage is the formation of an inner cavity where fluid starts to accumulate. The gradual enlargement of the inner cavity leads to an embryo finally consisting of a fluid filled cavity containing a cell mass, the inner cell mass, at one pole and being lined by a tightly sealed outer cell layer, the trophectoderm. A good-quality blastocyst consists of a tight well-defined inner cell mass, a regular paved trophectoderm spanning the cavity and a fluid-filled cavity expanding the blastocyst to the boundaries of the zona pellucida. These features will be helpful indicators in evaluating the freezability of a blastocyst.

The question we have to pose now is which type of blastocyst pre-freeze morphology is related to good cryosurvival. It is amazing that the criteria set in the pioneering works still hold today and are ready to be rediscovered. The most striking morphological change the blastocyst is undergoing when being cryopreserved is the collapse of the cavity in the hyper-osmotic cryoprotective media. A key feature to evaluate blastocysts before freezing and after thawing is exactly the degree of expansion before freezing and of re-expansion of this cavity in the hours following thawing. In 1985, Fehilly et al. described in a clear way the characteristics a blastocyst should correspond to in order to withstand freezing with success. Blastocysts are classified as follows: stage A, the blastocoel just appearing as one or more vacuoles; stage B, a distinct cavity equal to one quarter or one half the volume of the embryo, the diameter of the embryo being unchanged; stage C, a single cavity occupying most of the volume of the embryo, and the zona pellucida becoming thinner as the embryo expands; and stage D, the blastocyst fully expanded for 24 hours or more. It was demonstrated that stage C and D blastocysts gave the best results and expansion is therefore used as one of the main criteria when storing blastocysts. Most blastocysts expand on day 5 or 6 following insemination. Further morphological criteria that have relevance to survival and implantation of expanded blastocysts are the presence of a clear inner cell mass and trophoblast layer, the presence of one cavity and the absence of necrotic cells. An additional feature to judge blastocyst quality for freezing is the severity of contraction of blastocysts upon addition of the cryoprotective medium. The severity of contraction of the blastocyst as a sign of the reaction to the hyperosmotic extracellular environment was positively linked with the incidence of survival (Cohen et al., 1986).

Experience with outcome of human blastocyst freezing was first described by English IVF-groups who were pioneers to consider human blastocyst freezing (Cohen et al., 1985; Fehilly et al., 1985; Cohen et al., 1986) Although these teams were performing blastocyst culture and freezing already in the eighties the interest was fading because of the few embryos reaching the blastocyst stage in the media available at that time. It is only with the implementation of co-culture and the use of sequential embryo culture media for day 5 embryo culture in the IVF-routine programme in the nineties that the interest in human blastocyst freezing has become re-awakened (Menézo et al., 1995; 1998; Gardner et al., 1998; Shoukir et al., 1998; Wiemer et al., 1998; Behr et al., 1999; Schoolcraft et al., 1999). Pregnancy rates after transfer of frozen-thawed blastocysts look promising being around 20% per transfer. Although there may be fewer embryos reaching this stage and hence fewer embryos to freeze, their ability to implant may compare positively to that of earlier stage embryos (Menézo et al., 1992, 1993, 1995; Kaufman et al., 1995).

2.2. Embryo cleavage capacity and freezing tolerance

2.2.1. Pre-freeze cleavage capacity

The rate with which an embryo reaches a stage of division in the biologically clocked expected time range can also influence survival and implantation. On-time developing embryos did not survive better than slowly growing embryos but the incidence of implantation was higher with faster developing embryos (Cohen et al., 1986). Embryos showing 5-10 cells on day 3 following insemination had a higher implantation potential than embryos reaching these stages only at day 4. Pregnancy was established only with day 3 5-10-cell embryos. Similarly blastocysts expanding on day 5 did not survive cryopreservation significantly better than blastocysts expanded on day 6 but, the incidence of implantation with blastocysts expanding on day 5 was twice as high as with those expanding on day 6 (Cohen et al., 1986).

Lassalle et al. (1985) reported that survival was higher for synchronously dividing embryos (2, 4, 8-cell) compared to asynchronously dividing embryos (3, 5, 6, 7-cell). A low survival rate was observed in particular for embryos frozen at the three-cell stage or after the six-cell stage (Testart et al., 1986). Nucleated cells survive better than non-nucleated cells. It was however not possible to ascertain whether non-nucleated cells were observed during mitotic division or whether they were degenerating.

2.2.2. Post-thaw recovery

The idea is to define pre-transfer parameters that evaluate the embryo's ability to grow apart from maternal influences.

In the evaluation of post-thaw survival of PN zygotes cleavage capacity within a 24h post-thaw in vitro culture period has been traditionally used as a viability index to assess survival by cleavage rather than by morphological appearance only. Fugger et al. (1988) and Veeck et al. (1993) reported that PN zygotes, that survive the actual cryopreservation process have the same potential to cleave at the same rate as fresh PN zygotes if they were selected at random and frozen when the pronuclei are visible. Faster-growing frozen-thawed PN zygotes which resulted in 7-8-cell embryos in 24h after thawing did not result in pregnancies.

For multicellular embryos, the common strategy has been to evaluate post-thaw morphological appearance of the embryo and to base the decision of transfer on this assessment. Post-thaw morphology is described as the proportion of embryos having at least 50% of their blastomeres intact. Evaluation and selection of embryos to replace after thawing have mostly dealt with the fraction of survived blastomeres in each embryo. Very little attention has been paid to viability, defined as the embryo's ability to resume cleavage. In a study by Van der Elst et al. (1997) it was investigated whether it was possible to increase the success of a multicellular embryo cryopreservation programme by selecting embryos for transfer not only on the basis of their morphological appearance after thawing but also on their capacity to show sustained cleavage during a 24h culture period following thawing. The results showed that cleavage capacity is a powerful additional selection criterion for viable embryos with potential for implantation. The delivery and birth rates were significantly higher in transfers exclusively of cleaved embryos compared to mixed transfers of embryos with and without no further cleavage (11.2% and 6.5% versus 1.9% and 0.6%, respectively). Transfers exclusively of embryos with no further cleavage after thawing did not lead to any deliveries. These results indicate that the potential to cleave further in vitro may reflect metabolic survival of frozen-thawed embryos. Confirmation for this approach was also given by Ziebe et al. (1998), showing that resumption of mitosis during post-thaw culture is a key parameter in selecting the right embryos for transfer. Transfers including at least one cleaved embryo had significantly higher pregnancy and implantation rates than transfers of non-cleaved embryos.

Further embryological evidence to sustain the freezing-thawing policy to transfer only embryos showing cleavage potential is presented in a study by Laverge et al. (1998). Using fluorescent in situ hybridization for investigating numerical abnormalities for chromosomes X,Y and 1 in blastomeres of embryos surviving cryopreservation but not cleaving further, it was shown that a large proportion (approximately 80%) of these embryos that do not cleave further in culture after freezing and thawing carry chromosomal abnormalities. It gives furthermore reassurance to learn that the transfer decision based on post-thaw cleavage potential helps to prevent the transfer of genetically abnormal embryos.

2.3. Oocyte quality: follicular stimulation

Follicular stimulation in the oocyte retrieval cycle may leave a permanent imprint on the cryopreserved embryo. There is general agreement that the use of GnRH analogues (GnRHa) in combination with HMG for ovarian stimulation permits a strict scheduling of oocyte retrievals, decreases significantly the cancellation rate and increases the number of oocytes per retrieval and makes more embryos available for transfer and freezing. Therefore theoretically the cumulative pregnancy rate over the fresh and frozen cycles should be increased above that obtained with non-agonist ovarian stimulation. In literature there is still controversy about the estimated benefit of the use of GnRHa for the outcome of frozen embryo transfers. One opinion defends that using GnRHa in combination with HMG produces absolutely higher numbers of embryos for freezing but that the proportion of good quality embryos is reduced, leading to an adverse effect on survival and implantation rates of supernumerary frozen-thawed embryos (Cohen et al., 1986; Guérin et al., 1989; Testart et al. 1989a,b; Smitz et al., 1992; Van der Elst et al., 1996). The opposite opinion defends that the use of GnRHa has no influence on the viability of embryos after thawing (Macnamee et al., 1989; Mandelbaum., 1990; Demoulin et al., 1991; Fugger et al., 1991; Oehninger et al., 1992; Benshushan et al., 1993). According to Oehninger et al. (1992) the use of GnRHa produces pre-embryos of equal viability for development after cryopreservation at the PN stage than a similar gonadotrophin stimulation without GnRHa. Hartshorne et al. (1990) found no difference in the proportion of embryos in which all blastomeres were viable after thawing comparing a clomiphene citrate-HMG with a GnRHa-HMG stimulation regimen. When looking more carefully into the data of these studies, pronuclear stage embryos or very good quality cleaved embryos without fragments were frozen or rather short desensitizing protocols were used in the studies finding no adverse effect.

That cycle stimulation features can have indeed an influence on embryo quality was recently demonstrated in a comparison of different GnRHa regimens for down-regulation (buserelin versus nafarelin). The conclusion was drawn that differences exist in IVF-cycles down-regulated with buserelin or nafarelin which might affect embryo quality and outcome of frozen-thawed embryo transfers (Simberg et al., 1998). In frozen-thawed embryo transfer cycles, implantation rate was 21.1% in the nafarelin group versus 10.6% in the buserelin group.

More data supporting the influence of ovarian stimulation on the outcome of frozen-thawed embryo transfers is given by the data of Toner et al. (1991a and b), Lin et al. (1995) and Karlström et al. (1997), revealing that frozen sibling embryos from a prior successful ART cycle are more likely to result in pregnancy. Successful implantation of frozen sibling embryos is influenced by the outcome of the cycle in which they were created. This is interpreted that the outcome of the original IVF cycle has a predictive value for subsequent cryocycles. This reflects the influence of the quality of the cohort of oocytes not only on the outcome of the fresh but as well as of the frozen cycle coupled to it.

A limited number of data are available on the influence of sperm quality on the outcome of frozen-thawed embryo tranfer. After fertilization, semen parameters had no effect on the quality or cryosurvival of embryos in either IVF or ICSI (Moilanen et al., 1999).

2.4. Culture conditions

2.4.1. Co-culture

Before the introduction of sequential culture media for human blastocyst culture, co-culture was a popular means of increasing the efficiency of human blastocyst culture programmes. Not only did co-culture promote embryo development till day 5 and increase blastocyst cell numbers compared to culture in classical media, but also the ability of co-cultured blastocysts to withstand freezing was higher than without co-culture (Menézo et al., 1992; 1993; 1995; Menézo and Ben Khalifa, 1995).

2.4.2. Medium supplements

Bernart et al. (1987) investigated the freezing-thawing behaviour and cell membrane ultrastructure of mouse embryos cultured in serum-free versus serum-containing medium. A drastic decrease in the number of microvilli on the cell surface and a high degree of damage after freezing and thawing was observed in embryos cultured in serum-free medium. It was suggested that the smaller cell surface and the resulting lower permeability of the cell membrane in the embryos with few microvilli are due to sub-optimal culture conditions and was the main reason for decreased embryo viability after freezing and thawing.

In a study by Warnes et al. (1997) embryo freezability was compared for human embryos cultured and frozen in medium supplemented with normal serum albumin or with own patient serum. The implantation rate per frozen-thawed embryo transferred was significantly lower when normal serum albumin was included in the medium for culture and freezing compared to inclusion of patients' serum in culture and freezing medium.

2.4.3. Oxidative stress

Tarin and Trounson (1993) demonstrated that embryo tolerance to freezing-thawing stress can be modified by incubating murine embryos in the presence of stimulators or inhibitors of membrane lipid peroxidation. Since permeating cryoprotective agents need to enter cells to exert their cryoprotective action, perturbations of the membrane like peroxidation changing fluidity, structure and function may reduce permeability to water and to permeating cryoprotectants, decreasing embryo tolerance to freezing stress. In particular, the presence of a stimulator of lipid peroxidation (ferrous sulphate; $FeSO_4$) decreased the percentage of intact embryos recovered after thawing, whereas apotransferrin and ascorbate known as inhibitors of lipid peroxidation increased the number of intact embryos and blastocysts on day 5. Ascorbate also increased the number of hatching blastocysts and implantation sites after transfer to foster mice.

2.5. Embryo micromanipulation

The evolution of IVF is accompanied by increasingly sophisticated manipulations involving microscopically monitored techniques of assisted fertilization and embryo biopsy. These technological developments involve deliberate damage to the zona pellucida. Intracytoplasmic sperm injection (ICSI) allows to obtain successfully fertilization, embryo development and pregnancy in cases of severe male infertility which were formerly untreatable, but the zona and oolemma are punctured. The introduction of preimplantation genetic diagnosis (PGD) is interesting for couples at risk of transmitting genetic disease but a large gap has to be made in the zona pellucida for blastomere removal.

Pre-clinical studies using mouse embryos showed that the creation of artificial gaps in the zona of embryos does not appear to compromise development after cryopreservation (Depypere et al., 1991; Garrisi et al., 1992). Partial damage to the zona pellucida such as by zona drilling is compatible with rates of survival and development to the blastocyst stage which were not different to those observed in zona-intact control embryos. Even when creating a large hole in the ZP combined with limited blastomere removal, cryosurvival of mouse embryos was not jeopardized (Wilton et al., 1989; Krzyminska and O'Neill, 1991; Liu et al., 1993; Thompson et al., 1995; Ludwig et al., 1998).

In the early days of microsurgical assisted fertilization in the human, fertilization was attempted by subzonal insemination (SUZI) where a few spermatozoa were brought into the perivitelline space through perforations in the zona. Chang et al. (1991) reported a clinical pregnancy from the transfer of frozen-thawed zygotes with holes in their zonae because of SUZI. Further analysis was provided by Obasaju et al. (1994) showing that SUZI embryos with perforated zonae can be frozen and thawed with the same degree of confidence as zona-intact embryos but that implantation rates were lower than that obtained with fresh SUZI embryos or frozen-thawed zona-intact embryos. It was not clear whether reduced implantation was due to the zona effect or different morphological embryo quality between SUZI-frozen and intact frozen embryos.

From the team introducing ICSI as a more elegant and efficacious technique of assisted fertilization compared to SUZI came the first publication reporting on establishment of pregnancy after transfer of frozen-thawed embryos obtained after ICSI (Van Steirteghem et al., 1994). A similar embryo survival rate was found for frozen-thawed multicellular embryos resulting either from IVF or ICSI but the clinical ongoing pregnancy rate was lower for ICSI-derived embryos. A similar conclusion was reached by Simon et al. (1998) equally for multicellular embryos and by Macas et al. (1998) for zygotes arising from ICSI where the implantation potential of cryopreserved ICSI-zygotes or -embryos was lower than for IVF-derived ones. The timing of zygote freezing or difference in pre-freezing quality between IVF- and ICSI-derived multicellular embryos but not the partial zona damage by ICSI was considered to be the principal reason for the lower developmental potential of frozen-thawed ICSI-derived embryos.

Micromanipulation can influence cryopreservation potential of a cohort of oocytes quantitatively as demonstrated by Carrillo et al. (1998). The site and technique of oolemma rupture inside the ICSI needle influences fertilization rate and oocyte damage: improvement of the ICSI-technique can thus lead to an increased number of patients with cryopreserved embryos. Rupturing the oolemma by aspiration lowered oocyte damage rates and improved fertilization rates and embryo availability for storage.

So far, limited data are available on the success of cryopreservation after human embryo biopsy for PGD (Joris et al., 1999; Magli et al., 1999). In the approach by Joris et al. (1999) using human embryos derived from abnormal fertilization it was investigated whether zona drilling, eventually combined with removal of one or two blastomeres influences cryosurvival and further development after cryopreservation. It was found that cryosurvival for zona-drilled embryos was impaired and that this effect was more pronounced when zona drilling was combined with blastomere removal.

2.6. Embryo genetics

Viability of frozen-thawed mouse embryos is affected by genotype (Schmidt et al., 1985). The post-thaw survival of embryos collected from mice of various genotypes as measured by in vitro development varied significantly among mouse stocks and strains. Moreover, embryo viability post-freezing was affected by mating an incrossed male to a genotypically different female, suggesting a genetic influence of the male on the ability of embryos to survive cryopreservation.

Spearow and Bradford (1983) showed that genetic variability had great influence on induced ovulation in several strains of mice and determined that genotype contributed as much as 65% of the variation in numbers of ova collected in response to exogenous gonadotrophins. But, increased yield of embryos does not necessarily mean a favour for cryopreservation as documented further by Schmidt et al. (1985). An inbred strain producing greater numbers of 4-to-8-cell freezable embryos also produced more degenerate embryos having fragmented or lysed blastomeres, making them inferior for freeze storage. Previously it had been shown that Pregnant mare's serum gonadotrophin (PMSG) injection may cause certain strains to ovulate premature and inadvertently leading to the production of abnormal embryos (McLaren, 1962). It was suggested that this difference in sensitivity of strains to PMSG may have been the result of a genetically-based increased sensitivity to PMSG.

In the daily practice of the IVF-clinic the highly different background of patients is perhaps a factor unconsciously neglected in the interpretation of embryo development and pregnancy outcome. The extreme patient-to-patient variation and otherwise within patient consistency of patterns of ovarian response and embryo development may at least in part, be a reflection of underlying built-in genetic codes.

3. Conclusion

There is overwhelming evidence that embryo quality influences the freezability of an embryo. The closest approximation of embryo quality is currently made by microscopic description of embryo morphology. Microscopic embryo morphology gives information on the stage of development and the appearance of the embryo in relation to an ideal expected pattern. The degree to which the ideal pattern is reached is a measure for the quality of the embryo and for its freezing tolerance. Although embryo morphology may be a good index of embryo quality, it is far from being a blueprint. Full embryonic history is not translated into the phenotype visible under the microscope. Embryos showing similar morphology may not necessarily have the same quality and can be the product of a different embryonic history. The quality of an embryo depends as well on intrinsic factors such as gamete quality and genetics but also on environmental factors such as culture conditions and manipulations inflicted upon the embryo. In fact, the embryo is a child of its time and will behave accordingly when facing stress. The possibility of the embryo to withstand the freezing process is therefore a reflection of its quality.

References

Ashwood-Smith MJ. 1986. The cryopreservation of human embryos. Hum Reprod 1:319- 332.

Balakier H, Cadesky K. 1997. The frequency and developmental capability of human embryos containing multinucleated blastomeres. Hum Reprod 12:800-804.

Balakier H, MacLusky NJ, Casper R. 1993. Characterisation of the first cell cycle in human zygotes: implications for cryopreservation. Fertil Steril 59:359-365.

Barg PE, Barad DH, Feichtinger W. 1990. Ultrarapid freezing (URF) of mouse and human pre-embryos: a modified approach. J In Vitro Fert Embryo Transf 7:355-357.

Behr B, Pool TB, Milki AA, Moore D, Gebhardt J, Dasig D. 1999. Preliminary clinical experience with human blastocyst development in vitro without co-culture. Hum Reprod 14:454-457.

Benshushan A, Ezra Y, Simon A, Mordel N, Lewin A, Laufer N. 1993. The effect of gonadotrophin-releasing hormone agonist on embryo quality and pregnancy rate following cryopreservation. Fertil Steril 59:1065-1069.

Bernart W, Rabe T, Schleiermacher E, Anton-Lamprecht I, Runnebaum B. 1987. Freezing-thawing behaviour and cell membrane ultrastructure of mouse embryos pre-cultured in B2-Menézo medium before cryopreservation. Hum Reprod 2:679-687.

Camus M, Van den Abbeel E, Van Waesberghe L, Wisanto A, Devroey P, Van Steirteghem AC. 1989. Human embryo viability after freezing with dimethylsulfoxide as a cryoprotectant. Fertil Steril 51:460-465.

Carrillo AJ, Atiee SH, Lane B, Pridham DD, Risch P, Silverman IH, Cook CL. 1998. Oolemma rupture inside the intracytoplasmic sperm injection needle significantly improves the fertilization rate and reduces oocyte damage. Fertil Steril 70:676-679.

Chang SY, Soong YK, Chang MY, Lin PW, Guu HF, Wang ML 1991. A clinical pregnancy after a simple method of zona cutting, cryopreservation, and zygote intrafallopian transfer. Fertil Steril 55:420-422.

Cohen J, DeVane GW, Elsner CW, Fehilly CB, Kort HI, Massey JB, Turner GT. 1988a. Cryopreservation of zygotes and early cleaved human embryos. Fertil Steril 49:283-289.

Cohen J, Wiemer KE, Wright G. 1988b. Prognostic value of morphologic characteristics of cryopreserved embryos: a study using videocinematography. Fertil Steril 49:827-834.

Cohen J, Simons RS, Fehilly CB, Edwards RG. 1986. Factors affecting survival and implantation of cryopreserved human embryos. J In Vitro Fertil Embryo Transfer 3:46-52.

Cohen J, Simons RF, Fehilly CB, Fishel SB, Edwards RG, Hewitt J, Rowland GF, Steptoe PC,Webster, JM. 1985. Birth after replacement of hatching blastocyst cryopreserved at expanded blastocyst stage. Lancet 1(8429):647.

Demoulin A, Jouan C, Gerday C, Dubois M. 1991. Pregnancy rates after transfer of embryos obtained from different stimulation protocols and frozen at either pronucleate or multicellular stages. Hum Reprod 6:799-804.

Depypere HT, Carroll JC, Vandekerckhove D, Matthews CD. 1991. Normal survival and in vitro development after cryopreservation of zona-drilled embryos in mice. Hum Reprod 6:432-435.

Dobrinsky JR. 1997. Cryopreservation of pig embryos. J Reprod Fertil Suppl 52:301-312.

Fehilly CB, Cohen J, Simons RF, Fishel SB, Edwards RG. 1985. Cryopreservation of cleaving embryos and expanded blastocysts in the human: a comparative study. Fertil Steril 44:638-644.

Feichtinger W, Hochfellner C, Ferstl U. 1991. Clinical experience with ultra-rapid freezing of embryos. Hum Reprod 6:735-736.

Fugger EF, Bustillo M, Katz LP, Dorfmann AD, Bender SD, Schulman JD. 1988. Embryonic development from fresh and cryopreserved sibling pronucleate human zygotes. Fertil Steril 50:273-278.

Fugger EF, Bustillo M, Dorfmann AD, Schulman JD. 1991. Human preimplantation embryo cryopreservation: selected aspects. Hum Reprod 6:131-135.

Gardner DK, Schoolcraft WB, Wagley L, Schlenker T, Stevens J, Hesla J. 1998. A prospective randomized trial of blastocyst culture and transfer in in vitro fertilization. Hum Reprod 13:3434-3440.

Garrisi GJ, Talansky BE, Sapira V, Gordon JW, Navot D. 1992. An intact zona pellucida is not necessary for successful mouse embryo cryopreservation. Fertil Steril 57:677-681.

Gordts S, Roziers P, Campo R, Noto V. 1990. Survival and pregnancy outcome after ultrarapid freezing of human embryos. Fertil Steril 53:469-472.

Guérin JF, Lornage J, Boulieu D, Mathieu C, Pinatel MC. 1989. Stratégie du transfert des embryons congélés. Contr Fert Sex 17:771-773.

Hardy K, Winston RML, Handyside AH. 1993. Binucleate blastomeres in preimplantation embryos in vitro: failure of cytokinesis during early cleavage. J Reprod Fertil 98:549-558.

Hartshorne GM, Wick K, Elder K, Dyson H. 1990. Effect of cell number at freezing upon survival and viability of cleaving embryos generated from stimulated IVF cycles. Hum Reprod 5:857-861.

Joris H, Van den Abbeel E, De Vos A, Van Steirteghem A. 1999. Reduced survival after human embryo biopsy and subsequent cryopreservation. Hum Reprod 14:2833-2837.

Karlström PO, Bergh T, Forsberg AS, Sandkvist U, Wikland M. 1997. Prognostic factors for the success rate of embryo freezing. Hum Reprod 12:1263-1266.

Kaufman RA, Menézo Y, Hazout A, Nicollet B, Dumont M, Servy EJ. 1995. Cocultured blastocyst cryopreservation: experience of more than 500 transfer cycles. Fertil Steril 64:1125-1129.

Krzyminska U, O'Neill CO. 1991. The effects of cryopreservation and thawing on the development in vitro and in vivo of biopsied 8-cell mouse embryos. Hum Reprod 6:832-835.

Lai AC, Lin BP, Chang CC, Tsai HD, Hwang VW, Lo HY. 1996. Pregnancies after transfer of ultrarapidly frozen human embryos. J Assist Reprod Genet 13:625-628.

Lane M, Schoolcraft WB, Gardner DK. 1999. Vitrification of mouse and human blastocysts using a novel cryoloop container-less technique. Fertil Steril 72:1073-1078.

Lasalle B, Testart J, Renard JP. 1985. Human embryo features that influence the success of cryopreservation with the use of 1,2 propanediol. Fertil Steril 44:645-651.

Laverge H, Van der Elst J, De Sutter P, Verschraegen-Spae MR, De Paepe A, Dhont M. 1998. Fluorescent in-situ hybridization on human embryos showing cleavage arrest after freezing and thawing. Hum Reprod 13:425-429.

Leibo SP. 1977. Fundamental cryobiology of mouse ova and embryos. In Eliott K, Whelan J. eds. The Freezing of Mammalian Embryos. Ciba Foundation Symposium. Elsevier, Amsterdam, no.52, pp. 69-92.

Leibo SP. 1981. Preservation of ova and embryos by freezing. In Brackett EG, Seidel GE, Seidel SM. eds. New Technologies in Animal Breeding. Academic Press, New York, pp 127-139.

Leibo SP, Mazur P, Jackowski SC. 1974. Factors affecting survival of mouse embryos during freezing and thawing. Exp Cell Research 89:79-88.

Leibo SP. 1986. Cryobiology: Preservation of mammalian embryos. In Evans JW, Hollaender A. eds. Genetic Engineering of Animals: An Agricultural Perspective. Plenum Press, New York, pp. 251-272.

Leibo SP, Mazur P. 1978. Methods for the preservation of mammalian embryos by freezing. In Daniel JC. ed. Methods in Mammalian Reproduction. Academic Press, New York, pp. 179-201.

Lin YP, Cassidenti DL, Chacon RR, Soubra SS, Rosen GF, Yee B. 1995. Successful implantation of frozen sibling embryos is influenced by the outcome of the cycle from which they were derived. Fertil Steril 63:262-267.

Liu J, Van den Abbeel E, Van Steirteghem AC. 1993. The in vitro and in vivo developmental potential of frozen and non-frozen biopsied 8-cell mouse embryos. Hum Reprod 8:1481-1486.

Ludwig M, Muschalla H, Al-Hasani S, Diedrich K. 1998. The effect of multiple cryopreservation procedures and blastomere biopsy on the in vitro development of mouse embryos. Hum Reprod 13:3165-3168.

Macas E, Imthurn B, Borsos M, Rosselli M, Maurer-Majot E, Keller PJ. 1998. Impairment of the developmental potential of frozen-thawed human zygotes obtained after intracytoplasmic sperm injection. Fertil Steril 69:630-635.

Macnamee MC, Howles CM, Edwards RG, Taylor PJ, Elder KT. 1989. Short term luteinizing hormone-releasing hormone agonist treatment: prospective trial of a novel ovarian stimulation regimen for in vitro fertilization. Fertil Steril 52:264-269.

Magli MC, Gianaroli L, Fortini D, Ferraretti AP, Munné S. 1999. Impact of blastomere biopsy and cryopreservation techniques on human embryo viability. Hum Reprod 14:770-773.

Mahmoudzadeh AR, Van Soom A, Bols P, Ysebaert MT, de Kruif A. 1995. Optimization of a simple vitrification procedure for bovine embryos produced in vitro: effect of developmental stage, two-step addition of cryoprotectant and sucrose dilution on embryonic survival. J Reprod Fertil 103:33-39.

Mandelbaum J. 1990. La congélation des embryons humains. Contr Fert Sex 18:341-353.

Mandelbaum J, Junca AM, Plachot M, Alnot MO, Salat-Baroux J, Alvarez S, Tibi C, Cohen J, Debache C, Tesquier L. 1988. Cryopreservation of human embryos and oocytes. Hum Reprod 3:117-119.

Massip A, Mermillod P, Dinnyes A. 1995. Morphology and biochemistry of in vitro produced bovine embryos: implications for their cryopreservation. Hum Reprod 10:3004-3011.

Mauri AL, Petersen CG, Baruffi RLR, Ferreira RC, Franco Jr JG. 1999. Clinical experience with ultrarapid cryopreservation of human embryos resulting from intracytoplasmic sperm injection. Hum Reprod 14:2362-2366.

Mazur P. 1970. Cryobiology: The freezing of biological systems. Science 168:939-949.

Mazur P. 1977. The role of intracellular freezing in the death of cells cooled at supraoptimal rates. Cryobiology 14:251-272.

Mazur P. 1984. Freezing of living cells: mechanisms and implications. Am J Physiol 247: C125-C142.

Mazur, P. 1990. Equilibrium, quasi-equilibrium, and nonequilibrium freezing of mammalian embryos. Cell Biophysics 17:53-92.

Mazur P. 1988. Stopping biological time. The freezing of living cells. Ann N Y Acad Sci 541:514-531.

McLaren A. 1962. The relation between natural fecundity and response to follicle stimulating hormone. J Endocrinol 25:137-144.

Menézo YJ, Nicollet B, Dumont M, Hazout A, Janny L. 1993. Factors affecting human blastocyst formation in vitro and freezing at the blastocyst stage. Acta Eur Fertil 24:207-213.

Menézo YJ, Sakkas D, Janny L. 1995. Co-culture of the early human embryo: factors affecting human blastocyst formation in vitro. Microsc Res Tech 32:50-56.

Menézo YJ, Ben Khalifa M. 1995. Cytogenetic and cryobiology of human cocultured embryos: a 3-year experience. J Assist Reprod Genet 12:35-40.

Menézo Y, Veiga A, Benkhalifa M. 1998. Improved methods for blastocyst formation and culture. Hum Reprod 13 Suppl 4:256-265.

Menézo Y, Nicollet B, Herbaut N, Andre D. 1992. Freezing cocultured human blastocysts. Fertil Steril 58:977-980.

Mohr LR., Trounson A, Freemann L. 1985. Deep-freezing and transfer of human embryos. J In Vitro Fert Embryo Transf 2:1-10.

Moilanen JM, Tulppala M, Reima I, Hovatta O. 1999. Fertilization, embryo quality, and cryosurvival in in vitro fertilization and intracytoplasmic sperm injection cycles. J Assist Reprod Genet 16:17-23.

Obasaju MF, Venier W, Blake M, Merola G, Ying Yu-Kang. 1994. Comparative study of the survival rate of frozen-thawed embryos with perforated zonae from microsurgical fertilization and frozen-thawed embryos with intact zonae. Fertil Steril 61:886-890.

Oehninger S, Toner JP, Veeck LL, Brzyski RG, Acosta AA, Muasher SJ. 1992. Performance of cryopreserved pre-embryos obtained in in vitro fertilization cycles with or without a gonadotropin-releasing hormone agonist. Fertil Steril 57:620-625.

Pelinck MJ, De Vos M, Dekens M, Van der Elst J, De Sutter P, Dhont M. 1998. Embryos cultured in vitro with multinucleated blastomeres have poor implantation potential in human in vitro fertilization and intracytoplasmic sperm injection. Hum Reprod 13:960-963.

Pfaff RT, Liu J, Gao D, Peter AT, Li T-K and Critser JK.1998. Water and DMSO membrane permeability characteristics of in vivo- and in vitro-derived and cultured murine oocytes and embryos. Mol Hum Reprod 4:51-59.

Queenan JT, Veeck L, Muasher SJ. 1995. Clinical and laboratory aspects of cryopreservation. Sem Reprod Endocrinol 13:64-71.

Rall WF, Fahy GM. 1985. Ice-free cryopreservation of mouse embryos at -196°C by vitrification. Nature 313:573-575.

Saito H, Ishida GM, Kaneko T, Kawachiya S, Ohta N, Takahashi T, Saito T, Hiroi M. 2000. Application of vitrification to human embryo freezing. Gynecol Obstet Invest 49:145-149.

Schalkoff ME, Oskowitz SP, Powers RD. 1993. A multifactorial analysis of the pregnancy outcome in a successful embryo cryopreservation program. Fertil Steril 59:1070-1074.

Schmidt PM, Hansen CT, Wildt DE. 1985. Viability of frozen-thawed mouse embryos is affected by genotype. Biol Reprod 32:507-514.

Schoolcraft WB, Gardner DK, Lane M, Schlenker T, Hamilton F, Meldrum DR. 1999. Blastocyst culture and transfer: analysis of results and parameters affecting outcome in two in vitro fertilization programs. Fertil Steril 72:604-609.

Shoukir Y, Chardonnens D, Campane A, Bischof P, Sakkas D. 1998. The rate of development and time of transfer play different roles in influencing the viability of human blastocysts. Hum Reprod 13:676-681.

Simberg N, Tulppala M, Husa LM, Tiitinen A. 1998. Comparison of buserelin and nafarelin in IVF cycles and in subsequent frozen-thawed embryo transfer cycles. Acta Obstet Gynaecol Scand 77:854-859.

Simon A, Holzer H, Hurwitz A, Revel A, Zentner BS, Lossos F, Laufer N 1998. Comparison of cryopreservation outcome following intracytoplasmic sperm injection and conventional in vitro fertilization. J Assist Reprod Genet 15:431-437.

Smitz J, Ron-El R, Tarlatzis BC. 1992. The use of gonadotrophin releasing hormone agonists for in vitro fertilization and other assisted procreation techniques: experience from three centres. Hum Reprod 7:49-66.

Spearow JL, Bradford GE. 1983. Genetic variation in spontaneous ovulation rate and LH receptor induction in mice. J Reprod Fertil 69:529-537.

Tarin JJ, Trounson A. 1993. Effects of stimulation or inhibition of lipid peroxidation on freezing-thawing of mouse embryos. Biol Reprod 49:1362-1368.

Testart J, Lassalle B, Belaisch-Allart J, Forman R, Hazout A, Volante M, Frydman R. 1987a. Human embryo viability related to freezing and thawing procedures. Am J Obstet Gynecol 157:168-171.

Testart J, Lassalle B, Forman R, Gazengel A, Belaisch -Allart J, Hazout A, Rainhorn JD, Frydman R. 1987b. Factors influencing the success rate of human embryo freezing in an in vitro fertilization and embryo transfer program. Fertil Steril 48:107-112.

Testart J, Belaisch-Allart J, Forman R, Gazengel A, Strubb N, Hazout A, Frydman R. 1989a. Influence of different stimulation treatments on oocyte characteristics and in vitro fertilizing ability. Hum Reprod 4:192-197.

Testart J, Forman R, Belaisch-Allart J, Volante M, Hazout A, Strubb N, Frydman R. 1989b. Embryo quality and uterine receptivity in in vitro fertilization cycles with or without agonists of gonadotrophin-releasing hormone. Hum Reprod 4:198-201.

Testart J, Lasalle B, Belaisch-Allart J, Hazout A, Forman R, Rainhorn JD, Frydman R. 1986. High pregnancy rate after early human embryo freezing. Fertil Steril 46:268-272.

Thompson LA, Srikantharajah A, Hamilton MP, Templeton A. 1995. A comparison of the effects of different biopsy strategies on the post-thaw survival of 8-cell-stage mouse embryos: implications for preimplantation diagnosis. Hum Reprod 10:659-663.

Toner JP, Brzyski RG, Oehninger S, Veeck LL, Simonetti S, Muasher SJ. 1991a. Combined impact of the number of pre-ovulatory oocytes and cryopreservation on IVF outcome. Hum Reprod 6:284-289.

Toner JP, Veeck LL, Acosta AA, Muasher SJ. 1991b. Predictive value of pregnancy during original in vitro fertilization cycle on implantation and pregnancy in subsequent cryothaw cycles. Fertil Steril 56:505-508.

Trounson A, Jones G. 1993. Controversies in assisted reproduction. Freezing of embryos: early versus late stages. J Assist Reprod Genet 10:179-181.

Trounson A, Mohr L. 1983. Human pregnancy following cryopreservation, thawing and transfer of an eight-cell embryo. Nature 305:707-709.

Trounson A, Sjöblom P. 1988. Cleavage and development of human embryos in vitro after ultrarapid freezing and thawing. Fertil Steril 50:373-376.

Trounson A. 1990. Cryopreservation. Brit Med Bull 46:695-708.

Trounson A, Peura A, Kirby C. 1987. Ultrarapid freezing: a new low-cost and effective method of embryo cryopreservation. Fertil Steril 48:843-850.

Van den Abbeel E, Van der Elst J, Van Waesberghe L, Camus M, Devroey P, Khan I, Smitz J, Staessen C, Wisanto A, Van Steirteghem AC. 1988. Hyperstimulation: The need for cryopreservation of embryos. Hum Reprod 3 Suppl 2:53-57.

Van den Abbeel E, Camus M, Van Waesberghe L, Devroey P, Van Steirteghem A. 1997a. Viability of partially damaged human embryos after cryopreservation. Hum Reprod 12:2006-2010.

Van den Abbeel E, Camus M, Van Waesberghe L, Devroey P, Van Steirteghem AC. 1997b. A randomized comparison of the cryopreservation of one-cell human embryos with a slow controlled-rate cooling procedure or a rapid cooling procedure by direct plunging into liquid nitrogen. Hum Reprod 12:1554-1560.

Van der Elst J, Van den Abbeel E, Vitrier S, Camus M, Devroey P, Van Steirteghem AC. 1997. Selective transfer of cryopreserved human embryos with further cleavage after thawing increases delivery and implantation rates. Hum Reprod 12:1513-1521.

Van der Elst J, Van den Abbeel E, Camus M, Smitz J, Devroey P, Van Steirteghem A. 1996. Long-term evaluation of implantation of fresh and cryopreserved human embryos following ovarian stimulation with buserelin acetate-HMG or clomiphene citrate-HMG. Hum Reprod 11:2097-2106.

Van der Elst J, Camus M, Van den Abbeel E, Maes R, Devroey P, Van Steirteghem AC. 1995. Prospective randomized study on the cryopreservation of human embryos with dimethylsulfoxide or 1,2-propanediol protocols. Fertil Steril 63:92-100.

Van Steirteghem AC, Van den Abbeel E, Camus M, Van Waesberghe L, Braeckmans P, Khan I, Nijs M, Smitz J, Staessen C, Wisanto A, Devroey P. 1987. Cryopreservation of human embryos obtained after gamete intra-Fallopian transfer and/or in vitro fertilization. Hum Reprod 2:593-598.

Van Steirteghem A, Van den Abbeel E, Camus M, Devroey P. 1992. Cryopreservation of human embryos. Baillière's Clinical Obstet Gyn 6:313-325.

Van Steirteghem A, Van der Elst J, Van den Abbeel E, Joris H, Camus M, Devroey P. 1994. Cryopreservation of supernumerary multicellular human embryos obtained after intracytoplasmic sperm injection. Fertil Steril 62:775-780.

Van Steirteghem A, Van den Abbeel E. 1993. Controversies in assisted reproduction. Freezing of embryos: early versus late stages. J Assist Reprod Genet 10:185-186.

Veeck LL, Amundson CH, Brothman LJ, DeScisciolo C, Maloney K, Muasher S, Jones Jr HW. 1993. Significantly enhanced pregnancy rates per cycle through cryopreservation and thaw of pronuclear stage oocytes. Fertil Steril 59:1202-1207.

Veeck L. 1993. Controversies in assisted reproduction. Freezing of pre-embryos: early versus late stages. J Assist Reprod Genet 10:181-185.

Warnes GM, Payne D, Jeffrey R, Hourigan L, Kirby C, Kerin J, Matthews C. 1997. Reduced pregnancy rates following the transfer of human embryos frozen or thawed in culture media supplemented with normal serum albumin. Hum Reprod 12:1525-1530.

Whittingham DG, Leibo SP, Mazur P. 1972. Survival of mouse embryos frozen to -196°C and -269°C. Science 178:414.

Wiemer KE, Cohen J, Tucker MJ, Godke RA. 1998. The application of co-culture in assisted reproduction:10 years of experience with human embryos. Hum Reprod 13 Suppl 4:226-238.

Wilton LJ, Shaw JM, Trounson AO. 1989. Successful single-cell biopsy and cryopreservation of preimplantation embryos. Fertil Steril 51:513-517.

Wright G, Wiker S, Elsner C, Kort H, Massey J, Mitchell D, Toledo A and Cohen J. 1990. Observations on the morphology of pronuclei and nucleoli in human zygotes and implications for cryopreservation. Hum Reprod 5:109-115.

Yokota Y, Sato S, Yokota M, Ishikawa Y, Makita M, Asada T, Araki Y. 2000. Successful pregnancy following blastocyst vitrification: case report. Hum Reprod 15:1802-1803.

Ziebe S, Bech B, Petersen K, Mikkelsen AL, Gabrielsen A, Nyboe Andersen A. 1998. Resumption of mitosis during post-thaw culture: a key parameter in selecting the right embryos for transfer. Hum Reprod 13:178-181.

Chapter 5

FACTORS INFLUENCING CRYOPRESERVATION OF DOMESTIC ANIMAL EMBRYOS

Alban Massip[1] and S. P. Leibo[2]
[1] *Université Catholique de Louvain, Unité des Sciences Vétérinaires, Place Croix du Sud 3, B-1348 Louvain-la-Neuve, Belgium.*
[2] *Audubon Center for Research of Endangered Species; Department of Biological Sciences, University of New Orleans, New Orleans, LA* 70148-2960, *USA.*

1. Introduction

The ultimate test of the developmental potential of an embryo, after its transfer into a suitable recipient, is the birth of a normal young animal. Yet, the ability of an embryo to survive the stresses of freezing can be considered as an initial "screening test". Embryos that have been derived in vivo are usually selected on the basis of their morphology. In general, there appears to be a good association, if not a correlation, between embryo "quality", assigned on the basis of morphology, and the likelihood of pregnancy (Wright, 1985). This correspondence even holds for embryos after they have been cryopreserved (Niemann et al., 1982; Kennedy et al., 1983; Leibo, 1986). For example, although assigning a "grade" to an embryo is a rather subjective judgement, embryos can be assigned numerical grades on the basis of their morphology. The pregnancy results shown in Figure 1 illustrate that the likelihood of pregnancy can be estimated from the morphological grade of an embryo assigned to it either as a fresh embryo, or for a cryopreserved embryo prior to its being frozen. In general, poorer quality embryos are less likely to yield a pregnancy, and better quality

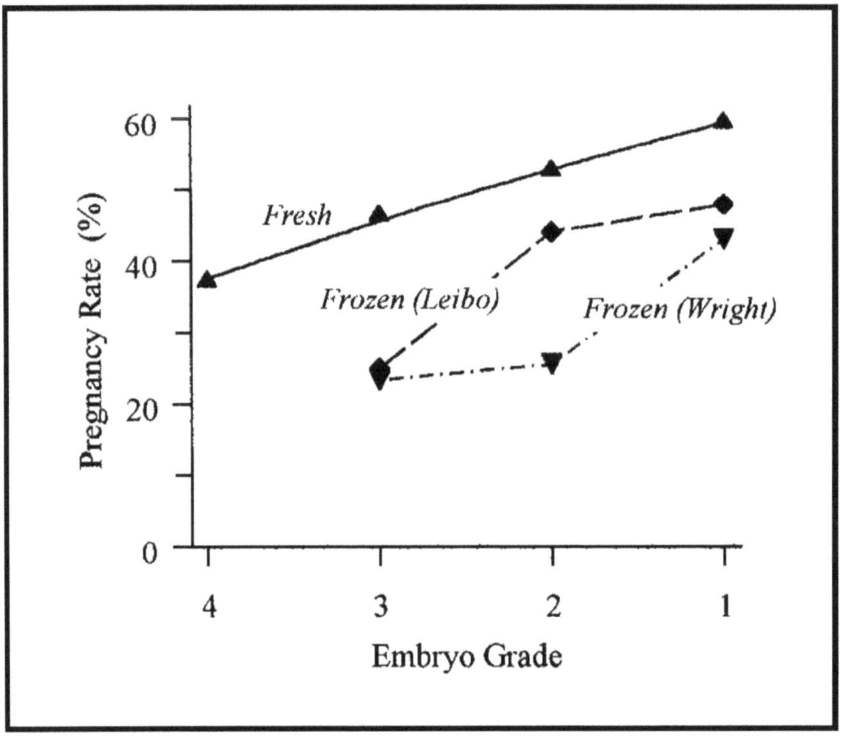

Figure 1: *Bovine pregnancies as a function of embryo grade. As they were collected, embryos were assigned a numerical grade as to their quality. Pregnancies were determined by rectal palpation 60 days after transfer. The data for fresh and for those labelled frozen embryos are those of Leibo (1986) and for the other frozen embryos are those of Wright (1985).*

embryos are more likely. Those results also indicate that, for a embryo of a given grade, freezing does reduce the likelihood of pregnancy. Some embryos classified as "good" or "excellent" may result in pregnancy if transferred as fresh embryos, but not if transferred after having been frozen. Embryos with identical morphologies do not necessarily have the same likelihood of producing a pregnancy. For many years, investigators have sought to derive in vitro methods to estimate the probability of a given embryo developing into a full-term foetus. Recently, metabolic tests have been developed to predict this probability (Donnay et al., 1999). There is an increasing demand for in vitro-produced bovine embryos to be used not only for the production of live calves, but also for production of transgenic animals and for their clones. Several reviews have been published describing the influence on various methods of embryo production on the resultant offspring (Hasler et al., 1995; van Wagtendonk et al., 2000; Farin and Farin, 1995; Farin et al., 2001).

Unfortunately, embryos produced in vitro are less viable than their in vivo-derived counterparts, and their cryopreservation is often less successful. Initially, the objective of in vitro embryo production was simply to increase the number of blastocysts, with relatively little attention given to their quality. It soon became evident, however, that embryo quality was of primary importance for success of micromanipulation of bovine embryos. Differences between the two kinds of embryos (those that develop in vivo and those produced in vitro) become most evident when they are subjected to various stresses, although reasons for these differences have not really been explained. Under certain conditions, morphological differences between in vitro-produced and in vivo-derived embryos are quite obvious, as illustrated by the embryos shown in Figure 2.

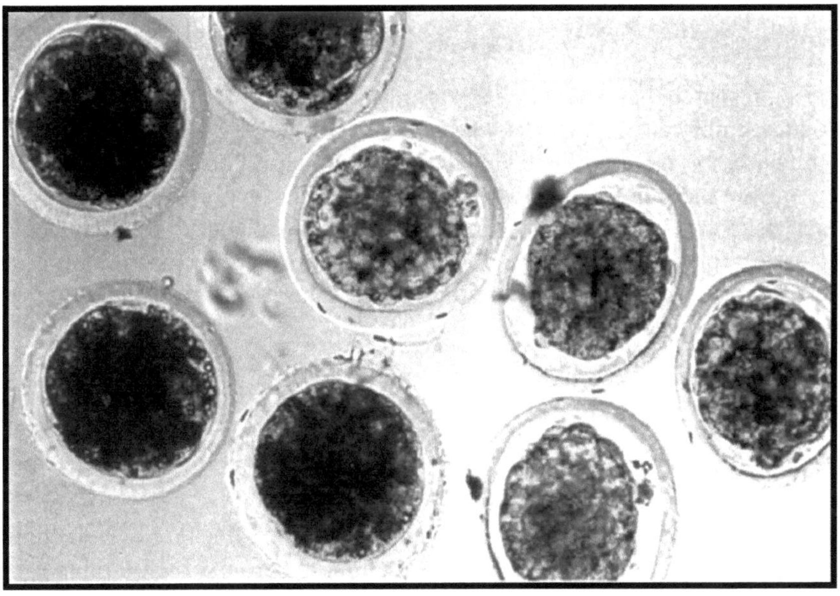

Figure 2: *Micrographs of bovine morulae. The four embryos on the left were produced in vitro by culture in the presence of bovine serum using methods described by Xu et al. (1992); those on the right were collected by non-surgical flushing of a superovulated cow 6 days after artificial insemination.*

When produced in the presence of serum, in vitro-produced embryos (IVP embryos) appear darker then their in vivo-derived counterparts, and have a much smaller perivitelline space, suggesting that compaction of adjoining cells within in vitro-produced embryos is less dense. We have speculated that the darker appearance and more diffuse packing reflect differences in lipid contents and composition of the two categories of embryos (Leibo and Loskutoff, 1993; Leibo et al., 1996). In this chapter, we will first discuss factors known to affect freezing tolerance of embryos, and then we will consider how deleterious effects of stress on embryos can be circumvented. We will conclude with a brief summary of recent developments in this field.

2. Factors affecting freezing tolerance

2.1. In vitro production systems

Recent research has shown that various aspects of the methods used to produce embryos in vitro may affect the morphology and physiology of resultant embryos, the kinetics of their development, the properties of their cell membranes through lipid per-oxidation, and finally the sensitivity of embryos to cryopreservation. In fact, embryos are influenced by each step of the in vitro procedure. Thus, there are as many different types of embryos as there are production systems. Such embryos cannot be differentiated on the basis of their morphology.

2.2. Culture conditions

The development of bovine embryos in vitro may be affected by the medium in which the oocytes were allowed to mature and from which the embryos were derived (Rose and Bavister, 1992). The culture medium in which IVP embryos are produced not only influences their development, but also exerts a significant effect on their survival after freezing (Leibo and Loskutoff, 1993). For example, Holm et al. (1994) achieved a higher post-thaw survival rate when embryos had been cultured in CR1 medium (58%) compared with SOF (8%), regardless of the gaseous atmosphere. Using differential staining to distinguish between cells of the inner cell mass (ICM) and those of the trophectoderm, Van Soom et al. (1996) compared embryos produced in TCM199 medium with those produced in B2 medium. When embryos were cultured in B2 medium, more blastocysts formed and they exhibited earlier blastulation. However, the ICM developed at a later developmental stage in embryos cultured in B2 medium compared to those cultured in TCM199 medium, and the number of cells of which the ICM was comprised was more variable in the B2-cultured embryos. The lower ratio of ICM-cells to trophectoderm cells observed for embryos grown in B2 medium may partly explain their higher sensitivity to freezing, since freezing may destroy some cells of an embryo without killing the entire embryo.

Other morphological defects of IVP embryos may also affect their survival after cryopreservation. These defects may include incomplete compaction, incomplete cell-to-cell tight junctions, and abnormal mitochondria; IVP embryos may also exhibit metabolic characteristics different from those of in vivo-derived embryos (Massip et al., 1995). For example, they show an earlier activation of glucose metabolism and a higher rate of aerobic glycolysis (Donnay et al., 1999). Together with the uncoupling of oxidative phosphorylation observed after cryopreservation due to mitochondrial damage, this increased glucose metabolism may induce a higher production of reactive oxygen species (ROS) known to be cytotoxic (Wentzel et al. 1999). Part of the toxic effect of ROS can be explained by lipid peroxidation that may alter physical and permeability properties of membranes. Similarly, IVP embryos have an altered lipid metabolism (Abd El Razek et al., 2000) that translates into lower density due to higher lipid / protein ratio caused by increased storage of triglycerides. This modified lipid content may explain a higher sensitivity to degradation by ROS generated after freezing-thawing.

2.3. Serum and lipids

When serum is added to basic culture medium, embryos may accumulate large quantities of lipids, mainly in their trophoblast cells (Ferguson and Leese, 1999). These intracellular lipids may be responsible for the low tolerance of in vitro-produced embryos to freezing and for their lower buoyant density (Pollard et al., 1993). Using a method first described for porcine embryos (Nagashima et al., 1994; 1999), partial removal of lipids by centrifugation and micromanipulation of bovine embryos has been attempted (Leibo et al., 1995). The result is that delipidation of one-cell stage bovine embryos is compatible with their normal development, and has a beneficial effect on their tolerance to freezing and thawing when frozen at the blastocyst stage (Ushijima et al., 1999). This procedure, however, alters the developmental potential of such blastocysts, suggesting that maternally inherited lipid stores may interfere with metabolic recovery after thawing (Diez et al., 1996). Charpigny et al. (2000) showed that in vitro produced embryos incorporate many more fatty acids into triglycerides than do in vivo-derived embryos. This suggests that the process of embryo culture itself may modify basal lipid metabolism and could be partly responsible for an excessive accumulation of lipids into embryonic cells, with possible effects on membranes. Non-optimum culture conditions of embryos may be responsible for the so-called "large calf" or "large lamb" syndrome (Thompson et al., 1995; van Wagtendonk et al., 2000).

It is common to supplement medium for the culture of IVP embryos with proteins or sera of various sources. However, bovine IVP embryos can be cultured with or without any added protein. As shown by the results in Figure 3, the total number of cells / blastocyst and the fraction of the total comprised of inner cell mass cells appears to be independent of specific protein supplements (Semple et al., 1993). That is, blastocysts produced in the absence of added protein consisted of the same number of total cells and of ICM-cells as those produced in media supplemented with various proteins. Moreover, the presence of serum in the medium may also influence the sensitivity of embryos to cryopreservation (Semple et al., 1995). When IVP bovine embryos are cultured in serum-free medium and then frozen under optimum conditions of cooling and warming, close to 100% will develop and hatch in culture (Hochi et al., 1996). Even IVP morulae, the embryonic stage known to be especially sensitive to chilling injury, exhibit relatively high survival and 40% hatch in culture. Recently, it has been shown that the survival of frozen-thawed bovine zygotes can be improved by allowing oocytes to mature and to be fertilized in the presence of linoleic acid plus albumin (Hochi et al., 1999).

The presence of serum supplementation in culture media may also influence the chemical composition of embryos. Sata et al. (1999) compared the total fatty acid composition of bovine embryos cultured in serum-free and serum-containing medium during early embryonic development. Embryos cultured in serum-supplemented medium had a rather dark morphology and a different fatty acid composition than those cultured in serum-free medium (See table 1 and 2 in Sata et al. 1999). Moreover, the lipid composition varied with the embryonic stage of development, thus modifying the membranes. However there was no correlation between the composition of serum used for culture and that of blastocysts cultured in its presence. Some fatty acids that are present in low concentrations in tissues are more abundant in oocytes and embryos.

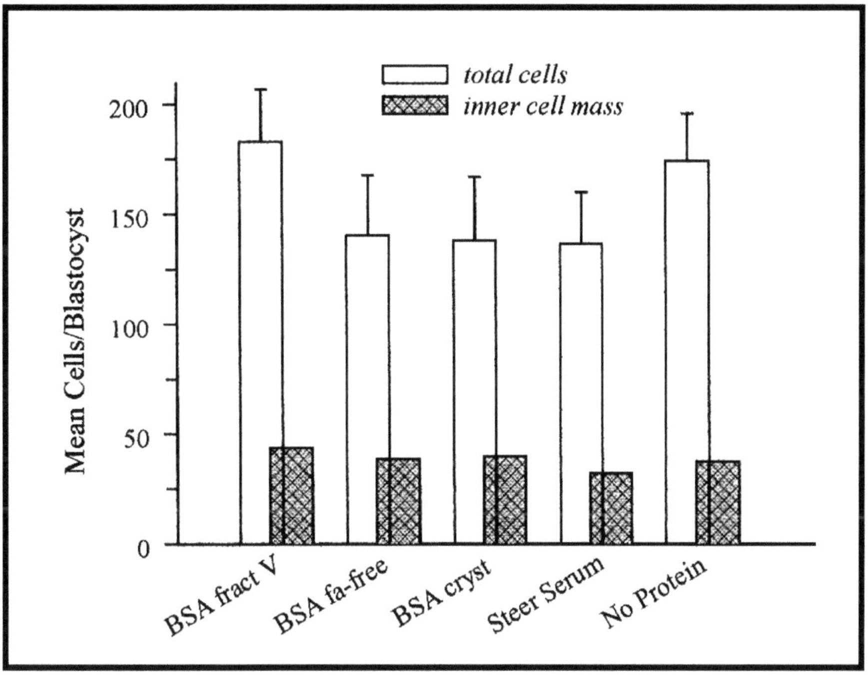

Figure 3: *The mean number of total cells in bovine in vitro-produced blastocysts cultured for ≈190 hours after fertilization in TCM199 medium supplemented with each of the indicated proteins, or with no added protein. By differential staining, the mean number of cells within the inner cell mass was also determined. The data are those of Semple et al., 1995.*

2.4. Kinetics of development

The rate at which embryos develop in vitro may also affect their response to freezing. Of those embryos that are at the 2- to 4-cell stage at 30 hours after insemination of the oocytes from which they are derived, 70% reach the blastocyst stage, whereas only 35% of those that cleave later develop into blastocysts (Plante and King, 1992; Grisart et al., 1994). Those embryos that develop at a faster rate appear to be more tolerant to cooling (Dinnyes et al., 1999). McKiernan and Bavister (1994) showed that hamster embryos which develop faster in vitro are twice as likely to develop into viable foetuses as the slower ones, even when blastocysts from both groups were transferred in equal proportions. Those authors suggested that analysis of the rate of embryonic development may be useful as a non-invasive indicator of embryo viability.

2.5. Oil used in microdrop culture

In many laboratories, embryos are cultured under paraffin or silicone oil and each new batch of oil is tested before use. However, the experience of one lab and its problems with oils used to overlay culture media are instructive. Blastocysts produced in medium under both types of oil were vitrified. None of the blastocysts produced under silicone oil survived the vitrification procedure whereas 59% of those cultured in the same medium but under paraffin oil did survive. In this case, not only was the rate of embryo production influenced by the type of oil overlay, but their sensitivity to cryopreservation was significantly affected (Van Soom et al., 2001). The problem was attributed to changes in fatty acid proportions of the oil that could cause membrane damage by lipid peroxidation. Perturbations in the fluidity and structure of the plasma membrane may be caused by lipid peroxidation. This, in turn, may reduce membrane permeability to water and to permeating cryoprotectants, as well as decreasing the tolerance of embryos to stresses resulting from cryopreservation. This tolerance can be modified by inhibitors or stimulators of lipid peroxidation (Tarin and Trounson, 1993).

2.6. Post-thaw recovery and treatment

Embryos classified as "freezable" must be able to recover their viability quickly after thawing so as to maintain synchrony between donor and recipients. The rate at which embryos resume development after cryopreservation may be a criterion of an embryo's viability. In our laboratory, alterations of cell membranes resulting from various vitrification methods (Donnay et al., 1998) have been visualized by double staining with bis-benzimide and propidium iodide (Kaidi et al., 1999). The speed of recovery after thawing depends not only on integrity of the embryo itself but also on the nature of the post-warming culture medium. Often, bovine embryos have been co-cultured with bovine oviductal or buffalo rat liver cells (Massip et al., 1993; Kaidi et al., 1998; Rizos et al., 2001).

3. Cryobiological Aspects

3.1. Chilling sensitivity

Initially, the same methods that are commonly used to cryopreserve in vivo-derived embryos were also used for in vitro-produced embryos. However, survival rates of IVP embryos were significantly lower than those for the in vivo-derived embryos, were more variable, and seemed to be highly dependent on the embryonic stage of development. Therefore, modifications were sought to adapt methods for these embryos that seemed especially sensitive to cryopreservation. It was soon learned that IVP-derived morulae were chilling sensitive (Pollard and Leibo, 1994; Leibo et al., 1996). Exposing them to temperatures below 10°C decreased their developmental capacity, and exposure to 0°C even for a few minutes drastically reduced their development into blastocysts. Therefore, it was soon realized that one way to circumvent chilling injury would be to cool the IVP embryos at very high rates through the injurious temperature zone of +10° to -5°C so as to "outrace" the injury. One way to achieve this was to use a procedure pioneered by Rall and Fahy (1985), a method referred to as "vitrification." The principle of the method is to use concentrated solutions of cryoprotection agents (CPAs) and very high cooling rates, conditions that cause the solution in which embryos are suspended to form a glass, or vitrify.

3.2. Vitrification

Numerous studies of vitrification of embryos of domestic species have been published. Recent examples of vitrification of IVP bovine embryos include papers by Bautista and Kanagawa (1998), Kaidi et al. (1998); Ohboshi et al. (1997), Pugh et al. (2000), Saha and Suzuki (1997). One especially novel method is the Open Pulled Straw method of Vajta et al. (1997, 1998; summarized in Vajta, 2000). There have also been comparative studies of the efficacy of conventional slow freezing and rapid methods of cryopreservation, including field trials to determine ultimate pregnancy rates achieved with each method, although these were with in vivo-derived embryos (van Wagtendonk et al., 1997; 2000). Despite the successes achieved with vitrification, the method has not been widely accepted by embryo transfer practitioners, as discussed recently by Vajta (2000).

In our opinion, successful cryopreservation of IVP embryos is not a problem of methodology, but rather of the ability of embryos to survive the stresses of the entire treatment. With standard methods of equilibrium cooling, these stresses include: [1] exposure to CPAs, [2] exposure to non-physiological temperatures, i.e. below +30°C, [3] phase change of the suspending solution from a liquid into a crystalline solid, [4] drastic increase in the concentration of the solute as water is removed in the form of ice. During warming and thawing, these changes are reversed. With methods of non-equilibrium cooling, such as vitrification, the primary stresses are: [1] exposure to extremely concentrated solutions of CPAs, and [2] abrupt change in the viscosity of the solution as the temperature is decreased. With vitrification, the CPA may consist of a 6 to 8 Molar solution of one or more non-physiological compounds, such as ethylene or propylene glycol, methanol, or dimethyl sulfoxide, plus saccharides and macromolecules.

3.3. Cryoprotectant solutions

It has long been recognized that many low molecular weight compounds can act to protect embryos against freezing damage. Therefore, many investigations have been performed to compare the efficacy of different CPAs for in vitro-derived bovine embryos. One very large study found that the post-thaw survival of IVP-derived blastocysts frozen in either 1.4 M glycerol (543 embryos) or in 1.5 M ethylene glycol (588 embryos) was very similar, yielding respective hatching rates in culture of 69% and 62%. (Hasler et al., 1997).

Provided that the embryos are highly permeable to the CPA being used and the CPA is non-toxic, it can be used at various concentrations (ethylene glycol, for example; Bautista and Kanagawa, 1998; Sommerfeld and Niemann, 1999). Under such conditions, survival is equivalent whatever the freezing method. Freezing media may also contain proteins, sera or other biological macromolecules, or any of a variety of synthetic macromolecules, such as polyethylene glycol, polyvinylpyrrolidone, or Ficoll. One problem with the use of naturally-occurring macromolecules, e.g. serum albumin, is that they may act as a carrier for many substances, such as steroids, fatty acids, citrate. Other biological macromolecules, such as hyaluronate or glycosamino-glycans, have also been used as serum substitutes for freezing of embryos (Palasz et al., 1993). Some of these have also been used for embryo culture.

3.4. Osmotic behaviour in cryoprotectant solutions

With our method of vitrification (Donnay et al., 1998), embryos are exposed to high concentrations of cryoprotectants before the vitrification solution itself. Depending on the permeability of the embryos to the CPA, the temperature of exposure, and the CPA concentration, the embryos may be subjected to osmotic shock either during addition and/or during removal of the cryoprotectants. We have found that those IVP bovine embryos that respond osmotically when subjected to various hypertonic solutions are those that will ultimately hatch in culture (Kaidi et al., 2000). Because this test is easy to perform, it might have a certain predictive value to estimate the chances of an embryo surviving cryopreservation.

4. Possible methods to Improve Freezing Resistance

In cattle, some 60 to 70 genes controlling about 15 physiological functions have been investigated (reviewed by Niemann and Wrenzycki, 2000). The majority of these showed either embryonic or maternal and embryonic genomic activity. Expression of genes of importance like connexins or heat shock protein (HSP70) were nil in vitro. The connexin 43 gene is crucial for maintenance of compaction. Here again, the basic culture medium (SOF or TCM199) may exert profound effects on levels of expression of the majority of these genes.

4.1. Provide the embryo with anti-stress protection

To survive cryopreservation, the embryo must be able to resist oxidative stress. To do this, it must possess protective mechanisms, such as those described by Guerin (1994). These include various enzymatic and non-enzymatic antioxidants. These systems may be present in the embryo itself or otherwise must be added to the extra-embryonic environment. Enzymatic protection against per-oxidations is provided by superoxide dismutase (SOD), catalase, glutathione peroxidase, and glutathione reductase. Glutathione plays an important role in embryo development and cysteamine added to the maturation medium increases the endogenous pool of glutathione (Hashimoto et al., 2000). As a consequence, more embryos develop faster and reach the blastocyst stage on day 6 of culture, which is a convenient stage for freezing (de Matos et al., 1995; 1996). Cysteamine and hypotaurine are precursors of taurine, but in the culture system used (co-culture with Vero cells, Guyader-Joly et al., 1998), cysteamine had no effect on embryo development. In contrast, the addition of hypotaurine to the culture medium significantly improved blastocyst production and quality.

4.2. Modification of embryo lipid content and composition

Phospholipids (lecithin) have been added to the freezing medium to try to minimize membrane damage (Guyader-Joly et al., 1999). Ayoagi et al. (1996) added a lipid-rich fraction of BSA (Albumax) to the cryoprotectant solution in which bovine embryos were frozen, and achieved pregnancy rates of 50% after direct transfer of the cryopreserved embryos. As mentioned above, Hochi et al. (1999) improved the tolerance to freezing of bovine zygotes by culturing them in Hepes-buffered TCM199 medium supplemented with linoleic acid-albumin (LAA) from bovine serum. The post-thaw cleavage rate of LAA-treated zygotes was significantly higher than that of untreated zygotes.

5. Conclusion

From this review, we conclude that it ought to be possible to produce embryos by in vitro methods of maturation, fertilization and culture that have been rendered considerably more resistant to the stresses of cryopreservation than those produced by most current methods. Optimally, the goal should be to produce such an embryo that is extremely tolerant of cryopreservation in a simple medium, without serum (Yamashita et al., 1999), but supplemented with anti-oxidants such as glutathione, ß-mercaptoethanol and/or trolox (Feugang et al., 2000a, b), as well as other exogenous substances, like growth factors (Lim and Hansel, 2000) and perhaps some lipid concentrates.

References

Abd El Razek IM, Charpigny G, Kodja S, Marquant-Leguienne B, Mermillod P, Guyader-Joly C, Humblot P. 2000. Differences in lipid composition between in vivo- and in vitro-produced bovine embryos. Theriogenology 53:346.

Aoyagi Y, Konishi M, Takedomi T, Itakura T, Itoh T, Yazawa S. 1996. Effect of lipid-rich bovine serum albumin on direct transfer of frozen-thawed bovine embryos. Theriogenology 45:165.

Bautista JAN, Kanagawa H. 1998. Current status of vitrification of embryos and oocytes in domestic animals: ethylene glycol as an emerging cryoprotectant of choice. Japan J Vet Res 45:183-191.

Charpigny G, Marquant-Leguienne B, Abd El Razek I, Humblot P, Mermillod P. 2000. Differences in fatty acid incorporation into lipids between in vivo and in vitro produced embryos from ruminant species. 7th Intern Conference on Goats, Poitiers, France.

de Matos DG, Furnus CC, Moses DF, Baldassarre H. 1995. Effect of cysteamine on glutathione level and developmental capacity of bovine oocytes matured in vitro. Mol Reprod Dev 42:432-436.

de Matos DG, Furnus CC, Moses DF, Martinez AG, Matkovic M. 1996. Stimulation of glutathione synthesis of in vitro matured bovine oocytes and its effect on embryo development and freezability. Mol Reprod Dev 45:451-457.

Diez C, Le Bourhis Y, Heyman Y, Renard JP. 1996. Effect of partial lipid removal from in vitro produced bovine zygotes on further development in vitro and on the freezing tolerance of blastocysts. Theriogenology 45:166.

Dinnyes A, Lonergan P, Fair T, Boland MP, Yang X. 1999. Timing of first cleavage post-insemination affects cryosurvival of in vitro-produced bovine blastocysts. Mol Reprod Dev 53:318-324.

Donnay I, Auquier P, Kaidi S, Carolan C, Lonergan P, Mermillod P, Massip A. 1998. Vitrification of in vitro produced bovine blastocysts: methodological studies and developmental capacity. Anim Reprod Sci 52:93-104.

Donnay I, Partridge RJ, Leese H. 1999. Can embryo metabolism be used for selecting bovine embryos before transfer? Reprod Nutr Dev 39:523-533.

Farin PW, Crossier AE, Farin CE. 2001. Influence of in vitro systems on embryo survival and fetal development in cattle. Theriogenology 55:151-170.

Farin PW, Farin CE. 1995. Transfer of bovine embryos produced in vivo or in vitro: survival and fetal development. Biol Reprod 52:676-682.

Ferguson EM, Leese HJ. 1999. Triglyceride content of bovine oocytes and early embryos. J Reprod Fert 116:373-378.

Feugang JM, Dessy F, Donnay I. 2000a. Oxidative stress: preventive effect of β-mercaptoethanol and trolox. 15th Ann Conf European Assoc Embryo Trans. Santander, Spain. p 144.

Feugang JM, Moens A, Bernard S, Dessy F, Donnay I. 2000b. Induction of degenerateness and apoptosis in in vitro-produced bovine embryos by pro-oxidants agents. Theriogenology 53:352.

Grisart B, Massip A, Dessy F. 1994. Cinematographic analysis of bovine embryos development in serum-free oviduct conditioned medium. J Reprod Fert 101:257-264.

Guerin P, El Mouatassim S, Menezo Y. 2001. Oxidative stress and protection against reactive oxygen species in the pre-implantation embryo and its surroundings. Hum Reprod Update 7:175-189.

Guyader-Joly C, Guérin P, Renard JP, Guillaud J, Ponchon S, Ménézo Y. 1998. Precursors of taurine in female genital tract: effects on developmental capacity of bovine embryos produced in vitro. Amino Acids 15:27-42.

Guyader-Joly C, Ponchon S, Durand M, Heyman Y, Renard JP, Ménézo Y. 1999. Effect of lecithin on in vitro and in vivo survival of in vitro produced bovine blastocysts after cryopreservation. Theriogenology 52:1193-1202.

Hashimoto S, Minami N, Yamada M, Imai H. 2000. Excessive concentration of glucose during in vitro maturation impairs the developmental competence of bovine oocytes after in vitro fertilization: relevance to intracellular reactive oxygen species and glutathione contents. Mol Reprod Dev 56:520-526.

Hasler JF, Henderson WB, Hurtgen PJ, Jin QZ, McCauley AD, Mower SA, Neely B, Shuey LS, Stokes JE, Trimmer SA. 1995. Production, freezing and transfer of bovine IVF embryos and subsequent calving results. Theriogenology 43:141-152.

Hasler JF, Hurtgen PJ, Jin ZQ, Stokes JE. 1997. Survival of IVF-derived bovine embryos frozen in glycerol or ethylene glycol. Theriogenology 48:563-579.

Hochi S, Kanamori A, Sugisawa K, Kimura K, Hanada A. 1999. Effect of linoleic acid-albumin in the IVM and IVF media on survival of frozen-thawed pronuclear bovine zygotes. J Mamm Ova Res 16:19-22.

Hochi S, Semple E, Leibo SP. 1996. Effect of cooling and warming rates during cryopreservation on survival of in vitro-produced bovine embryos. Theriogenology 46:837-847.

Holm P, Smith S, Callesen H. 1994. Post-thaw in vitro viability of bovine blastocysts produced in vitro in different media and gas atmospheres. Proc. 10th Europ Emb Trans Assoc, Lyon, France, p 182.

Kaidi S, Donnay I, Lambert P, Dessy F, Massip A. 2000. Osmotic behavior of in vitro produced bovine blastocysts in cryoprotectant solutions as a potential predictive test of survival. Cryobiology 41:106-115.

Kaidi S, Donnay I, Van Langendonckt A, Dessy F, Massip A. 1998. Comparison of two co-culture systems to assess the survival of in vitro produced bovine blastocysts after vitrification. Anim Reprod Sci 52:39-50.

Kaidi S, Van Langendonckt A, Massip A, Dessy F, Donnay I. 1999. Cellular alteration after dilution of cryoprotective solutions used for the vitrification of in vitro-produced bovine embryos. Theriogenology 52:515-525.

Kennedy LG, Boland M, Gordon I. 1983. The effect of embryo quality at freezing on subsequent development of thawed cow embryos. Theriogenology 19:823-832.

Leibo SP, Loskutoff NM. 1993. Cryobiology of in vitro derived embryos. Theriogenology 39:81-94.

Leibo SP, Martino A, Kobayashi S, Pollard JW. 1996. Stage-dependent sensitivity of oocytes and embryos to low temperatures. Anim Reprod Sci 42:45-53.

Leibo SP, Pollard JW, Martino A. 1995. Chilling and freezing sensitivity of "re-assembled" in vitro-derived bovine embryos. Theriogenology 43:264.

Leibo SP. 1986. Cryobiology: Preservation of mammalian embryos. In: Genetic Engineering of Animals: An Agricultural Perspective. (Evans JW, Hollaender A, eds). Plenum Press, NY, pp 251-272.

Lim JM, Hansel W. 2000. Exogenous substances affecting development of in vitro derived bovine embryos before and after embryonic genome activation. Theriogenology 53:1081-1091.

Massip A, Mermillod P, Dinnyes A. 1995. Morphology and biochemistry of in vitro bovine embryos: implications for their cryopreservation. Hum Reprod 10:3004-3011.

Massip A, Mermillod P, Wils C, Dessy F. 1993. Effects of dilution procedure and culture conditions after thawing on survival of frozen bovine blastocysts produced in vitro. J Reprod Fertil 97:65-69.

McKiernan SH, Bavister BD 1994. Timing of development is a critical parameter for predicting successful embryogenesis. Hum Reprod 9:2123-2129.

Nagashima H, Cameron RDA, Kuwayama M, Young M, Beebe L, Blackshaw AW, Nottle MB. 1999. Survival of porcine delipated oocytes and embryos after cryopreservation by freezing or vitrification. J Reprod Devel 45:167-176.

Nagashima H, Kashiwasaki N, Ashman RD, Grupen CG, Seamark RF, Nottle MB. 1994. Removal of cytoplasmic lipid enhances the tolerance of porcine embryos to chilling. Biol Reprod 51:618-622.

Niemann H, Schilling E, Sacher B, Smidt D. 1982. Effect of bovine embryo quality on survival rate after fast freezing and thawing. In: Embryo Transfer in Mammals. 2nd Int Congr (Mérieux C, Bonneau M, eds). Fondation Marcel Mérieux, Lyon, France, p 313.

Niemann H, Wrenzycki C. 2000. Alteration of expression of developmentally important genes in preimplantation bovine embryos by in vitro culture conditions: implications for subsequent development. Theriogenology 53:21-34.

Ohboshi S, Fujihara N, Yoshida T, Tomogane H. 1997. Usefulness of polyethylene glycol for cryopreservation by vitrification of in vitro-derived bovine blastocysts. Anim Reprod Sci 48:27-36

Palasz A, Alkemade S, Mapletoft RJ. 1993. The use of sodium hyaluronate in freezing media for bovine and murine embryos. Cryobiology 30:172-178.

Plante L, King WA. 1992. Effect of time to first cleavage on hatching rate of bovine embryo in vitro Theriogenology 37:274 (Abstract).

Pollard JW, Leibo SP. 1994 Chilling sensitivity of mammalian embryos. Theriogenology 41:101-107.

Pollard JW, Plante C, Songsasen N, Leibo SP. 1993. Correlation of sensitivity to chilling and freezing with buoyant density of mammalian embryos. Cryobiology 30:631.

Pugh PA, Tervit HR, Niemann H. 2000. Effects of vitrification medium composition on the survival of bovine in vitro produced embryos, following in straw-dilution, in vitro and in vivo following transfer. Anim Reprod Sci 58:9-22.
Rall WF, Fahy GM. 1985. Ice-free cryopreservation of mouse embryos at -196ºC by vitrification. Nature 313:573-575.
Rizos D, Ward F, Boland MP, Lonergan P. 2001. Effect of culture system on the yield and quality of bovine blastocysts as assessed by survival after vitrification. Theriogenology 56:1-16.
Rose TA, Bavister BD. 1992 Effect of oocyte maturation medium on in vitro development of in vitro fertilized bovine embryos. Mol Reprod Dev 21:72-77.
Saha S, Suzuki T. 1997. Vitrification of in vitro produced bovine embryos at different ages using one- and three-step addition of cryoprotective additives. Reprod Fertil Dev 9:741-746.
Sata R, Tsujii H, Abe H, Yamashita S, Hoshi H. 1999. Fatty acid composition of bovine embryos cultured in serum-free and in serum-containing medium during early embryonic development. J Reprod Dev 45:97-103.
Semple E, Loskutoff N, Leibo SP, Betteridge KJ. 1993. Effects of culture medium and maturation time on in vitro development of bovine oocytes into blastocysts. Theriogenology 39:285.
Semple ME, Betteridge KJ, Leibo SP. 1995. Cryopreservation of in vitro-derived bovine embryos produced in a serum-free culture system. Theriogenology 43:320.
Sommerfeld V, Niemann H. 1999. Cryopreservation of bovine in vitro produced embryos using ethylene glycol in controlled freezing or vitrification. Cryobiology 38:95-105.
Tarin JJ, Trounson AO. 1993. Effects of stimulation or inhibition of lipid peroxidation on freezing-thawing of mouse embryos. Biol Reprod 49:1362-1368.
Thompson JG, Gardner DK, Pugh PA, McMillan WH, Tervit HR. 1995. Lamb birth weight is affected by culture system utilized during in vitro pre-elongation development of ovine embryos. Biol Reprod 53:1385-1391.
Ushijima H, Yamakawa H, Nagashima H. 1999. Cryopreservation of bovine pre-morula-stage in vitro matured/in vitro fertilized embryos after delipidation and before use in nucleus transfer. Biol Reprod 60:534-539.
Vajta G, Booth PJ, Holm P, Greve T, Callesen H. 1997. Successful vitrification of early stage bovine in vitro-produced embryos with the Open Pulled Straw (OPS) method. CryoLetters 18:191-195.
Vajta G, Holm P, Kuwayama M, Booth PJ, Jacobsen H, Greve T, Callesen H. 1998. Open pulled straw (OPS) vitrification: a new way to reduce cryoinjuries of bovine ova and embryos. Mol Reprod Dev 51:53-58.
Vajta G. 2000. Vitrification of the oocytes and embryos of domestic animals. Anim Reprod Sci 60:357-364.
Van Soom A, Boerjan A, Ysebaert MT, De Kruif A. 1996. Cell allocation to inner cell mass and the trophectoderm in bovine embryos cultured in two different media. Mol Reprod Dev 45:171-182.

Van Soom A, Mahmoudzadeh AR, Christophe A, Ysebaert MT, de Kruif A. 2001. Silicone oil used in microdrop culture can influence bovine embryonic development and freezability in a negative way. Reprod Dom Anim 36:169-176

van Wagtendonk-de Leeuw AM, den Daas JHG, Rall WF. 1997. Field trial to compare pregnancy rates of bovine embryo cryopreservation methods: vitrification and one-step dilution versus slow freezing and three-step dilution. Theriogenology 48:1071-1084.

van Wagtendonk-de Leeuw AM, Mullaart E, de Roos APW, Merton JS, den Daas JHG, Kemp B, de Ruigh L. 2000. Effects of different reproduction techniques: AI, MOET or IVP on health and welfare of bovine offspring. Theriogenology 53:575-597.

Wentzel P, Welch N, Eriksson UJ. 1999. Increased lipid peroxidation, diminished cyclo-oxygenase expression, developmental damage and lowered prostaglandin E2 levels in rat embryos exposed to a diabetic environment. Diabetes 48:813-820.

Wright J. 1985. Commercial freezing of bovine embryos in straws. Theriogenology 23:17-29.

Xu KP, Yadav BR, Rorie RW, Plante L, Betteridge KJ, King WA. 1992. Development and viability of bovine embryos derived from oocytes matured and fertilized in vitro and co-cultured with bovine oviductal epithelial cells. J Reprod Fert 94:33-43.

Yamashita S, Abe H, Itoh T, Satoh T, Hoshi H. 1999. A serum-free culture system for efficient in vitro production of bovine blastocysts with improved viability after freezing and thawing. Cytotechnology 31:1-9.

Chapter 6

TIMING OF EMBRYO DEVELOPMENT

Barry Bavister
Department of Biological Sciences
University of New Orleans
200 Computer Center
New Orleans, LA 70148-2960, USA

1. Introduction

In vitro produced (IVP) embryos often exhibit wide variations in developmental competence and viability, considerably more than are exhibited by embryos that develop in vivo. These anomalies in IVP embryos may be due to heterogeneity of oocyte quality, suboptimal culture conditions, disturbances in gene expression, or most likely a combination of these factors (Ho et al., 1994; Roth et al., 1994; McKiernan and Bavister, 1998; Hasler, 1998; Schramm and Bavister, 1999; Doherty et al., 2000; Hyttel et al., 2000; Niemann and Wrenzycki, 2000; Wrenzycki et al., 2001). In research studies or in clinical applications with domesticated animals, cats, non-human primates and humans, oocytes used for IVP are usually collected from a heterogeneous cohort of ovarian follicles that include oocytes which normally might not be ovulated and/or are deficient in developmental competence. Moreover, although major improvements in culture media for oocyte maturation and embryo culture have been made in the last decade or so, IVP technology is still far from perfect (Bavister, 1995; Thompson and Peterson, 2000; Gardner et al., 2000; Vanroose et al., 2001). It is now clear that gene expression, including the critical transition from maternal to embryonic genome control, is subject to perturbations by culture media components, although the mechanism of this disturbance is not understood.

Thus, there are two principal concerns with respect to IVP technology. We need to further improve IVP procedures, especially culture media formulations, in order to raise the average quality of embryos; and also to improve methods for selecting the most viable embryos for transfer. For embryo selection, the dilemma is to evaluate embryos in ways that are rapid, simple, accurate and non-invasive. At present, the method that best fits all of these criteria is embryo development timing. While the timing of embryo development in vivo mostly proceeds according to a precise schedule in all species, IVP embryos commonly show heterogeneity in their development timing, concomitant with variability in developmental competence. Those embryos that develop in a timely manner have the highest developmental competence and viability post-transfer, as shown in studies with a variety of species, such as rodents, cattle, and primates including humans.

2. Timing of embryo development as an indicator of viability

2.1. Hamster embryos

Rodent embryos provide an ideal model for studying the timing of early development both in vivo and in vitro. They can be obtained in quite large numbers so that multifactorial experiments can be done; they are fairly homogeneous in quality because they are usually collected during or after fertilization in vivo; and most importantly, numerous embryos can be transferred into the same recipient female for evaluation of their viability. Some of these features are shared by pigs and rabbits, which have also been widely used for embryo research. Another major advantage of laboratory rodents is their very short gestation period, i.e., 3 weeks or less (only 16 days in the golden hamster), which allows rapid data collection on foetal development or production of offspring. The golden hamster also has a particular feature making it very useful for evaluation of embryo viability, which is the anatomical separation of the two uterine horns so that embryo migration from one to the other is prevented. Thus, different groups of embryos can be transferred into each uterine horn of the same female, allowing direct comparison to be made of embryo viability without the confounding effect of different recipients. This model has been used extensively in my laboratory.

In one study, pronucleate ova collected from mated, naturally-cycling (i.e., non-superovulated) hamsters were cultured on the stage of an inverted microscope for up to four days while time-lapse digital images were collected (Gonzales et al., 1995). A similar, cinematographic approach was also used effectively by Grisart et al. (1994). The great advantage of dynamic recording is that development timing data are obtained for each embryo independently of the others, so that a complete data set (mean timing ± sem) is produced from a single batch of cultured embryos. By replaying the stored images, timing data can be obtained relating any one developmental event to any other. Using this method with hamster embryos, it was found that the time interval from the second to the third cleavage divisions was inversely correlated with the probability of reaching the blastocyst stage and escaping from the zona pellucida (Gonzales et al., 1995). Of the embryos that took 750 min or less to complete the third cleavage division (i.e., the 4-cell to 8-cell transition), 92% developed into blastocysts and 69% escaped their zonae pellucidae, while for embryos that took up to 900 min for this transition, only 75% reached the blastocyst stage and 49% escaped their zonae. These differences are significant ($P \leq 0.002$ and $P \leq 0.03$ for blastocyst development and for zona escape, respectively). This study showed that by simply observing the time needed for embryos to progress from the 4-cell to the 8-cell stage we could predict which ones were the most competent to develop into blastocysts. Why the third cleavage division is so critical is unknown; although the major maternal embryo genome control transition occurs in hamsters at the 2-cell stage (Seshagiri et al., 1992), possibly some transcriptional activity occurring at that time affects subsequent development timing (see 'Genetic regulation' below).

Under in vivo conditions, almost all of the pronucleate ova from naturally-cycling females would develop and implant, as evidenced by the high correlation between the numbers of corpora lutea and of blastocysts ($r = 0.97$; Gonzales and Bavister, 1995). In view of this, the wide differences reported by Gonzales et al. (1995) in development timing in vitro with pronucleate ova fertilized in vivo that were illustrates two related key points. One is that embryos produced in vivo are heterogeneous but differences are not expressed due to the optimal environment provided by the female reproductive tract. The corollary of this is that embryo culture conditions are sub-optimal because the inherent heterogeneity is exhibited, as evidenced by the timing differences.

These points were further elaborated in a study on the viability of hamster embryos related to their development timing in vitro, using as a reference point the time normally taken to complete the third cleavage division by in vivo produced embryos, i.e., 58 h post-egg activation (PEA) (McKiernan and Bavister, 1994). Whereas by 58 h PEA in vivo virtually all embryos have cleaved to 8-cells, when fertilized ova were collected from the oviducts of mated superovulated females at approx. 10 h PEA and then cultured for 48 hr, 61% had reached 8-cells but 19% were still at 4-cells and 16% were at intermediate stages. Embryos in the latter two groups were still developmentally competent because almost all of them would reach 8-cells when cultured for a few hours longer. One-cell embryos collected 10 h PEA were cultured in medium HECM-6 for 48 h (20 embryos per drop), then evaluated for cleavage stage reached. Embryos still at the 4-cell stage by 58 h PEA were considered "slow" and embryos at the 8-cell stage were considered "fast", i.e., normal, according to the timetable for in vivo development. All embryos were then cultured in fresh medium for an additional 20 h (=78 h PEA); fast cleaving embryos produced more blastocysts than slow cleaving embryos (74% vs. 13%; Figure 1). Some of this difference might be explained by slow embryos being derived from oocytes that were ovulated and hence fertilized later than fast embryos. However, the slow cleaving embryos did not catch up to the fast embryos in terms of blastocyst production even when culture was continued for another 4 or 8 h (82 and 86 h PEA, respectively), approx. the time needed for completion of an additional cell cycle. Thus, the slower cleaving embryos were on average less developmentally competent under in vitro culture conditions.

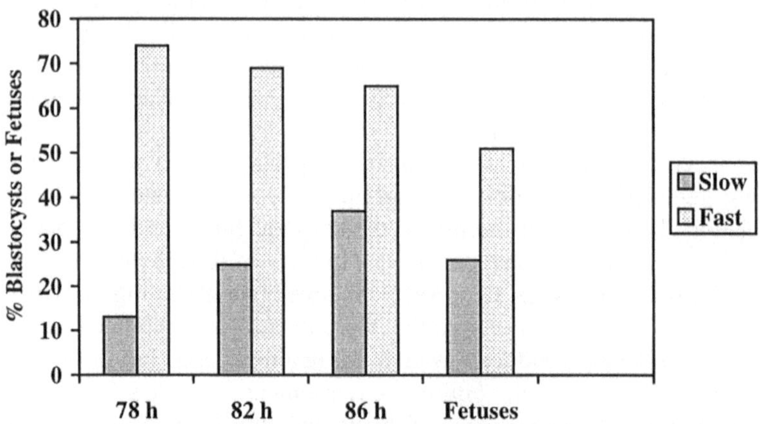

Figure 1: Differences in developmental competence and viability of slow- vs. fast-cleaving hamster embryos.

Values for % blastocysts within each time interval (h PEA) are significantly different ($P<0.0001$); foetal development values are also significantly different ($P=0.02$). Data from McKiernan and Bavister, 1994.

In a separate experiment, morulae and blastocysts that developed by 72 h of culture (82 h PEA) from slow cleaving vs. fast cleaving fertilized ova were transferred into opposite uterine horns of pseudopregnant recipients. The proportions and numbers of morulae and blastocysts transferred were the same across the two conditions (slow and fast). The proportions of embryos that developed into near-term foetuses (day 14 of pregnancy) were double for the fast-cleaving vs. the slow-cleaving embryos (51% vs. 26%, respectively; Figure 1). This study illustrates the heterogeneity of in vivo fertilized ova referred to in the Introduction. It also showed that (i) slower developing embryos, while they could reach the morula and blastocyst stages, could not do so within the same culture time period as faster developing embryos (i.e., those that were already 8-cells in vitro at 58 h PEA); (ii) even with prolonged culture, embryos that were still 4-cells at 58 h PEA never developed into morulae or blastocysts at the same frequency as the faster cleaving embryos did without additional culture; and (iii) blastocysts derived from culturing slow cleaving embryos were much less viable compared with blastocysts from fast-cleaving embryos as judged by production of near-term foetuses. This last point, while emphasizing the critical importance of embryo development timing for viability, also illustrates that the viability of embryos cannot be accurately evaluated by appearance alone, because only well-formed morulae and blastocysts were transferred in this study.

2.2. Cattle embryos

The key role of cleavage timing for selecting the most competent embryos and for revealing physiological differences in apparently identical embryos is also well illustrated in studies with cattle. Fast cleaving IVP embryos reached morulae and blastocysts at significantly higher frequencies than slower cleaving embryos (Grisart et al., 1994; Van Soom et al., 1997a). In addition, the faster cleaving embryos showed a lower incidence of fragmentation (Van Soom et al., 1997a). This study did not find any correlation between the timing of the first cleavage division with timing of blastocyst formation, in accordance with a study on hamster embryos (Gonzales et al., 1995). However, the timing of blastocyst development was correlated with cell numbers in the blastocysts: there were significantly fewer total and ICM-cells in day 8 vs. day 6 or day 7 blastocysts (Van Soom et al., 1997a). This is remarkable because one would expect that embryos cultured for 24 h longer would have more blastocyst cells instead of less.

That the reverse was found is a telling indictment of the quality of slower developing embryos. Valuable data on the timing of developmental events in vivo were provided by Van Soom et al. (1997b). The timing of compaction, blastocyst formation and hatching as well as the corresponding cell numbers were examined, including the minimum number of ICM-cells in blastocysts. Parenthetically, in contrast, the absence of comparable data on human embryo development in vivo is an impediment to our ability to evaluate timing as a factor in blastocyst production in vitro, as well as hindering the improvement of culture media for supporting normal blastocyst development.

Another study with cattle involved more than 6,000 embryo transfers over a period of 5 years (Hasler, 1998). Regardless of whether grades 1 or 2 blastocysts were transferred (and in a few cases, grade 3), pregnancy rates were very high with IVP fresh embryos (54%), although they were even higher with in vivo produced blastocysts (67%). The results with IVP blastocysts are remarkable considering that the embryos were derived from IVM oocytes and that only one embryo was transferred per recipient (we should perhaps ask why similar success could not be obtained with human IVP embryos). Pregnancy rates were significantly reduced when cattle blastocysts that took 8 days to develop in vitro were compared with those for day 7 blastocysts (43% vs. 54%, respectively; Hasler, 1998). In a small subset (107 transfers total), cryopreserved day 7 blastocysts gave 42% pregnancies vs. only 20% for frozen day 8 blastocysts. Thus, it appears that not only does cryopreservation of IVP cattle blastocysts reduce their subsequent viability (42% vs. 54% for frozen and fresh day 7 blastocysts, respectively), but cryopreservation has a disproportionately adverse effect on day 8 vs. day 7 blastocysts (20% vs. 43% blastocysts, respectively). There is clearly a need to investigate the reasons for these differences, with respect to both the timing and the cryopreservation differentials.

3. Genetic regulation of embryo development timing

The timing of embryo development in mice is regulated by the Ped gene (Goldbard and Warner, 1982; Brownell and Warner, 1988; Warner et al., 1998; Cao et al., 1999). The mouse Ped gene product is an MHC class 1b protein called Qa-2 (Cao et al., 1999). Some haplotypes of the Ped gene were associated with fast and others with slow development (Goldbard and Warner, 1982). In a remarkable demonstration of the functional importance of this gene, when the Ped gene protein Qa-2 was "painted" onto mouse embryos, the rate (speed) of cleavage was increased (McElhinny et al., 2000). How the activity of this gene controls the speed of embryo development is not known. Embryos of other species including humans may also have Ped gene analogs that could account for some of variability in embryo development timing and associated differences in viability. Several candidate genes have been proposed in day 3 human embryos (Cao et al., 1999). Other proteins may also play a role in regulating the timing of embryo development. Transcription-enhancing ("high mobility") proteins appear to be constitutive components of mouse embryo chromatin. When mouse pronucleate ova were injected with antisense oligonucleotides targeting these proteins, the second and subsequent cleavage divisions were delayed while RNA and protein synthesis were reduced (Mohamed et al., 2001). It would be most interesting to investigate whether these proteins affect activation of the embryonic genome (Latham, 1999). Genome activation in mouse embryos is necessary for the switch in metabolic pathways which occurs at compaction or early cavitation (Martin and Leese, 1999). Thus, alterations in the timing of genome activation, caused by defective oocyte maturation or culture imperfections, could have serious consequences for the embryo's ability to provide sufficient energy for differentiation events. It is interesting in this context that failure of embryo development in vitro often occurs at the morula-blastocyst transition (e.g., Roth et al., 1994; Pinyopummintr and Bavister, 1991; Seshagiri and Hearn, 1992; Schramm and Bavister, 1996).

4. Timing of events during fertilization

Heterogeneity of embryo developmental competence, sometimes manifested in the timing of embryo cleavage as described earlier, may originate in the oocyte and/or during fertilization. Such anomalies could be exacerbated by culture conditions used for IVM or IVF. Both maternal and paternal effects on the timing of embryo development have been documented (Shire and Whitten, 1980a, b; Hillary et al., 1990). Comizzoli et al. (2000) demonstrated that the timing of pronuclear formation in bovine oocytes was not affected by IVF with spermatozoa that were associated with high or low blastocyst development frequencies, and the proportions of oocytes that were fertilized were not different. However, in the high blastocyst-forming experimental group, the onset of S-phase was earlier and the duration of S-phase was longer (>30% of the fertilization cell cycle vs. <20% with low blastocyst-forming spermatozoa). The use of high blastocyst-forming spermatozoa was also associated with earlier (approx. 3 hr) first cleavage division. Thus, in this study timing of fertilization events was strongly associated with embryo development competence in vitro. This result appears to contradict other studies that did not find any differences in the timing of the first cleavage division (e.g., van Soom et al., 1997a). However, in the latter study, the spermatozoa from a single bull were used, as is common practice in IVP studies to reduce variability, but this means that any male-related effects on the timing of fertilization would not of course be seen. Consistent with the study by Comizzoli et al. (2000), differences in the timing of first cleavage in human IVP embryos have been reported, and even used as an indicator of developmental competence and viability (Shoukir et al., 1997; Sakkas et al., 1998).

Bovine zygotes that completed their first cell division early (24 to 30 hpi) had a higher likelihood of reaching the 8-cell stage and further than slower-cleaving zygotes (40-62 hpi), and were more likely to be male (Yadav et al.,1993). These authors concluded that either this difference in cleavage timing is due to differential expression of X- and Y-bearing spermatozoa during fertilization, i.e., well before the commonly accepted time of "embryonic genome expression" (Telford et al., 1990; Memili and First, 2000), or else that the oocyte processes these two classes of spermatozoa differently.

5. Culture media effects on timing of embryo development

The foregoing clearly illustrates that there are inherent differences in developmental competence and viability in embryos produced in vivo or in vitro, and that these differences are exacerbated by IVP, with consequences for pregnancy outcomes, both in animals and in humans. However, it is also known that different culture media formulations can profoundly affect embryo development and pregnancy rates. For example, changes in the amino acid composition of a culture medium altered the proportion of bovine IVP embryos that reached the 8-cell stage in a timely manner, i.e., at 72 hpi (Pinyopummintr and Bavister, 1996). The overall range of 8-cell development by this time was almost 2-fold. In general, treatments that produced fewer 8-cell embryos by 72 hpi also resulted in a lower yield of blastocysts. Because embryo culture after 72 hpi was done in the same medium regardless of initial culture treatment, then differences in blastocyst development were due to amino acid effects during the 1- to 8-cell culture period. Thus, the timing or frequency of 8-cell development by 72 hpi was correlated with the ability of those embryos to develop into blastocysts.

Even relatively simple changes in the formulation of culture media can have striking effects on embryonic gene expression. For example, lowering the concentration of sodium chloride by 40 mM substantially increased total mRNA synthesis and dramatically improved the amounts of specific mRNAs, e.g., for growth factors and their receptors, in mouse embryos (Ho et al., 1994). Another study examined the expression of the maternally-imprinted H19 gene in mouse embryos cultured in two very different media, Whitten's medium as used for embryos of inbred mouse strains, and KSOM+AA which was developed for supporting embryos of a wide range of strains including those that typically blocked in culture. The expression of H19 was much more like the in vivo profile in embryos cultured in KSOM+AA than in Whitten's medium, which caused a loss of methylation of this gene (Doherty et al., 2000). In a study with IVP bovine embryos, the relative abundance of specific gene products was markedly different between two culture media (TCM-199 and SOF) compared with in vivo produced embryos (Wrenzycki et al., 2001). In view of these striking examples of the effects of culture media on gene expression in preimplantation embryos, it would be interesting to examine if the expression of the Ped gene, which regulates embryo cleavage timing in mouse and possibly human embryos (Warner et al., 1998; Cao et al., 1999), is influenced by different culture media conditions. If this were the case, then culture media formulations that enhanced Ped gene activity should improve embryo quality, because of the linkage between the timing of cleavage and viability (see 'Timing of embryo development'). Indeed, examination of mRNA profiles as a rapid, quantifiable criterion for embryo quality could be used as a tool for the development of new culture media formulations, as suggested by Niemann and Wrenzycki (2000).

The studies cited above, as well as others, indicate that suboptimal culture conditions can alter gene expression of preimplantation embryos and thus affect subsequent embryo development. It remains to be shown if differential activity of some of these genes might alter embryo development timing. It would also be helpful to examine if the timing of embryo cleavage, perhaps especially the third division, is correlated with the degree or onset of embryonic gene activation. Human IVP embryos exhibit a high incidence of failure of embryonic genome activation (Tesarik et al., 1988). Defects in embryonic genome activation can have serious consequences for embryo development, some of which may not be manifested until late in gestation (Latham, 1999; Niemann and Wrenzycki, 2000).

6. Timing of human embryo development

Human IVP embryos are known for their high frequency of genetic anomalies (Winston et al., 1991; Munne and Cohen, 1998; Munne et al., 1998), which might disturb the timing of embryo development. As mentioned previously, there is some evidence for a Ped gene analog in human embryos (Cao et al., 1999). Thus, development timing might serve as a non-invasive screening method for the presence of major genetic anomalies, especially if there is a connection between genome activation and development timing, but this remains to be seen. However, for various reasons, it is much more difficult to conduct studies on the timing of human preimplantation embryos in vitro than with animals. Difficulties in recording human embryo development timing in vitro have been discussed (Bavister, 1995). As already mentioned, there is no database on preimplantation embryo development in vivo for humans, which precludes using normal development timing to select embryos for transfer, as was done in some animal studies (McKiernan and Bavister, 1994; Hasler, 1998).

Almost twenty years ago, it was reported that non-human primate IVP embryos that took longer to reach the 8-cell stage were less developmentally competent (Bavister et al., 1983). Several early studies with human embryos attempted to use cleavage timing as an indicator of embryo quality, but for practical reasons (discussed by Bavister, 1995) the timing data were not sufficiently precise to validate this approach. For example, attempts to use cleavage timing to grade embryo quality were compromised by the wide variation of times used for examining embryos, a fact that was recognized by the investigators (Cummins et al., 1986). A more complex, cumulative scoring system was devised that incorporated embryo morphology, blastomere numbers (which are related to cleavage timing) and number of embryos transferred (Steer et al., 1992), but again problems inherent in examination of embryo development at specific times compromised this worthy attempt.

However, in a carefully designed study, cleavage timing was used successfully as a selection criterion for transfer of human embryos (Racowsky et al., 2000). This study eliminated much of the variability inherent in previous studies by using a narrow time window in the morning of day 3 (mean time 67 h post-insemination) for examination of embryos, when it was expected that competent embryos would be at the 8-cell stage. Embryos were selected for transfer based on the numbers of normal-looking 8-cell embryos observed at this time. When 1 to 3 or more acceptable embryos were found, either embryo transfer was performed on day 3, or transfer was postponed until day 5 when embryos had developed in culture to the blastocyst stage. There was no difference in ongoing pregnancy rates (% of transfers) between these two procedures (40-55%; Table 1). However, when there were no embryos on day 3 that met the timing and morphological criteria, an acceptable pregnancy rate (33%) was still obtained with day 3 embryo transfers but there were no ongoing pregnancies after day 5 embryo transfers. The lack of pregnancies in the latter group was correlated with significantly lower values for the numbers and percentages of day 5 blastocysts and expanded blastocysts, compared with the group that had 3 or more quality 8-cell embryos on day 3. These observations suggest that blastocysts which develop from slow-cleaving embryos are non-viable, consistent with data obtained in a study with hamster embryos (McKiernan and Bavister, 1994). The study by Racowsky et al. (2000) is an outstanding example of how the application of information largely derived from experimental studies with animal models can benefit clinical outcomes. In addition, this study avoided a problem inherent in several other studies comparing 'day 3' and 'day 5' embryo transfer outcomes, in which embryos were transferred on day 3 that were at various stages of cleavage and which therefore, most likely, were widely different in their viability. Some of these studies reported an improvement in pregnancy rates with blastocyst transfers compared with day 3 embryos, which is not surprising when slow-cleaving embryos that had an increased likelihood of being non-viable were selected for transfer (Table 1). On the other hand, the blastocysts that formed by day 5 are likely to be the most viable ones, but not necessarily more viable than day 3 8-cell embryos that are carefully selected by cleavage timing.

Table 1. Comparison of pregnancy rates with day 3 or day 5 embryo transfers*

No. 8-cell embryos on day 3[a]	Day of embryo transfer	No. patients with foetal heartbeat / No. transfers (CPR %)[b]
0	3	9 / 27 (33)
	5	0 / 11 (0)
1 or 2	3	33 / 77 (43)
	5	19 / 47 (40)
≥ 3	3	62 / 117 (53)
	5	46 / 83 (55)

* *Data from Racowsky et al. (2000)*

[a] *Embryo cleavage examined in the morning of day 3 (mean time 67 hpi)*

[b] *CPR = clinical pregnancy rate; single and multiple gestations combined*

Shoukir et al. (1997) used timing of the first cleavage division to evaluate embryo viability and to provide a non-invasive selection method. In a study involving 143 treatment cycles, fertilized ova were examined within 25 h post-insemination (hpi) to see if they had cleaved, then embryos were transferred to patients the next day. When 3 embryos were transferred per patient, embryos that had reached 2-cells by 25 hpi produced significantly higher clinical pregnancy rates compared with those that had cleaved later (33% vs. 15%, respectively). An even more striking difference in pregnancy rates was found by Sakkas et al. (1998) who compared ICSI embryos that cleaved within 27 hpi vs. those that did not (26% vs. 3% pregnancy rates, respectively, in 88 cycles). These authors found no post-fertilization effect of males on cleavage timing or pregnancy rates, suggesting that male effects on embryo development reported in other studies may be more related to differences in egg penetration or fertilization. In contrast to these studies with human embryos, Van Soom et al. (1997a) found no correlation between the timing of the first cleavage division in bovine IVP embryos and the timing of blastocyst formation or with blastocyst cell numbers (but see 'Fertilization' above). Similarly, no correlation was found between blastocyst development and the timing of either the first or the second cleavage division in cultured hamster embryos (Gonzales et al., 1995). Possibly, differences in culture media formulations, which can certainly affect cleavage development and blastocyst timing (see 'Culture media effects' above), might partly account for the differences reported in these studies, while species differences may also be involved.

7. Summary and Conclusions

It is clear from the foregoing that the timing of embryo development, both through cleavage stages and for blastocyst formation, is a critically important aspect of IVP technologies. The timing of development is firmly linked to embryo developmental competence and to viability post-transfer, in several species examined, including rodents, cattle and humans. Defects in embryogenesis that may be very subtle, i.e., not detectable by simple morphological inspections, can be expressed as a measurable slowing of the normal time-course through particular cleavage stages, and/or by the length of time to reach the blastocyst stage. It is now abundantly clear that morphologically normal-looking blastocysts may be non-viable if they exhibit prolonged development timing. In light of this, the study of development timing should be incorporated into any IVP protocol aimed at producing normal embryos. Examination of development timing is simple, rapid and requires no elaborate equipment other than a standard microscope.

The genetic control of embryo development timing is a subject in need of further exploration. More studies on the Ped gene and its analogs in different species may be especially rewarding. Additional research on the timing and extent of genome activation in IVP embryos could be very valuable and help explain why such embryos are on average inferior to their in vivo produced counterparts. Improvement in culture media formulations for supporting normal embryo development in vitro could be aided by using as a marker the activity of specific genes associated with development timing. In addition, this approach would increase basic understanding about the regulation of embryogenesis.

References

Bavister BD, Boatman DE, Leibfried L, Loose M, Vernon MW. 1983. Fertilization and cleavage of rhesus monkey oocytes in vitro. Biol Reprod 28:983-999.

Bavister BD. 1995. Culture of preimplantation embryos: facts and artifacts. Hum Reprod Update 1:91-148.

Brownell MS, Warner CM. 1988. Ped gene expression by embryos cultured in vitro Biol Reprod 39:806-811.

Cao W, Brenner CA, Alikani M, Cohen J, Warner CM. 1999. Search for a human homologue of the mouse Ped gene. Mol Hum Reprod 5:541-547.

Comizzoli P, Marquant-LeGuienne B, Heyman Y, Renard JP. 2000. Onset of the first S-phase is determined by a paternal effect during the GI-phase in bovine zygotes. Biol Reprod 62:1677-1684.

Cummins JM, Wilson, LM, Breen TM, Hennessey JF. 1986. A formula for scoring human embryo growth rates in IVF: value in predicting pregnancy and in comparison with other estimates of embryo quality. J In Vitro Fert Embryo Transfer 3:284-295.

Doherty AS, Mann MR, Tremblay KD, Bartolomei MS, Schultz RM. 2000. Differential effects of culture on imprinted H19 expression in the preimplantation mouse embryo. Biol Reprod 62:1526-1535.

Gardner DK, Pool TB, Lane M. 2000. Embryo nutrition and energy metabolism and its relationship to embryo growth, differentiation, and viability. Semin Reprod Med 18:205-218.

Goldbard SB, Warner CM. 1982. Genes affect the timing of early mouse embryo development. Biol Reprod 27:419-424.

Gonzales DS, Bavister BD. 1995. Zona pellucida escape by hamster blastocysts in vitro is delayed and morphologically different compared with zona escape in vivo. Biol Reprod 52:470-480.

Gonzales DS, Pinheiro JC, Bavister BD. 1995. Prediction of the developmental potential of hamster embryos in vitro by precise timing of the third cell cycle. J Reprod Fertil 105:1-8.

Grisart B, Massip A, Dessy F. 1994. Cinematographic analysis of bovine embryo development in serum-free oviduct-conditioned medium. J Reprod Fertil 101:257-264.

Hasler JF. 1998. The current status of oocyte recovery, in vitro embryo production, and embryo transfer in domestic animals, with an emphasis on the bovine. J Anim Sci 76:52-74.

Hillary FL, Parrish JJ, First NL. 1990. Bull specific effect on fertilization and embryo development in vitro. Theriogenology 33:249.

Ho Y, Doherty AS, Schultz RM. 1994. Mouse preimplantation embryo development in vitro: effect of sodium concentration in culture media on RNA synthesis and accumulation and gene expression. Mol Reprod Dev 38:31-141.

Ho Y, Wigglesworth K, Eppig, JJ Schultz RM. 1995. Preimplantation development of mouse embryos in KSOM: augmentation by amino acids and analysis of gene expression. Mol Reprod Dev 41:232-238.

Hyttel P, Viuff D, Laurincik J, Schmidt M, Thomsen PD, Avery B, Callesen H, Rath D, Niemann H, Rosenkranz C, Schellander K, Ochs RL, Greve T. 2000. Risks of in vitro production of cattle and swine embryos: aberrations in chromosome numbers, ribosomal RNA gene activation and perinatal physiology. Hum Reprod 15:87-97.

Latham KE. 1999. Mechanisms and control of embryonic genome activation in mammalian embryos. Int Rev Cytol 193:71-124.

Martin KL, Leese HJ. 1999. Role of developmental factors in the switch from pyruvate to glucose as the major exogenous energy substrate in the preimplantation mouse embryo. Reprod Fertil Dev 11:425-433.

McElhinny AS, Exley GE, Warner CM. 2000. Painting Qa-2 onto Ped slow preimplantation embryos increases the rate of cleavage. Am J Reprod Immunol 44:52-58.

McKiernan SH, Bavister BD. 1994. Timing of development is a critical parameter for predicting successful embryogenesis. Hum Reprod 9:2123-2129.

McKiernan SH, Bavister BD. 1998. Gonadotropin stimulation of donor females decreases post-implantation viability of cultured 1-cell hamster embryos. Hum Reprod 13:724-729.

Memili E, First NL. 2000. Zygotic and embryonic gene expression in cow: a review of timing and mechanisms of early gene expression as compared with other species. Zygote 8:87-96.

Mohamed OA, Bustin M, Clarke HJ. 2001. High-mobility group proteins 14 and 17 maintain the timing of early embryonic development in the mouse. Dev Biol 229:237-249.

Munne S, Marquez C, Reing A, Garrisi J, Alikani M. 1998. Chromosome abnormalities in embryos obtained after conventional in vitro fertilization and intracytoplasmic sperm injection. Fertil Steril 69:904-908.

Munne S, Cohen J. 1998. Chromosome abnormalities in human embryos. Hum Reprod Update 4:842-855.

Niemann H, Wrenzycki C. 2000. Alterations of expression of developmentally important genes in preimplantation bovine embryos by in vitro culture conditions: implications for subsequent development. Theriogenology 53:21-34.

Pinyopummintr T, Bavister BD. 1991. In vitro-matured/in vitro-fertilized bovine oocytes can develop into morulae/blastocysts in chemically-defined, protein-free culture media. Biol Reprod 45:736-742.

Pinyopummintr T, Bavister BD. 1996. Effects of amino acids on in vitro development of cleavage stage bovine embryos into blastocysts. Reprod Fertil Dev 8:835-841.

Racowsky C, Jackson KV, Cekleniak NA, Fox JH, Hornstein MD, Ginsburg ES. 2000. The number of eight-cell embryos is a key determinant for selecting day 3 or day 5 transfer. Fertil Steril 73:558-564.

Roth TL, Swanson WF, Wildt DE. 1994. Developmental competence of domestic cat embryos fertilized in vitro versus in vitro. Biol Reprod 51:441-451.

Sakkas D, Shoukir Y, Chardonnens D, Bianchi PG, Campana A. 1998. Early cleavage of human embryos to the two-cell stage after intracytoplasmic sperm injection as an indicator of embryo viability. Hum Reprod 13:182-187.

Schramm RD, Bavister BD. 1996. Development of in vitro fertilized primate embryos into blastocysts in chemically-defined, protein-free culture medium. Hum Reprod 11:1690-1697.

Schramm RD, Bavister BD. 1999. Onset of nucleolar and extranucleolar transcription and expression of fibrillarin in macaque embryos developing in vitro. Biol Reprod 60:721-728.

Seshagiri PB, Hearn JP. 1992. Protein-free culture media that support in vitro development of Rhesus monkey blastocysts. ARTA 3:225-232.

Seshagiri PB, McKenzie DI, Bavister BD, Williamson JL, Aiken JM. 1992. Golden hamster embryonic genome activation occurs at the 2-cell stage: correlation with major developmental changes. Molec Reprod Dev 32:229-235.

Shire JG M, Whitten WK. 1980a. Genetic variation in the timing of first cleavage in mice: effect of maternal genotype. Biol Reprod 23:369-376.

Shire JGM, Whitten WK. 1980b. Genetic variation in the timing of first cleavage in mice: effect of paternal genotype. Biol Reprod 23:363-368.

Shoukir Y, Campana A, Farley T, Sakkas D. 1997. Early cleavage of in vitro fertilized human embryos to the 2-cell stage: a novel indicator of embryo quality and viability. Hum Reprod 12:1531-1536.

Steer CV, Mills CL, Tan SL, Campbell S, Edwards RG. 1992. The cumulative embryo score: a predictive embryo scoring technique to select the optimal number of embryos to transfer in an in vitro fertilization and embryo transfer programme. Hum Reprod 7:117-119.

Telford NA, Watson AJ, Schultz GA. 1990. Transition from maternal to embryonic control in early mammalian development: a comparison of several species. Molec Reprod Dev 26:90-100.

Tesarik J, Kopecny V, Plachot M. Mandelbaum J. 1988. Early morphological signs of embryonic gene expression in human preimplantation development as revealed by quantitative electron microscopy. Devel Biol 128:15-20.

Thompson JG, Peterson AJ. 2000. Bovine embryo culture in vitro: new developments and post-transfer consequences. Hum Reprod 15:59-67.

Van Soom A, Ysebaert MT, de Kruif A. 1997a. Relationship between timing of development, morula morphology, and cell allocation to inner cell mass and trophectoderm in in vitro-produced bovine embryos. Mol Reprod Dev 47:47-56.

Van Soom A, Boerjan ML, Bols PE, Vanroose G, Lein A, Coryn M, de Kruif A. 1997b. Timing of compaction and inner cell allocation in bovine embryos produced in vivo after superovulation. Biol Reprod 57:1041-1049.

Vanroose G, Van Soom A, de Kruif A. 2001. From co-culture to defined medium: state of the art and practical considerations. Reprod Domest Anim, 36:25-28.

Warner CM, Cao W, Exley GE, McElhinny AS, Alikani M, Cohen J, Scott RT, Brenner CA. 1998. Genetic regulation of egg and embryo survival. Hum Reprod 13:178-190.

Winston NJ, Braude PR, Pickering SJ, George MA, Cant A, Currie J, Johnson MH. 1991. The incidence of abnormal morphology and nucleocytoplasmic ratios in 2-, 3- and 5-day human pre-embryos. Hum Reprod 6:17-24.

Wrenzycki C, Herrmann D, Keskintepe L, Martins AJ, Sirisathien S, Brackett B, Niemann H. 2001. Effects of culture system and protein supplementation on mRNA expression in pre-implantation bovine embryos. Hum Reprod 16:893-901.

Yadav BR, King WA, Betteridge KJ. 1993. Relationships between the completion of first cleavage and the chromosomal complement, sex and developmental rates of bovine embryos generated in vitro. Mol Reprod Dev 36:434-439.

Chapter 7

BIOPSY AND PREIMPLANTATION DIAGNOSIS OF HUMAN EMBRYOS

Deborah Taylor and Maria Tachataki
Institute of Reproductive Biology, Faculty of Medicine, Imperial College, Hammersmith Hospital, Du Cane Road London W11 0NN, UK

1. Introduction

Currently, in the human, prenatal diagnosis of inherited disease involves recovering cells of foetal origin by chorion villus sampling (CVS) or amniocentesis between 8 and 20 weeks of gestation, prior to cytogenetic, biochemical or molecular analysis. If the foetus is diagnosed as affected, therapeutic termination can be offered. The accessibility of preimplantation embryos following IVF has raised the possibility of preimplantation diagnosis. This involves the removal of a cell or cells from a preimplantation embryo for genetic analysis. Once the diagnosis is complete only the unaffected embryos are selected for transfer to the uterus (Handyside and Delhanty, 1997). In this way, many first and second trimester abortions can be avoided. The techniques for the removal of cells from embryos during their preimplantation development are well established. Although primarily these techniques are used for clinical purposes, this technology also provides a means of obtaining material from viable oocytes and embryos which allow their viability or quality to be assessed prior to their implantation. This chapter will detail all aspects of embryo biopsy and preimplantation genetic diagnosis and discuss how these techniques can provide a means to evaluate mammalian embryo quality.

2. Preimplantation genetic diagnosis (PGD)

Preimplantation genetic diagnosis (PGD) allows the diagnosis of a genetic disorder in an embryo prior to its implantation in the uterus. It is of benefit to couples at risk of passing on a genetic disease to their offspring. Conventional prenatal diagnosis, carried out between 12 - 16 weeks of pregnancy by chorionic villus sampling (CVS) or amniocentesis carry a 1% risk of miscarriage. This may result in the loss of an unaffected pregnancy. If however, the foetus is affected, couples then have to consider terminating the pregnancy. Because only unaffected embryos are transferred back to the uterus following PGD, it avoids the need for termination. This technique was pioneered at the Hammersmith Hospital in the late 1980's (Handyside et al., 1989) and there are currently over 100 unaffected children born following PGD.

3. Obtaining oocytes and embryos

3.1. Human

The diagnosis is carried out on cells obtained from the embryo during its preimplantation development. Couples must first undergo an IVF treatment cycle to obtain the oocytes and embryos from which the cells will be removed. During a typical IVF treatment cycle, oocytes are collected from the female partner, which are then inseminated later that day using the male partner sperm (day 0). Normally fertilized embryos exhibit 2 pronuclei on the following morning (day 1) approximately 16-18 hours post insemination (pi). On day 2 pi embryos are usually around the 4-cell stage. Following a further 24 hours in culture each cell undergoes a further cleavage division to produce an 8-cell embryo. Embryos reach the differentiated blastocyst stage on day 5/6 pi. Most IVF units will transfer the embryos back to the uterus on day 2 or day 3 pi, however attempts have been made to establish routine blastocyst transfers (Cruz et al., 1999; Gardner et al., 2000).

3.2. Rabbit

Superovulation protocols for mouse, rabbit and rat are well established and also require the use of gonadotrophins (Kennelly and Foote, 1965; Foote and Simkin, 1993). Rabbits require six doses of follicle stimulating hormone (FSH) followed by an ovulating dose of luteinizing hormone (LH) or human chorionic gonadotrophin (hCG). The superovulation of 12 week old females can be further increased by priming animals with progesterone to produce something in the region of 30 oocytes.

3.3. Rat and mouse

Protocols for superovulation of the mouse and rat are similar (Zarrow and Wilson, 1961). The adult female is initially given an injection of pregnant mare's serum gonadotrophin (PMSG) (Folligon: Intervet Norge, Norway) 48 hours prior to an injection of human chorionic gonadotrophin (hCG) (Profasi: Serono, Switzerland) to provide supernumary oocytes for analysis.

4. Methods of oocyte and embryo biopsy

There are three main methods used for the removal of cells for the diagnosis. Each method requires the use of specialized micromanipulation equipment, allowing the operator to undertake the biopsy procedure with the accuracy required to work on such minute samples. Each biopsy is classified according to the stage of development on which it is performed, as summarized in Figure 1. Each procedure has its advantages and disadvantages, however unlike the polar body biopsy and the cleavage stage biopsy, the blastocyst biopsy is yet to be used clinically.

Stage of embryo development	Embryo biopsy procedure	Diagrammatic representation
Polar body biopsy	**Removal of the polar body from metaphase II unfertilized oocyte (day 0), and fertilized pronucleate zygote stage (day 1) (optional)**	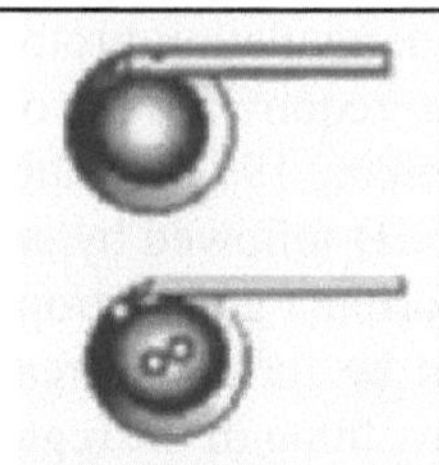
Cleavage stage biopsy	**Removal of up to a maximum of 2 blastomeres from a day 3 cleavage stage (4-12 cell stage)**	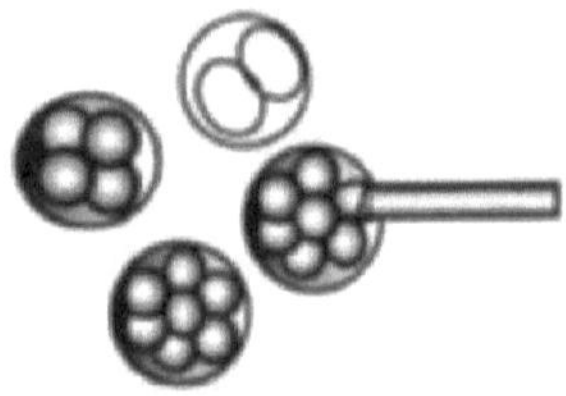
Blastocyst stage biopsy	**Removal of approximately 10 trophectoderm (TE) cells from a differentiated day 5/6 blastocyst**	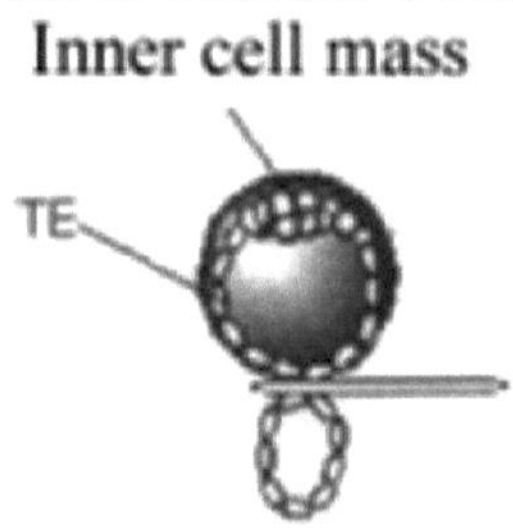

Figure 1: Schematic summary of biopsy procedures

4.1. Polar Body Biopsy

Although the polar bodies contain chromosomes derived from the oocyte itself, they are not required for the subsequent development of the oocyte or embryo. The removal of both the first and second polar bodies has therefore been used to obtain genetic information regarding the oocyte or embryo from which it came. A polar body biopsy is performed by using gentle suction to immobilize the unfertilized oocyte on to the holding pipette. A hole is created in the zona pellucida mechanically using a sharp sampling pipette. The pipette is introduced into the perivitelline space above the polar body, which is then gently aspirated out. The polar body is then used for the genetic analysis of choice, and is typically used for diagnosis of age related aneuploidies (Verlinsky et al., 1999) or detection of translocations carried by the female partner (Munne et al., 1998). Diagnosis of single gene defects and chromosomal errors can be made after analysis of the first polar body, or both the first and second polar bodies.

Although the polar body biopsy is a semi-invasive procedure, it has the advantage of not reducing the cellular mass of the embryo, nor does it interfere with the embryo's subsequent development (Verlinsky et al., 1995). The major limitation of this technique is that only defects carried by the mother can be diagnosed. For dominant diseases, a diagnosis can only be made if it is the mother who carries the defective gene. For recessive conditions, when both parents carry a mutant allele, only the maternal allele can be detected and hence the carrier status of the embryo cannot be assessed. Importantly polar body biopsy cannot be used for the determination of sex and therefore can not identify male embryos, which are at risk of carrying severe X-linked disorders.

4.2. Cleavage stage embryo biopsy

Cleavage stage embryo biopsies are usually performed in the morning of day 3 post insemination. At this stage a maximum number of embryos should have reached the 6-10 cell stage, before the start of cell aggregation and embryo compaction. The embryo is immobilized on a holding pipette, a hole is drilled into the zona pellucida by applying a stream of acidic solution through a fine micropipette, or using laser technology. A second micropipette (the sampling pipette) is inserted through the resulting hole in the zona pellucida and one or two cells are removed by gentle suction. Experiments in the mouse show that biopsy of one or two cells at the 8-cell stage and one cell at the 4-cell stage has no deleterious effects on blastocyst formation, embryo implantation and postimplantation development after transfer to pseudopregnant recipients (Monk and Handyside, 1988; Wilton and Trounson, 1989). Similarly, Hardy et al. (1990) demonstrated in the human that cleavage stage biopsy was not detrimental to subsequent embryo development to the blastocyst stage. Although the cell numbers per embryo and the uptake of glucose and pyruvate are reduced, it is only in proportion to the reduced cellular mass following removal of 1/8 or 2/8 cells (Hardy et al., 1990). This work also shows that the biopsy procedure does not increase the proportion of morphologically abnormal blastocysts. A similar study, showed that cleavage stage biopsy at the 4-cell stage (day 2 post insemination) does not affect the proportion of human embryos developing to the blastocyst stage (Tarin et al., 1992).

When the cleavage stage biopsy is performed, the individual blastomeres of the embryo are totipotent and are capable of contributing to the foetus and extra embryonic tissues (Papaioannou and Ebert, 1986). Genetic analysis from blastomeres should therefore represent the true genotype of the future baby. More than 160 children have now been born following cleavage stage biopsy and PGD (ESHRE PGD consortium steering committee, 2000) with no report of congenital abnormalities. Although we are not yet able to rule out the possibility of subsequent developmental defects, this seems highly unlikely.

4.3. Blastocyst stage biopsy

At the blastocyst stage, the mammalian embryo is differentiated into two distinct cell lineages, the trophectoderm (TE) layer which goes on to form the placenta, and the inner cell mass (ICM), which forms the embryo proper and some extraembryonic cell lineages. The removal of a number of trophectoderm cells was first described in 1968 (Gardner and Edwards, 1968) and was performed on rabbit blastocysts. The biopsied embryos were transferred into female recipients after determination of sex and implantation was observed. A small hole was made into the zona pellucida and the trophectoderm (TE) cells were cut out using two sharp pipettes. A similar approach has been used in the smaller mouse blastocyst: a small slit was made in the zona pellucida, away from the inner cell mass (ICM) cells. Trophectoderm cells, herniating through the hole, either spontaneously or after induction by pressure with a glass needle, were pulled away from the blastocyst by gentle suction with a second micropipette (Monk et al., 1988). The same method has been used successfully for research purposes on human blastocysts, however unlike the mouse, blastocysts were allowed to hatch before removal of trophectoderm cells (Dokras et al., 1990; Muggleton Harris and Findlay, 1991; Muggleton Harris et al., 1993).

The main advantage of blastocyst biopsy is the increased number of cells that can be removed for analysis. Duplicate or triplicate samples can be examined, providing a more reliable diagnosis. Another important advantage is that TE-Cells contribute only to placental tissues after implantation. The foetus is derived exclusively from the ICM, therefore removal of a small number of the TE-Cells should not affect the normal development of the foetus. Because it affects the tissue destined to form the placenta, removal of trophectoderm cells has the potential to compromise implantation. It was believed that aneuploid cells in the developing embryo were preferentially allocated to the trophectoderm layer, which would result in a greater risk of misdiagnosis. However recent FISH studies on human blastocysts show that this should not be the case, as the level of aneuploidy in the inner cell mass is equivalent to that found in the trophectoderm layer (Evsikov and Verlinsky, 1998). Only 17-40% of human embryos reach the blastocyst stage following *in vitro* culture (Hardy et al., 1989). After elimination of the affected embryos and embryos which fail to provide a diagnosis, analysis at blastocyst stage could result in an unacceptably high incidence of cycles with no embryo transfer. Due to the difficulties involved in obtaining a good number of blastocysts and the uncertainty associated with their transfer there has been no report of clinical PGD cases using blastocyst biopsy. Currently more IVF units are attempting to implement routine blastocyst transfers, and with the current availability of sequential media, this may promote the use of the blastocyst biopsy in the near future.

5. Methods of diagnosis

5.1. Metabolic assays

Approximately 200 of 1,500 defined monogenic genetic disorders result in enzyme deficiency. Around 60 of these diseases can be diagnosed by enzyme assays on cultured amniotic fluid cells or chorionic villus samples (Benson and Fensom, 1985) examples of which are given in Table 1.

Table 1: Examples of human genetic diseases caused by enzyme abnormalities

Genetic Disorder	Deficient Enzyme
Tay-Sachs Disease	Hexosaminidase
Fabry Disease	Galactosidase A
Gaucher Disease	β-glucosidase
M Leucodystrophy	Aryl Sulfatase A
Globoid Leucodystrophy	Gal-cerase
Nieman-Pick Disease	Sphyngomyelinase
Hurler Syndrome	α-L-iduronidase
Hunter Syndrome	Iduronate sulfatase
Lesch-Nyhan Syndrome	Hypoxanthine phosphoribosyl transferase
Severe Combined Immunodeficiency Disease	Adenosine deaminase
Severe haemolytic anaemia	Glucose-6-phosphate dehydrogenase

Such diseases are often caused by heterogeneous mutations within the gene, and separate strategies would be required for each family in order to detect the defect at the DNA level. In many cases, the inherited disease manifests itself early in childhood causing severe physical and mental problems, and often results in death. In principle, biochemical assays of specific enzymes could be used to diagnose specific disorders as well as sex determination. The use of biochemical microassays for PGD is dependent on a number of criteria, namely the availability of highly sensitive assays, the embryonic gene for the enzyme must be expressed at the time of the biopsy and the embryonic enzyme activity must be distinguishable from the oocyte derived, maternal activity. Furthermore use of biochemical microassay for PGD are restricted to cleavage stage and blastocyst biopsies only, as polar bodies only contain small amounts of oocyte derived activity.

5.1.1. Mouse metabolic assays

Attempts at sexing preimplantation embryos have been made using enzymatic measurements. Day 2 mouse embryos were split into two, one half was transferred into a pseudopregnant recipient and the other half used for enzymatic measurements. The ratio of the enzymatic activity of X-linked hypoxanthine phosphoribosyl transferase (HPRT) and somatic adenosine phosphoribosyl transferase (APRT) was calculated at different periods of time and was found to be 1:1 in female embryos and 1:2 in male embryos (Epstein et al., 1978). Using a similar approach, preimplantation diagnosis was performed in HPRT deficient mice (Hooper et al., 1987) as a model for sexing (Monk et al., 1987; Monk and Handyside, 1988) and Lesch-Nyhan syndrome diagnosis (Monk et al., 1988). HPRT activity was detected in single cells removed from embryos at the 8-cell stage or trophectoderm cells excised from blastocysts while the embryos remained in culture. After the transfer of selected embryos to pseudopregnant recipients, the ratio of X-linked HPRT and autosomal control APRT activities allowed both sex and HPRT status of the biopsied embryos to be accurately predicted (Monk and Handyside, 1988). Deficiency of the enzyme adenine deaminase (ADA) accounts for approximately 20% of cases of severe combined immunodeficiency disease (SCID) in humans. Benson and Monk (1988) developed a microassay for preimplantation diagnosis of this enzyme in a mouse model for SCID (Abbott et al., 1986). Use of α-amanitin blocked the activity of ADA beyond the 2-cell stage, indicating that the detectable levels of enzyme activity were not oocyte derived. They demonstrated that it was possible to obtain enzyme assays from individual cells removed from 8-cell mouse embryos and hence PGD based on a biochemical assay was possible.

5.1.2. Human metabolic assays

Comparable biochemical microassays have been performed on human oocytes and embryos. One study measured high levels of HPRT activity throughout all preimplantation stages of human development, and were unable to detect any embryo coded enzyme activity (Braude et al., 1989). The more sensitive high performance liquid chromatography (HPLC) assay was also used to detect HPRT enzyme activity in human embryos. A significant increase in activity was detected between days 2 and 4 pi, following activation of the embryonic genome on day 3 pi. Unlike the previous study this work provided some evidence of embryonic HPRT expression in the human (Leese et al., 1991). A biochemical microassay has also been designed for the analysis of Tay-Sachs and Sandhoff disease, which are caused by deficiency of the α- and β- chains of β-N-hexosaminidase, respectively. Once again the, high levels of enzyme activity measured in oocytes and throughout preimplantation development made it impossible to reliably distinguish between maternal and embryonic enzyme activity (Sermon et al., 1992).

The work carried out in the mouse shows that biochemical assays can be used for the diagnosis of disease during preimplantation development. However, there is a vast difference in the success between human and mice studies. This may simply be due to the fact that global embryonic gene activation occurs at the 2-cell stage in the mouse, but occurs later in the human, at the 4- to 8-cell stage. If the cells used for the assays are removed from an 8-cell embryo, a significant amount of oocyte derived products will be degraded by this stage in the mouse, but in the human embryo a significant amount may still be present.

5.2. The Polymerase Chain Reaction (PCR)

Methods for the detection of single gene defects for preimplantation genetic diagnosis are based on the polymerase chain reaction (PCR). PCR allows the two DNA molecules present in a single cell to be amplified to levels easily detected for analysis. Initially PCR was used for the diagnosis of embryo sex (Handyside et al., 1989), but it now has much wider applications (review, Wells and Sherlock, 1998). Despite the exquisite sensitivity of the single cell PCR protocols used for PGD, there are three problems associated with this procedure namely; failure of amplification, contamination and allele drop out (ADO). Failure of amplification can be a direct result of the limited starting material. This should not affect the diagnosis of an embryo, but can reduce the number of embryos available for transfer back to the uterus following the diagnosis. ADO is the failure to amplify a single allele in a heterozygous cell. This phenomenon affects each allele at random, and is specific to amplification from single cells. It is important to determine the incidence of ADO for each diagnostic approach, as it can potentially result in misdiagnosis. For clinical cases a number of steps can be used to avoid any misdiagnosis resulting from ADO. The level of ADO can be reduced but not eliminated, by increasing the PCR denaturing temperature (Ray and Handyside, 1996). Alternatively any ADO event can be recognized as such by performing a multiplex PCR, co-amplifying an informative marker which is linked to the disease gene (Kuliev et al., 1998; Xu et al., 1999). Multiplex

PCR can also be used to avoid misdiagnosis through contamination from a stray cell or piece of DNA. It is possible that contamination could result in a misdiagnosis. However, by co-amplifying a highly polymorphic site, DNA fingerprints can be used to ensure that the amplified product is derived only from the embryo and not from contaminating sperm, cumulus cells or any other external source. Conventional methods for detection of PCR products separated by gel electrophoresis, require a significant amount of DNA for ethidium bromide (EtBr) staining, silver staining or radiolabelling of primers or product. Furthermore, products of similar size are difficult to resolve for direct fragment analysis. With the arrival of fluorescent PCR technology (Hattori et al., 1992) not only is the sensitivity of the PCR vastly increased but fluorescent PCR allows the investigator to separate DNA molecules differing in size by a single base pair. Between 30 - 40 PCR cycles can be used to generate a product detectable by fluorescent DNA analysis, compared with > 50 cycles using conventional nested PCR protocols. This significantly reduces the time taken to obtain a diagnosis. For these reasons fluorescent PCR is quickly becoming the method of choice for contemporary diagnostic approaches for PGD (Findlay et al., 1996; Sermon et al., 1998). There are many different PCR based strategies employed for the diagnosis of a variety of mutations for PGD (Handyside and Delhanty, 1997; Wells and Sherlock, 1998), however PCR protocols do not usually provide information on chromosomal aneuploidies. One PCR based method has been used to analyse both chromosome number and structure. Comparative genome hybridization (CGH) allows every chromosome in a human blastomere to be identified and analysed for number and structural integrity (Voullaire et al., 2000; Wells et al., 1999). It is dependent on an initial PCR step, which amplifies the whole genome, degenerate oligonucleotide primer PCR (DOP-PCR) is the preferred method used to achieve this. PCR products from both the test and control templates are fluorescent labelled with different fluorochromes and hybridized to a control metaphase spread. Differences between hybridization of the control template and the test cell is then used to detect any chromosomal abnormalities. This method has been used on single human blastomeres and despite the procedure taking 3-4 days, it has the potential to be of great use for routine PGD analysis in the future.

5.3. Fluorescence in situ hybridization (FISH)

Karyotyping has been used to identify all the chromosomes in mammalian embryos, and has provided information on the existence of aneuploidies in the human preimplantation embryo (Angell et al., 1986; Plachot et al., 1987). However, due to the low number of available metaphase spreads and the difficulty in banding the condensed chromosomes obtained from human embryos, karyotyping is not used for clinical PGD cases. Although recent methods, usually involving the fusion of a blastomere with an intact or enucleated mammalian oocyte, can produce metaphase spreads with 90% success (Willadsen et al., 1999; Verlinsky and Evsikov, 1999) it is not feasible to use these approaches for clinical diagnoses. Fluorescent in situ hybridization (FISH) is less technically demanding and allows the detection of specific chromosomes in an intact interphase or metaphase nucleus. This method is used for the diagnosis of embryo sex for X-linked diseases, chromosomal aneuploidies such as Down's syndrome as well as major chromosomal aberrations such as translocations (Delhanty et al., 1994; Conn et al., 1999). FISH analysis of normally fertilized spare embryos from fertile patients undergoing PGD for X-linked disorders has revealed a surprisingly high incidence of chromosomal mosaicism in human cleavage stage embryos (Munne et al., 1994; Harper et al., 1995; Delhanty et al., 1997). A significant number of embryos have been shown to carry a proportion of blastomeres with abnormal chromosomal complements, and in the case of chaotic embryos, the majority of cells possess a different chromosomal aneuploidy. These aneuploidies could result from aberrant chromosome replication and/or segregation. The embryos normal cell-cycle check points should function to ensure the division of each blastomere into two genetically identical daughter cells (Murray, 1992). It is possible that the genes required for this control are only expressed following activation of the embryonic genome at the 4 to 8-cell stage in the human (Braude et al., 1988; Tesarik et al., 1988). This would explain why the initial cleavage divisions are prone to error, resulting in the high incidence of aneuploidies during human preimplantation development. Embryo aneuploidy may be a major contributor to implantation failure and hence limit IVF success rates. In an attempt to improve implantation rates, preimplantation genetic diagnosis of aneuploidies has been attempted on selected groups of patients, including patients of 36 years or older, patients with recurrent IVF failures and patients with recurrent miscarriage. The use of FISH probes for detection of up to 9 different chromosomes has been used to screen for aneuploidies in these patients groups, resulting in improved ongoing implantation rates (Gianaroli et al., 1999; Vidal et al., 2000).

5.4. Reverse transcription PCR (RT-PCR)

Most researchers agree one of the best ways of avoiding inaccuracies in the diagnosis is to preferentially select those embryos giving confirmation of diagnosis following removal of 2 cells. This is not always possible if the embryo has not developed sufficiently to allow 2 cells to be removed. It is important therefore, that other strategies for obtaining confirmation of diagnosis are explored. One such method is reverse transcription PCR (RT-PCR). Once a gene is expressed then each copy will produce multiple copies of itself in the form of mRNA. With multiple templates on which to base a diagnosis, the incidence of amplification failure, contamination and ADO should be significantly reduced. Furthermore as the introns have been spliced out, the amplification of a much larger part of the coding sequence can be used for the diagnosis, although the detection of rare mutations located in the introns is not possible. RT-PCR has previously been used for PGD of the autosomal dominant condition Marfan syndrome (Eldadah et al., 1995). The possibility of an RT-PCR based PGD protocol demands that the gene is expressed during human preimplantation development, and that the diagnosis is based on wholly embryonic gene products, not the maternal transcripts carried over from the oocyte. RT-PCR is a relatively simple procedure and has been used to detect a variety of gene transcripts expressed either throughout or at specific stages of human preimplantation development (Table 2 and 3 respectively). Hence it can be used to determine whether a gene of interest is expressed. Following amplification by RT-PCR, cDNA markers such as coding sequence polymorphisms, can be used to distinguishing between oocyte derived and embryonic genes and determine the time at which the embryonic gene is being expressed (Taylor et al., 1997). The stage of development at which degradation of the oocyte derived mRNA occurs can also be determined using quantitative RT-PCR (Taylor et al., 2001). Although the use of RT-PCR for PGD requires further evaluation, it has the potential to vastly improve the diagnosis of single gene defects in the future.

Table 2: Transcripts detected throughout human preimplantation development by RT-PCR

Specific gene transcript	**Reference**

ZFX	Ao et al., 1994
SRY	Fiddler et al., 1995
OCT 4	Abdel-Rahman et al., 1995
EGF	Chia et al., 1995
TGF-alpha	"
EGFR	"
Cyclin B1	Heikinheimo et al., 1995
Beta-actin	"
SCF	Sharkey et al., 1995
c-kit/ SCFR	"
c-fms	"
MPK	Daniels et al., 1995
LIFR	van Eijk et al., 1996
Gp130	"
IGF2	Lighten et al., 1997
IGF1R	"
IGF2R	"
PDGF	Österlund et al., 1996
HPRT	Taylor et al., 2001
APRT	"
ADA	Taylor et al., 1997
G6PD	"

Table 3: Transcripts detected by RT-PCR showing stage specific expression during human preimplantation development

Specific gene transcript	Stage when expressed	Reference
TNFRp80	Oocyte to 4-cell stage	Sharkey et al., 1995
TNF-alpha	4-cell to Morula stage	"
TNFRp60	6-cell to Blastocyst	"
IL-6	Blastocyst	"
IL-6R	Blastocyst	"
LIFR	Blastocyst	"
Insulin Receptor	Oocyte and 8-cell stage	Lighten et al., 1997
OCT 6	10-cell to Blastocyst	Abdel-Rahman et al., 1995
C-mos	Oocyte specific	Heikinheimo et al., 1995

6. Use of PGD technology for evaluation of mammalian embryo quality

The techniques required to perform a preimplantation genetic diagnosis are divided into two main procedures; the micromanipulation of oocytes and embryos and the method for the diagnosis. Each procedure can provide a means for evaluation of embryo quality. We have already discussed how embryo biopsy followed by FISH analysis can identify chromosomally abnormal and hence less viable embryos, which can be considered a means for assessing embryo quality. At present the choice of the 'best' embryos for embryo transfer following an IVF treatment cycle is usually made on gross morphology and cleavage rate. These are very limited markers for embryo health. Because the embryo biopsy does not affect the embryos subsequent development, an isolated cell can be used to determine the developmental potential of the embryo from which it came. This can be achieved simply by observing the blastomere in culture or by a more involved method of analysis.

Blastomeres biopsied from an 8-cell embryo can develop further in culture, by dividing and cavitating. The development of these isolated blastomeres correlates with the subsequent development of the embryo from which they are removed. Following the removal of a single blastomere from a cleavage stage embryo, the embryos can be frozen and the isolated blastomeres placed in culture. The isolated blastomeres which cavitate indicate which embryos are more likely to develop to the blastocyst stage, and these embryos can be selected for transfer. Although this type of analysis may be an improved method of embryo selection, it is still based on observations of morphology. Alternatively, PGD technology can be used to evaluate the genetic status of an embryo, whether it is through the detection of a transgene or a specific mutation. We know that human embryos will arrest during in vitro culture, and that the majority of arrest occurs at the 4 to 8-cell stage, the time at which embryos are selected for embryo transfer (Bolton et al., 1989). This developmental arrest also coincides with the timing of global embryonic gene activation in the human (Braude et al., 1988; Tesarik et al., 1988). The failure to activate the embryonic genome is related to embryonic developmental arrest (Artley et al., 1991). Embryo biopsy followed by RT-PCR on the isolated cell allows the investigator to identify in which embryos zygotic gene activation has taken place. The detection of any paternally derived transcript, even if this is simply a coding sequence polymorphism indicates that the embryonic genome has been switched on. By selecting embryos in which embryonic gene activation has occurred we can avoid a proportion of embryos destined to arrest. Coding sequence polymorphisms have already been used to show activation of the embryonic genes of glucose-6-phosphate dehydrogenase (G6PD) and adenosine deaminase (ADA) in the human (Taylor et al., 1997). Amplification of a specific gene transcript by RT-PCR may provide a more definitive marker for embryo quality. For instance the expression of IGF-1 is significantly associated with blastocyst formation in the mouse (Kowalik et al., 1999). RT-PCR analysis of IGF-1 in isolated blastomeres allows the embryos with the greater developmental potential to be identified. This is another example of how PGD technology can be used to select the embryos more likely to develop to the blastocyst stage. Once a marker for developmental potential is identified in other mammalian species including the human this strategy will be of great use for assessing embryo quality in the future.

References

Abbott CM, Skidmore CJ, Searle AG, Peters, J. 1986. Deficiency of adenosine deaminase in the wasted mouse. Proc Natl Acad Sci. USA 83:693-695.

Abdel-Rahman B, Fiddler M, Rappolee D, Pergament E. 1995. Expression of transcription regulating genes in human embryos. Hum Reprod 10:2787-2792.

Angell R, Templeton A, Aitkin R. 1986. Chromosome studies in human in vitro fertilization. Hum Genet 72:333-339.

Ao A, Erickson RP, Winston RML, Handyside AH. 1994. Transcription of paternal Y-linked genes in the human zygote as early as the pronucleate stage. Zygote 2:281-288.

Artley J, Braude P, Johnson MH. 1991. Gene activity and cleavage arrest in human preimplantation embryos. Hum Reprod 7:1014-1021.

Benson PF, Fensom AH. 1985. Genetic biochemical disorders. Oxford Monographs on Medical Genetics No. 12. Oxford: Oxford University Press.

Benson C, Monk M. 1988. Microassay for adenosine deaminase, the enzyme lacking in some forms of immunodeficiency, in mouse preimplantation embryos. Hum Reprod 3:1004-1009.

Bolton VN, Hawes SM, Taylor CT, Parsons JH. 1989. Development of spare human preimplantation embryos in vitro: an analysis of the correlations among gross morphology, cleavage rates and development to the blastocyst. J in vitro Fert Embryo Transf 6:30-35.

Braude P, Bolton V, Moore S. 1988. Human gene expression first occurs between the four- and eight-cell stages of preimplantation development. Nature 332:459-461.

Braude PR, Monk M, Pickering SJ, Caut A, Johnson MH. 1989. Measurement of HPRT activity in the human unfertilized oocyte and pre-embryo. Prenat Diagn 9:839-850.

Chia CM, Winston RM, Handyside AH. 1995. EGF, TGF-alpha and EGFR expression in human preimplantation embryos. Development 121:299-307.

Conn CM, Cozzi J, Harper JC, Winston RM, Delhanty, JD. 1999. Preimplantation genetic diagnosis for couples at high risk of Down syndrome pregnancy owing to parental translocation or mosaicism J Med Genet 36:45-50.

Cruz J, Dubey A, Patel J, Peak D, Hartog B, Gindoff P. 1999. Is blastocyst transfer useful as an alternative treatment for patients with multiple in vitro fertillization failures? Fertil Steril 72:218-220.

Daniels R, Kinis T, Serhal P, Monk M. 1995. Expression of myotonin protein kinase gene in preimplantation human embryos. Hum Mol Genet 4:389-393.

Delhanty JD, Handyside AH, Winston RM, Hughes M. 1994. Sex determination of preimplantation embryos. Lancet 343:549.

Delhanty JD, Harper JC, Ao A, Handyside AH, Winston RM. 1997. Multicolour FISH detects frequent chromosomal mosaicism and chaotic division in normal preimplantation embryos from fertile patients. Hum Genet 99:755-760.

Dokras A, Sargent IL, Ross C, Gardner RL, Barlow DH. 1990. Trophectoderm biopsy in human blastocysts Hum Reprod 5:821-825.

ESHRE consortium steering committee. 2000. ESHRE preimplantation genetic diagnosis (PGD) consortium: data collection II. Hum Reprod 15:2673-83.

Eldadah ZA, Grifo JA, Dietz HC. 1995. Marfan syndrome as a paradigm for transcript-targeted preimplantation diagnosis of heterozygous mutations Nat Med 1:798-803.

Epstein C J, Travis B, Tucker G, Smith S. 1978. The direct demonstration of an X-chromosome dosage effect prior to inactivation Basic Life Sci 12:261-267.

Evsikov S, Verlinsky Y. 1998. Mosaicism in the inner cell mass of human blastocysts Hum Reprod 13:3151-3155.

Fiddler M, Abdel Rahman B, Rappolee DA, Pergament E. 1995. Expression of SRY transcripts in preimplantation human embryos. Am J Med Genet 55:80-84.

Findlay I, Quirke P, Hall J, Rutherford A. 1996. A fluorescent PCR: a new technique for PGD of sex and single-gene defects. J Assist Reprod Genet 13:96-103.

Foote RH, Simkin ME. 1993. Use of gonadotropic releasing hormone for ovulating the rabbit model. Lab Anim Sci 43:383-385.

Gardner RL, Edwards RG. 1968. Control of the sex ratio at full term in the rabbit by transferring sexed blastocysts. Nature 218:346-349.

Gardner DK, Phil D, Lane M, Stevens J, Schlenker T, Schoolcraft W. 2000. Blastocyst score affects implantation and pregnancy outcome: towards a single blastocyst transfer. Fertil Steril 73:1155-1158.

Gianaroli L, Magli MC, Ferraretti AP, Munne S. 1999. Preimplantation diagnosis for aneuploidies in patients undergoing in vitro fertilization with a poor prognosis: identification of the categories for which it should be proposed. Fertil Steril 72:837-844.

Handyside AH, Delhanty JD. 1997. Preimplantation genetic diagnosis: strategies and surprises. Trends Genet 13:270-275.

Handyside AH, Pattinson JK, Penketh RJ, Delhanty JD, Winston RM, Tuddenham EG. 1989. Biopsy of human preimplantation embryos and sexing by DNA amplification. Lancet 1:347-349.

Hardy K, Handyside AH, Winston RM. 1989. The human blastocyst: cell number, death and allocation during late preimplantation development in vitro. Development 107:597-604.

Hardy K, Martin KL, Leese HJ, Winston RM, Handyside AH. 1990. Human preimplantation development in vitro is not adversely affected by biopsy at the 8-cell stage. Hum Reprod 5:708-714.

Harper JC, Coonen E, Handyside AH, Winston RM, Hopman AH, Delhanty JD.1995. Mosaicism of autosomes and sex chromosomes in morphologically normal, monospermic preimplantation human embryos. Prenat Diagn 15:41-9.

Hattori M, Yoshioka K, Sakaki Y.1992. High-sensitive fluorescent DNA sequencing and its application for detection and mass-screening of point mutations. Electrophoresis 13:560-565.

Heikinheimo O, Lanzendorf SE, Baka SG. 1995. Cell cycle genes c-mos and cyclin-B1 are expressed in a specific pattern in human oocytes and preimplantation embryos. Hum Reprod 10:699-707.

Hooper M, Hardy K, Handyside A, Hunter S, Monk M. 1987. HPRT deficient (Lesch-Nyhan) mouse embryos derived from germline colonisation by cultured cells. Nature 326 (6110):292-295.

Kennelly JJ, Toote RH. 1965. Superovulatory response of pre- and post-pubertal rabbits to commercially available gonadotrophins. J Reprod Fertil 9:177-88.

Kowalik A, Liu H-C, Mele C, Barmat L, Rosenwaks Z. 1999. Expression of the insulin-like growth factor-1 gene and its receptor in preimplantation mouse embryos; is it a marker of embryo viability? Mol Hum Reprod 5:861-865.

Kuliev A, Rechitsky S, Verlinsky O, Ivakhnenko V, Evsikov S, Wolf A, Angastinoitis M, Georghiou D, Kukharenko V, Strom C, Verlinsky Y. 1998. Preimplantation diagnosis of thalassemia. J Assist Reprod Genet 15:219-225.

Leese HJ, Humpherson PG, Hardy K, Hooper MA, Winston RM, Handyside AH. 1991. Profiles of hypoxanthine guanine phosphoribosyl transferase and adenine phosphoribosyl transferase activities measured in single pre-implantation human embryos by high-performance liquid chromatography. J Reprod Fertil 91:197-202.

Lighten A, Hardy H, Winston RM, Moore GE. 1997. Expression of mRNA for the insulin-like growth factors and their receptors in human preimplantation embryos. Mol Reprod Dev 47:134-139.

Marquez C, Cohen J, Munne S. 1998. Chromosome identification in human oocytes and polar bodies by spectral karyotyping. Cytogenet Cell Genet 81:254-8.

Monk M, Handyside AH, Hardy K, Whittingham D. 1987. Preimplantation diagnosis of deficiency of hypoxanthine phosphoribosyl transferase in a mouse model for Lesch-Nyhan syndrome. Lancet ii:423-425.

Monk M, Muggleton Harris AL, Rawlings E, Whittingham DG. 1988. Pre-implantation diagnosis of HPRT-deficient male and carrier female mouse embryos by trophectoderm biopsy. Hum Reprod 3:377-381.

Monk M, Handyside AH. 1988. Sexing of preimplantation mouse embryos by measurement of X-linked gene dosage in a single blastomere. J Reprod Fert 82:365-368.

Muggleton Harris AL, Findlay I. 1991. In vitro studies on 'spare' human preimplantation embryos in culture. Hum Reprod 6:85-92.

Muggleton-Harris AL, Glazier AM, Pickering SJ.1993. Biopsy of the human blastocyst and polymerase chain reaction (PCR) amplification of the beta-globin gene and a dinucleotide repeat motif from 2-6 trophectoderm cells. Hum Reprod 8:2197-2205.

Munne S, Scott R, Sable D, Cohen J. 1998. First pregnancies after pre-conception diagnosis of translocations of maternal origin. Fertil Stertil 69:675-81.

Munne S, Weier HU, Grifo J, Cohen J.1994. Chromosome mosaicism in human embryos. Biol Reprod 51:373-379.

Murray AW. 1992. Creative blocks: cell-cycle checkpoints and feedback controls. Nature 359:599-604.

Osterlund C, Wramsby H, Pousette A. 1996. Temporal expression of platelet-derived growth factor (PDGF)-A and its receptor in human preimplantation embryos. Mol Hum Reprod 2:507-512.

Papaioannou VE, Ebert KM. 1986. Experimental approaches to mammalian embryonic development (Rossant, J and Pedersen, R, Eds) Cambridge University Press, Cambridge 67-96.

Plachot M, de Grouchey J, Junca A-M, Mandelbaum J, Turleau C, Couillin P, Cohen J, Salat-Baroux J. 1987. From oocyte to embryo: a model, deduced from in vitro fertilization for natural selection against chromosome abnormalities. Ann Genet 30:22-32.

Ray PF, Handyside AH. 1996. Increasing the denaturation temperature during the first cycles of amplification reduces allele dropout from single cells for preimplantation genetic diagnosis. Mol Hum Reprod 2:213-218.

Sermon K, Lissens W, Tarlatzis B, Braude PR, Devroey P, Van Steirteghem A, Liebaers I. 1992. Beta-N-acetylhexosaminidase activity in human oocytes and preimplantation embryos. Hum Reprod 9:1278-1280.

Sermon K, De VA, Van de Velde H, Seneca S, Lissens W, Joris H, Vandervorst M, Van SA, Liebaers I. 1998. Fluorescent PCR and automated fragment analysis for the clinical application of preimplantation genetic diagnosis of myotonic dystrophy (Steinert's disease). Mol Hum Reprod 4:791-796.

Sharkey AM, Dellow K, Blayney M, Macnamee M, Charnock-Jones S, Smith SK. 1995. Stage specific expression of cytokine and receptor messenger ribonucleic acids in human preimplantation embryos. Biol Reprod 53:974-981.

Tarin JJ, Conaghan J, Winston RML, Handyside AH. 1992. Embryo biopsy on the 2nd day after insemination for preimplantation diagnosis: biopsy of a quarter of embryo retards cleavage. Fert Steril 58:970-976.

Tesarik J, Kopecny V, Plachot M, Mandelbaum J. 1988. Early morphological signs of embryonic genome expression in human preimplantation development as revealed by quantitative electron microscopy. Dev Biol 128:15-20.

Taylor DM, Ray PF, Ao A, Winston RM, Handyside AH. 1997. Paternal transcripts for glucose-6-phosphate dehydrogenase and adenosine deaminase are first detectable in the human preimplantation embryo at the three- to four-cell stage. Mol Reprod Dev 48:442-448.

Taylor DM, Handyside AH, Ray PF, Dibb NJ, Winston RM, Ao A. 2001. Quantitative measurement of transcript levels throughout human preimplantation development: analysis of hypoxanthine phosphoribosyl transferase. Mol Hum Reprod 7:147-154.

van Eijk MJ, Mandelbaum J, Salat-Baroux J, Belaisch-Allart J, Plachot M, Junca AM.1996. Expression of leukaemia inhibitory factor receptor subunits LIFR beta and gp130 in human oocytes and preimplantation embryos. Mol Hum Reprod 2:355-360.

Verlinsky Y, Cielak J, Freidine M, Ivakhnenko V, Wolf G, Kovalinskaya L, White M, Lifchez A, Kaplan B, Moise J, Valle J, Ginsberg N, Strom C, Kuliev A. 1995. Pregnancies following pre-conception diagnosis of common aneuploides by fluorescent in-situ hybridization. Hum Reprod 10:1923-1927.

Verlinsky Y, Cielak J, Ivakhnenko V, Evsikov S, Wolf G, White M, Lifchez A, Kaplan B, Moise J, Valle J, Ginsberg N, Strom C, Kuliev A. 1999. Prevention of age-related aneuploidies by polar body testing of oocytes. J Assist Reprod Genet 16:165-169.

Verlinsky Y, Evsikov S. 1999. A simplified and efficient method for obtaining metaphase chromosomes from individual human blastomeres. Fertil Steril 72:1127-33.

Vidal F, Rubio C, Simon C, Gimenez C, Minguez Y, Pellicer A, Santalo J, Remohi J, Egozcue J. 2000. Is there a place for preimplantation genetic diagnosis screening in recurrent miscarriage patients? J Reprod Fertil 55:143-146.

Voullaire L, Howard S, Williamson R, Wilton L. 2000. Chromosome analysis of blastomeres from human embryos by using comparative genomic hybridization. Human Genet 106:210-217.

Wells D, Sherlock JK. 1998. Strategies for preimplantation genetic diagnosis of single gene disorders by DNA amplification. Prenat Diagn 18:1389-401.

Wells D, Sherlock J, Handyside A, Delhanty J. 1999. Detailed chromosomal and molecular genetic analysis of single cells by whole genome amplification and comparative genomic hybridization. Nucleic Acid Res 27:1214-1218.

Willadsen S, Levron J, Munne S, Schimmel T, Marquez C, Scott R, Cohen J. 1999. Rapid visualization of metaphase chromosomes in single human blastomeres after fusion with in vitro matured bovine eggs. Hum Reprod 14:470-475.

Wilton LJ, Trounson AO. 1989. Biopsy of preimplantation mouse embryos: development of micromanipulated embryos and proliferation of single blastomeres in vitro. Biol Reprod 40:145-152.

Xu K, Shi Z, Veek L, Hughes M, Reosenwaks Z. 1999. First unaffected pregnancy using preimplantation genetic diagnosis of sickle cell anaemia. JAMA 281:1701-1706.

Zarrow MX, Wilson ED. 1961. The influences of age on superovulation in the immature rat and mouse. Endocrinology 69:851-855.

Chapter 8

BIOPSYING AND GENOTYPING CATTLE EMBRYOS

Erik Mullaart
Holland Genetics, P.O. Box 5073, 6802 EB Arnhem, The Netherlands

1. Introduction

Within the practice of cattle breeding embryo transfer (ET) technology is now well established. In vivo grown embryos can be generated via superovulation and flushing of the uterus at day 7 after insemination. Alternatively, in vitro grown embryos can be generated from immature oocytes after in vitro maturation, fertilization and culture for 7 days. The selection of transferable embryos is generally based on the morphology and their expected breeding value (Lindner and Wright, 1983; Chapter 2, Merton). However, the efficiency of breeding programmes would increase, if embryos could be selected based on their sex, the presence of particular economically important traits (i.e. milk, meat, fertility, etc.) and the absence of particular diseases. In addition, the use of this marker-assisted selection (MAS) of embryos will lead to a shorter generation interval and higher selection intensity, thereby increasing the genetic gain.

Due to the rapid development and progression within the field of molecular biology and genetics, more and more genes are discovered which are responsible for particular economically important traits and genetic defects (Georges, 1998). Theoretically, it is possible to analyse the DNA of embryos for the presence of economically important genes and genes related to diseases. Methods to analyse the DNA from a few embryonic cells, which are taken from the embryo (i.e. a biopsy), have recently become available.

In this chapter, different methods to take a biopsy from cattle embryos are described and their impact on pregnancy rates is assessed. Subsequently the available DNA tests presently used in cattle breeding programmes are discussed. Furthermore, possibilities for future application are presented.

2. Embryo biopsy methods

Several methods are available to take a biopsy of preimplantation bovine embryos. Two routine methods, i.e. aspiration and micro-section are described and some recently developed methods will be evaluated. Both methods are suitable for in vivo as well as for in vitro produced embryos.

2.1. The aspiration method

With this method, a holding pipette is used to stabilize the embryo while a second pipette is needed for the aspiration of embryonic cells using suction pressure. Both pipettes are connected to a micro-manipulator. The aspiration pipette, which is connected to a pump system is inserted through the zona into the embryo and a few cells are aspirated. A prerequisite for optimal results of this method is the absence of strong cell-cell interactions. This means that early cleavage stage embryos or morulae are the optimal embryonic stages for aspiration biopsies. Two to four cells can be collected from a morula without disturbing the further development of the embryo. Using the aspiration method a skilled lab technician can collect about 10 biopsies per hour.

The main advantage of this method is that it makes only a small incision in the zona, allowing the embryo to stay within the zona, which is an advantage during freezing (Thibier and Nibart, 1995). According to these authors, the relatively small insult has only a minor negative effect on the pregnancy rates. The disadvantage of this method is the fact that only a limited amount of material is collected. This can be a problem for further DNA testing, but also increases the possibility of loosing the biopsy during the subsequent transfer of the cells into the test tube. Another disadvantage of this method is the fact that the technique requires sufficient technical experience from the person who is performing the biopsy (Bredbacka, 2001). The aspiration biopsy method can also be used for blastocysts but this implies the use of a third micro-manipulator with a closed pipette serving as a “knife” to separate the cells collected by the aspiration (Geldhof et al., 2000).

2.2. The micro-section method

With this method, first described by Bondroli et al. (1989), a piece of the embryo is cut off using a microblade. In most cases the microblade is connected with a micromanipulator but the complete procedure can also be performed manually (Bredbacka, 1998). The biopsy contains about 10-20 cells, but since there is a large incision in the zona, the embryo will escape from the zona in most cases. The micro-section method is most often used for blastocysts. At this stage of development, the differentiation between the inner cell mass and trophoblast has taken place and therefore it is possible to take a few trophoblast cells without damaging the inner cell mass. Obviously, a biopsy taken from the inner cell mass might disturb the normal development of the embryo. The method can also be applied to morulae, but in this case it is important to take a small biopsy. In addition, biopsies from morulae more often fall apart into loose cells, which are easily lost during further processing (Carbonneau et al., 1997, Bredbacka, 2001). Using this method one skilled lab technician can perform a biopsy of some 15 embryos per hour (Bredbacka, 1998; personal observation).

2.3. The two methods compared

The aspiration and micro-section methods are the most popular methods for the performing of biopsies on cattle embryos. Each method has its own advantages and disadvantages which are summarized in Table 1.

Table 1: The advantages and disadvantages of the two different biopsy methods.

Biopsy method	Aspiration	Micro-section
Size of biopsy	Small (2-4 cells)	Large (10-20 cells)
Damage to the zona	Minimal	Considerable
Possibility of loosing biopt	High	Low
Possible with morulae	Yes	Yes
Possible with blastocysts	No	Yes
Trained personnel	Yes	Less necessary

2.4. New biopsy methods

Besides these two basic approaches, several alternative ways for obtaining embryo biopsies have been developed. For instance, Leoni et al. (2000) make a small incision with a micro-needle in the zona of an ovine embryo at the morula or blastocyst stage. After a culture period of several hours, a few cells pop out of the hole in the zona, allowing them to be excised by hand with a microblade under a stereo microscope. The advantage of this method is that the embryo remains within the zona. The disadvantage is that the embryos have to be cultured for several hours. However, despite the fact that embryonic survival in vitro was not affected by the biopsy procedure, a large drop in pregnancy rates was observed compared to control embryos (from 68.7% to 40.0% based on only 15-30 transfers) (Leoni et al., 2000). Because the biopsied embryos were also subjected to vitrification in this study, manipulated embryos might have a different sensitivity to the cryo-preservation procedures.

The group of Galli is using a modification of the micro-section method (Giovanna Lazzari, personal communication). Positive pressure is applied upon the embryo with the side of the blade until it bursts, and a small crack in the zona appears . It is then possible to cut a very small piece with a micro blade from the flattened embryo. They argue that it is much easier to cut a small piece from a flat, well-immobilized embryo than from a round rolling embryo. Using this method the embryo also remains within the zona (which has one crack and a small hole) and this biopsy method procedure appears to have little or no negative effect on the pregnancy rates.

Something very comparable has been described by Bredbacka (2001) as a method developed by Ryan. Precompacted morulae are being used. A very small incision is prepared in the zona pellucida by a vertical movement of the microblade. Then the medium is aspirated from the micro-drop in which the embryo is floating. Due to the decreased volume, the micro-drop flattens, as well as the embryo and the zona opens through the incision with several cells being released into the medium. As the aspirated medium is put back into the droplet, the zona closes and some of the extruded cells remain outside the zona and can be used for sampling.

3. Pregnancy rates

Obviously, it is very important that the biopsy method should not have a negative effect upon the pregnancy rate. Basically there is a conflict of interest here. For DNA testing the biopsy should be as large as possible (more material), while in order to keep the pregnancy rates high, the biopsy should be as small as possible. A very important factor determining the effect of the biopsy method on the pregnancy rate is the quality of the embryo. Grade 1 embryos (see Chapter 2) appear to be much less vulnerable for the biopsy procedure than grade 2 embryos. Thibier and Nibart (1995) obtained a pregnancy rate of 53% with grade 1 and of only 34% with grade 2 embryos after biopsy with a micro-blade. Similar data are reported by Hasler (in preparation), i.e. 61% for grade 1 and 45% for grade 2 embryos. When grade 2 embryos are subsequently being frozen, the pregnancy rates can decline even more. In table 2 and 3 the pregnancy rates are presented of several groups using micro-section and/or aspiration methods for taking biopsies from in vivo or in vitro embryos.

In general, there appears to be a drop in the pregnancy rate of about 0-10% for fresh and frozen in vivo produced embryos when aspiration biopsies are being performed (Table 2). When the biopsies are collected by micro-section, the drop in the pregnancy rate appears to be somewhat higher, especially when embryos are being frozen after having taken a biopsy and then being transferred (0-30%). For in vitro produced embryos, most pregnancy data are obtained after micro-section (Table 3).

Table 2: Pregnancy rates of in vivo cattle embryos after aspiration or micro-section biopsy compared with controls.

Biopsy method	Embryo	Percentage of animals pregnant after transfer of		Decrease in pregnancy rate	Reference
		control embryos	biopsied embryos		
Aspiration	Fresh	ND	56 %	-	Geldhof, 2000
		55-65 %	56 %	0-9 %	Ponsart, 2000
		53 %	46 %	7 %	Roschlau et al., 1997
		50-60 %	60 %	0	Shea et al., 1999
	Frozen	49 %	46 %	3 %	Garcia et al., 1997
		ND	45 %	-	Geldhof, 2000
		49 %	46%	3 %	Ponsart, 2000
		49 %	44 %	6 %	Roschlau et al., 1997
		40-50 %	41 %	0-10 %	Shea et al., 1999
		49 %	55 %	0 %	Thibier and Nibart, 1995
Micro-section	Fresh	70 %	62 %	8 %	Hasler, 2000
		ND	55 %	-	Jindra et al., 2000
		50-60 %	49 %	0-10 %	Shea et al., 1999
	Frozen	>60%	56 %	> 4 %	Bousquet, (pers comm.)
		66 %	52 %	14 %	Hasler, 2000
		52 %	44 %	8 %	Otter et al., 2000
		40-50 %	23 %	20-30 %	Shea et al., 1999
		49 %	28 %	21 %	Thibier and Nibart, 1995

ND = not determined

Table 3: Pregnancy rates of in vitro produced cattle embryos after micro-section biopsy compared with controls.

Embryo	Percentage of pregnant animals after transfer of			
	Control embryos	Biopsied embryos	Decrease in pregnancy rate	Reference
Fresh	> 60 %	57 %	> 3 %	Carbonneau et al,. 1997
	56 %	49%	7%	Hasler, 2000
	48 %	42 %	6 %	Otter et al., 2000
Frozen	35 %	33 %	2 %	Otter et al., 2000

The decrease in pregnancy rate is comparable with that of in vivo produced embryos. Interestingly and contrary to the studies with in vivo produced embryos, one group found better results for micro-section biopsy (66 % pregnant) than for aspiration (45 % pregnant) (Carbonneau et al., 1997). This can be explained by the fact that for aspiration biopsies, morulae are being used. It is common knowledge that not all in vitro produced morulae are capable of reaching the blastocyst stage and this might be the reason for the lower pregnancy rates. Using blastocysts for biopsies gives the opportunity to select better quality embryos.

Interestingly, taking a biopsy from an already frozen embryo, also appears to be possible. Hasler et al. (in preparation) obtained promising data with this approach (pregnancy rates of 53%) but the numbers are still very small (36 transfers). This can be of practical value since, due to the sometimes large distances between the place of embryo collection and the place of biopsy, it might not always be possible, or very difficult, to work with fresh embryos.

4. DNA tests performed on the biopsy

After the biopsy has been taken, several DNA-based tests can be performed. But first the DNA has to be isolated from the biopsied cells. The most commonly used method for this procedure is to incubate the cells with proteinase K at 55° C. The genetic material is released from the cells by the enzymatic action. Proteinase K is subsequently inactivated at 95° C to prevent destruction of DNA Polymerase, which is added in the following steps of the PCR-reaction (Bredbacka et al., 1995). This procedure has the advantage that it is reliable, quick (10 min) and simple. Furthermore, sperm cells that might be attached to the zona are not lysed by proteinase K and therefore sperm DNA cannot contaminate the embryonic DNA.

4.1. Determination of embryonic sex

The first developed and now most common assay that is performed on an embryo biopsy is the determination of sex. Knowing the sex of an embryo allows for a more efficient use in breeding programmes but can also present a major advantage for the farmer, who is preferably interested in female replacement animals. With cattle, like with all mammalian species, the male has one copy of the Y-chromosome and one copy of the X-chromosome. Female animals have two copies of the X-chromosome. All sex determinations on biopsies are therefore based on the detection of the presence of the Y-chromosome. When PCR is being used to detect male specific DNA, Y-chromosome specific primers are used to amplify a part of the Y-chromosome. In the simplest form of this assay (Bredbacka et al., 1995), the next step is to check if amplification has occurred. For this specific assay, amplification will only take place when the Y-chromosome is present. Consequently, since the fluorescent DNA stain is present in the amplification mix, Y-positive samples (from male embryos) will fluoresce pink under UV light (Bredbacka et al., 1995).

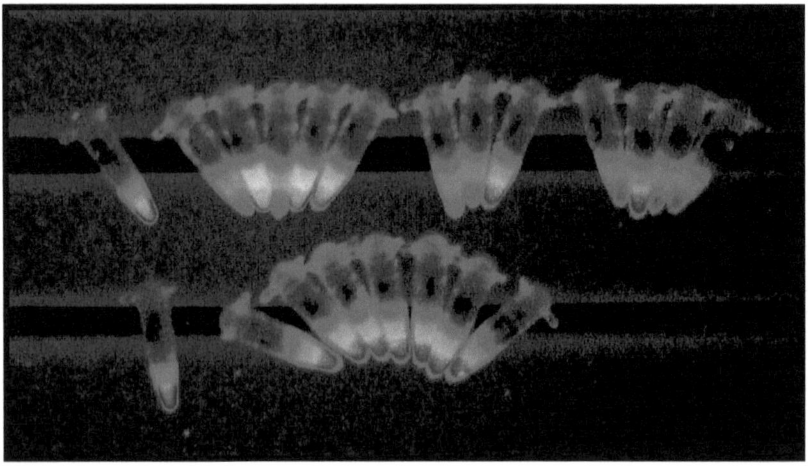

Figure 1: *Example of the sex test on embryo biopsies. In male biopsies containing the Y-chromosome the amplified DNA is clearly visible as a pink staining. The female biopsies lacking the Y-chromosome are blank since no amplified DNA is present.*

Samples, which do not fluoresce, are Y-negative and thus female embryos. An example of this test is shown in Figure 1.

The advantage of this test is that it is very simple and does not require gel electrophoresis. Moreover, it is not necessary to open the reaction tube after PCR, which is greatly reducing the risk of contamination of the laboratory with PCR product (see also section 4.5.).

The disadvantage is the fact that embryos are determined female based upon the absence of any PCR reaction product. When the biopsy of a male embryo is lost during transfer into the reaction tube, a similar result is obtained and the embryo is identified as being female. The error rate of this test is about 8%, where all errors reported until now are males falsely reported (by the PCR test) as females.

The problem of males being wrongly identified as females, can be solved by PCR amplification of part of an autosomal chromosome (Herr and Reed, 1991). This implies that the resulting PCR products need to be analysed by gel electrophoresis to determine the presence of the different bands. The analysis of the extra chromosome will be visible as an extra band on the gel. If the control band is visible it is an indication that the biopsy was present and has been amplified. This adaptation makes the test more reliable.

The efficiency (percentage result) of the gel based method is about 90-95%, while the efficiency of the non-gel based method is 100%. Since no amplification automatically equals a female embryo, there is always a result, but not necessarily a correct one. The gel based methods have a higher (about 96-99%) accuracy than the non-gel based methods (92-95%).

4.2. Sex/BLAD test

Bovine leukocyte adhesion deficiency (BLAD) is an autosomal recessive disease described in 1992 by Shuster. If both alleles of the BLAD gene are present, the animal is affected and will die. If only one allele has been mutated the animals are healthy but are so-called carrier. Hence the disease is still present in the population. Therefore, screening embryos for their BLAD status could be useful when the father or the mother is carrier. Hochman et al. (1996) described the use of this test on embryo biopsies and recently a sex / BLAD test has been developed by Bredbacka (Finnzymes, personal communication). After releasing the DNA from the biopsy, PCR amplification is performed with primers specific for the Y-chromosome (sex) and for the BLAD gene. Subsequently, a restriction enzyme digestion is effected using an enzyme that cuts the BLAD DNA fragment when it does not contain the BLAD mutation. However, a DNA fragment containing the BLAD mutation is not cut by this restriction enzyme.

Subsequently, the entire mixture is analysed by gel electrophoresis. When the embryo is male, the Y-chromosome band will be visible, while in case of a female embryo no band is observed at this position in the gel. When the embryo is BLAD free, the restriction enzyme cuts both alleles and only one band remains visible on the gel. If the embryo is a BLAD carrier one allele is cut, while the other is not cut. On gel, two bands with different lengths can be observed. If the animal is homozygous for the BLAD mutation neither of the two alleles are cut and only one band remains visible on gel, being longer than the normal (none mutated) band. An example of such a gel with different samples is shown in Figure 2.

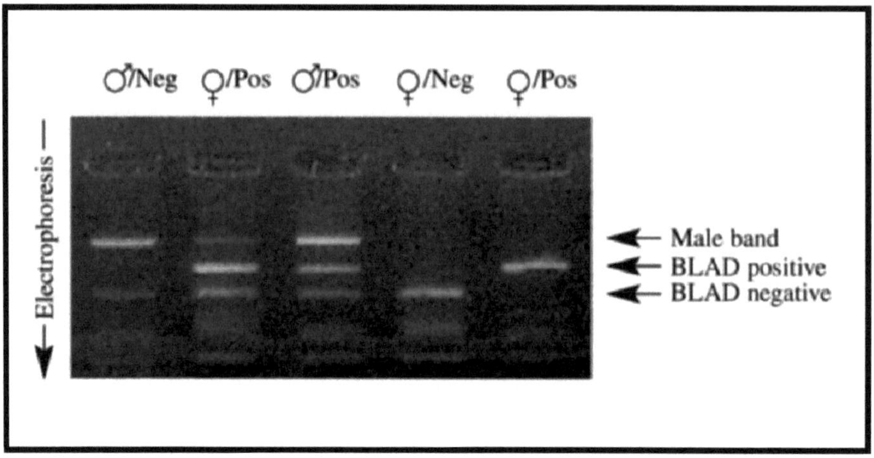

Figure 2: Example of the BLAD/sex test as analysed on an agarose gel. The presence or absence of the Y band indicates the sex. The BLAD status can be deduced from the presence or absence of two bands below the Y band. When the BLAD mutation is absent the fragments become smaller. A homozygous BLAD mutation results in one large band (lane 5), homozygous BLAD free in one smaller band (lanes 1 and 4) and heterozygous in two bands, one large and one small (lanes 2 and 3).

Due to the fact that an extra step, i.e. restriction enzyme digestion, and gel electrophoresis has to be performed, this test is somewhat more difficult. Especially the interpretation of the bands requires some experience. The reliability of this test (especially the BLAD test) is not known yet.

4.3. Other tests

Until now little information is available on the use of other DNA tests for embryo biopsies. Hochman et al. (1995) reported about the analysis of casein, mitochondrial sequences, a micro-satellite (D9S1) and BLAD-factor on a single biopsy after pre-amplification (see also section 4.4.). Finnzymes (Finnland) is now working on the development of a coat colour (red-factor) test in combination with a sex test. Recently, also Joerg et al. (2001) has reported about a coat colour assay on embryo biopsies. Theoretically any disease or trait of which the gene is known can be analysed by PCR on embryo biopsies. However, until now no reports of the use of these tests on embryo biopsies have been published.

4.4. Multiple tests

With the present increase in the discovery of genes and QTLs (Quantitative Trait Loci, which code for milk production, meat quality, etc) found for different traits and diseases in cattle (Georges, 1998), there is a growing demand for procedures allowing multiple tests to be performed on a single embryo biopsy. However, a single biopsy is estimated to contain between 1 and 20 cells and therefore contains enough DNA for only a few PCR reactions. This small number of cells normally obtained after embryo biopsy is the major obstacle for large-scale DNA analysis. It is of course possible to perform multiplex PCR reaction, which increases the number of loci analysed per PCR reaction (perform both sex and BLAD test for instance), but there are limits (e.g. multiplex reaction of more than 3-4 genes are generally not very robust).

A possibility to solve this problem is to increase the amount of DNA material before the specific PCRs are performed. This can be done biochemically (by total genome amplification (TGA)) or biologically (by culturing the cells from the biopsy).

When total genome amplification is performed, the idea is to amplify the entire genome several times to generate enough DNA to perform all the different tests. Several protocols have been developed, with varying success rates (Zhang et al., 1992). Generally the genome is amplified by a factor 30 - 100, but it remains uncertain whether the entire genome is amplified. Since the amplification process happens at random, particular parts might be over-amplified while others are not amplified at all. This will give problems when the gene of interest lies in a non-amplified region (locus drop-out). In addition, it is quite often found that the quality of the amplification of one copy of the chromosome is less than the other copy. This might lead to allele drop-out: one allele which is located on the less amplified chromosome is lost during further analysis. This results in originally heterozygous samples which are then incorrectly called homozygous (Wells et al., 1999). Another potential risk with total genome amplification is the risk of contamination. Since during the pre-amplification minute amounts of DNA from the biopsy are amplified multiple times, any small contamination is also being amplified several times.

Several groups have been using total genome amplification to analyse small biopsies from tumours (e,g, Ao et al., 1998) or to analyse single cells (Sun et al., 1995). For embryo biopsies it is not used very often. Hochman et al.

(1996) reported about the Primer-extension Preamplification (PEP) and the subsequent analysis of up to 40 different tests. In their hands the success rate was about 95% as tested with PCR amplification (after TGA) of a locus that should be present in all biopsies. Until now however, no follow-up has been seen by this group. Some preliminary data from Malhomme et al. (2000), indicated that after TGA it is possible to test several different (up to 7) Single Nucleotide Polymorphism (SNP) markers.

Chrenek et al. (2001) used PEP followed by specific PCR for sex, kappa-casein, growth hormone and prolactin, with an average efficiency of 90%.

Another method to increase the amount of DNA is to multiply the cells of the biopsy by means of cell culture techniques. Adequate culture techniques should result in a harvest of thousands to millions of cells, which should be sufficient to perform all the different tests. Until now however, no reliable cell culture protocols have been developed to make this possible. The major obstacle appears to be the attachment of the embryonic cells to the surface of the culture dish, which seems to be necessary for outgrowth of the biopsy (Moens, personal communication). Success rate is in the order of 40%, which is still too low for large-scale use.

4.5. Some practical considerations when PCR tests on embryo biopsies are performed

Basically, these PCR-based tests are simple and easy to perform (simple protocols). One of the most serious problems however, is the contamination with tiny amounts of foreign DNA. Since the methods are often very sensitive, contamination with tiny amounts of DNA is sufficient to cause large problems. During a standard PCR reaction of 30 cycles, the starting material is theoretically multiplied by a factor 1×10^9. Potential sources of contamination are products from previous amplifications, cells adhering to the zona (i.e. sperm cells or cumulus cells), bovine DNA that might be present in FCS and BSA. It is therefore advisable to use buffers which do not contain any material of animal origin. Furthermore, it is very important to include proper controls (both positive and negative) in the PCR test. In addition, it is imperative to use different rooms for PCR-analysis and for the taking biopsies and running the gels. Especially the act of running a gel is a serious source of contamination since the tubes contain large amounts of the end product and when they are opened, the air will be contaminated. Be careful that the different rooms are also not connected with each other via the air conditioning system. To reduce this risk of contamination with human material (skin cells, hairs from the laboratory technicians) , all tests work with primers specific for cattle. One other point of concern is the fact that embryos often contain haploid or mixoploid cells (Viuff et al., 1999). When these cells (especially the haploid ones) are used for the analysis, the results can be incorrect. Especially when only a few cells are collected by aspiration biopsy, this can be a serious problem. Until now no clear solutions are available to circumvent this problem.

5. Conclusion

Taking biopsies by aspiration or micro-section is now more or less a standard procedure for cattle embryos, however new methods are becoming available. Generally, a drop in pregnancy rates of about 10-15% is observed, but this depends on the quality of the embryos. Good quality embryos can withstand more easily the different insults (biopsy, freezing) while moderate quality embryos show a larger drop in pregnancy rates. In addition, skill and practice are required to take a biopsy without damaging the embryo. It is clear that whenever possible, freezing should be circumvented to limit embryo degeneration. Alternative freezing methods (e.g. the OPS vitrification method from Vajta et al., 1998) could yield an interesting solution to this problem. At the moment simple tests like the sex test and possibly some others can be performed within a few hours. The more complicated tests, like those requiring pre-amplification and multiple testing, will take at least 1-3 days and embryos need to be frozen after biopsy. However, developments in this area of amplification and mutation detection are very fast, which might make it possible to perform those tests within a few hours in the near future. The other option would be to take a biopsy of 1-2 cells at day 4 and a transfer at day 7, which leaves about 3 days for the analysis. This requires, however, a very strict schedule and well-organized logistics.

The most frequently used test until now is the sex test, both gel based and non-gel based. The accuracy of these tests is high, 91-97% for the tube and 95-98% for the gel based assay respectively. In the meantime, several groups (Roschlau et al., 1997; Hasler, 2000; Otter et al., 2000; Geldhof et al., 2000, etc.) have proven that this biopsy sexing method can be used within a commercial breeding programme. Although other tests are theoretically possible, only the BLAD test is now used on a routine basis, since genes for several diseases and traits are known. Perhaps in some cases the cost to develop and perform such a test on embryo biopsies is higher than the benefits. On the other hand, performing multiple tests on a single embryo biopsy seems very attractive, allowing accurate selection of the best embryos before transplantation. In this way only disease-free embryos with specific traits are transferred which might dramatically reduce the costs of recipients. It is clear that such a selection procedure is only possible when enough embryos can be produced to select from. It is therefore important to search further for reliable methods to perform multiple tests on a single embryo biopsy.

If it would be possible to perform multiple tests on one biopsy, the value of a particular embryo (containing all the right gene variants) might increase dramatically. It would be very desirable then to obtain a calf from that particular "good" embryo. However, after biopsy and freezing, the pregnancy rates are only about 40%. This pregnancy rate might be increased by alternative ways of biopsy and/or freezing. One alternative is making more embryos from the one “top” embryo by splitting or cloning techniques.

This added value to a particular embryo, will increase the genetics’ progress in the breeding programme. The generation interval will be shortened since it will not longer be necessary to wait for phenotypic data to select the best ones, while choosing from high numbers of embryos will increase the selection intensity (selection from comparable high numbers of calves is very expensive). Besides its use within the breeding programme, embryo biopsy and analysis might also lead to special products to be commercialized. For example female, BLAD free embryos or male embryos with specific production characteristics could be offered for sale.

Acknowledgement

I would like to thank Peter Bredbacka for his critical reading of the manuscript and making useful suggestions for improvement.

References

Ao A, Wells D, Handyside AH, Winston RM, Delhanty JD. 1998. Preimplantation genetic diagnosis of inherited cancer: familial adenomatous polyposis coli. J Assist Reprod Genet 15:140-144.

Bredbacka P, Kankaanpää A, Peoppo J. 1995. PCR-sexing of bovine embryos: A simplified protocol. Theriogenology 44: 167-176.

Bredbacka P. 1998. Recent developments in embryo sexing and its field application. In: 14^e Reunion AETA pp 105-111.

Bredbacka P. 2001. Progress on methods of gene detection in preimplantation embryos. Theriogenology 55:23-34.

Carbonneau G. Morin N, Durocher J, Bousquet D. 1997. Viability of bovine IVF embryo biopsied with micro-section or microaspiration technique for sexing. Theriogenology 49: 266 (abstract).

Chrenek P, Boulanger L, Heyman Y, Uhrin P, Laurincik J, Bulla J, Renard JP. 2001. Sexing and multiple genotype analysis from a single cell of bovine embryo. Theriogenology 47:268 (abstract).

Garcia JF, Nogueira MFG, Pupim F, Pantano T, Visintin JA. 1997. Pregnancy rates of blastomere biopsied bovine embryos frozen in ethylene glycol. Theriogenology 47:268 (abstract).

Geldhof A, Goossens L, Moyaert I, Cuelenaere M, Roschlau K, Roschlau D, Nivot A, Beckers JF and Ectors F. 2000. Sex determination in bovine embryos by micro-aspiration. Results of five years practice under farm conditions. In 16e Réunion A.E.T.E: pp 150 (abstract).

Georges M, 1998. Mapping genes underlying production traits in livestock. In: Clark J (ed) Animal Breeding: Technology for the 21st century. Harwood Academic Publishers: 77-101.

Hasler J, 2000. Practitioners forum; Field applications of PCR sexing. IETS, 2000 Maastricht.

Herr CM, Reed KC, 1991. Micromanipulation of bovine embryos for sex determination. Theriogenology 35:45-54.

Hochman D, Zaron Y, Dekel I, Feldmesser E, Madrano JF, Shani M and Ron M. 1996. Multiple genotype analysis and sexing of IVF bovine embryos. Theriogenology 46:1063-1075.

Jindra M, Lopatarova M, Krontorad P, Fialla M. 2000. Practical use of bovine embryo sexing by PCR under filed conditions in the Czech and Slovak republics. Theriogenology 53:481 (abstract).

Joerg HW, Ménétrey F, Graphodatskaya DA, Stanzinger GF. 2001 Genetic diagnosis of sex and coat colour for bovine embryo selected prior to implantation. Theriogenology 55:364 (abstract).

Leoni G, Ledda S, Bogliolo L, Naitana S. 2000. Novel approach to cell sampling from preimplantation ovine embryos and its potential use in embryonic genome analysis. J Reprod Fertil. 119:309-314.

Lindner GM and Wright RW. 1983. Bovine embryo morphology and evaluation. Theriogenology 20:407-416.

Malhomme O, Karim L, Moens A, Georges M, Dessy F. 2000. Detection of single nucleotide polymorphism markers on biopsies form in vitro produced bovine embryos. In 16[e] reunion AETA: p 180 (abstract).

Otter T, Dotinga F, Jong-Wubs S, Landman RR, Bredbecka P, Mullaart E. 2000. PCR sexing of day 7 bovine in vivo and in vitro embryos. In 16[e] Reunion A.E.T.A.

Ponsart Practitioners forum; Field applications of PCR sexing. IETS 2000 Maastricht.

Roschlau, K, Roschlau D, Roselius R, Dexne U, Michaelis U, Strhl R, Unicki P, Rink N. 1997. Over 5 years experience in sexing of bovine morulae and blastocysts during routine embryo transfer. Theriogenology 47: 273.

Shea BF. 1999. Determining the sex of bovine embryos using polymerase chain reaction results: a six-year retrospective study. Theriogenology 51:841-854.

Shuster DE, Kehrli MEJ, Ackerman MR, Gilbert RO. 1992. Identification and prevalence of a genetic defect that causes leukocyte adhesion deficiency in Holstein cattle. Proc Natl Acad Sci USA. 89:9225-9229.

Sun F, Arnheim N, Waterman MS. 1995. Whole genome amplification of single cells: mathematical analysis of PEP and tagged PCR. Nucleic Acids Res. 23:3034-3040.

Thibier M, Nibart M. 1995. The sexing of bovine embryos in the field. Theriogenology 43:71-80.

Vajta G, Holm P, Kuwayama M, Booth PJ, Jacobsen H, Greve T, Callesen H. 1998. Open Pulled Straw (OPS) vitrification: a new way to reduce cryoinjuries of bovine ova and embryos. Mol Reprod Dev. 51:53-58.

Viuff D, Rickords L, Offenberg H, Hyttel P, Avery B, Greve T, Olsaker I, Williams JL, Callesen H, Thomsen PD. 1999. A high proportion of bovine blastocysts produced in vitro are mixoploid. Biol Reprod. 60:1273-1278.

Wells D, Sherlock J, Handyside A, Delhanty J. 1999. Detailed chromosomal and molecular genetic analysis of single cells by whole genome amplification and comparative genomic hybridisation. Nucleic Acids Res. 27:1214-1218.

Zhang L, Cui X, Schmitt K, Hubert R, Navidi W, Arnheim N. 1992. Whole genome amplification from single cell: implications for genetic analysis. Proc. Natl. Acad Sci USA 89:5847-5851.

Chapter 9

MULTIPHOTON MICROSCOPY FOR IMAGING MAMMALIAN EMBRYOS

Jayne M. Squirrell
Laboratory of Molecular Biology,
University of Wisconsin-Madison,1525 Linden Dr., Madison, WI 53717, USA

1. Introduction

It has been said that "you can see a lot just by looking". This is certainly true in the field of embryology. Indeed, morphology is one, if not the major, criterion for assessing embryo quality in both human and domestic animal reproductive fields. Microscopy has been an important tool for improving our ability to observe and investigate mammalian embryo development. Much has been learned by electron microscopy (Stern et al., 1971; Longo, 1973; Panigel et al., 1975; Calarco et al., 1976; Cran et al., 1980; Hillman and Tasca, 1983; Cran, 1985; Hyttel and Niemann, 1990; Oura et al., 1990; 1994; Van Blerkom and Runner, 1990; Gallicano et al., 1992; Plante and King, 1994; Sathananthan, 1994) and fluorescence microscopy (Maro et al., 1984; Van Blerkom et al., 1984; Maro et al., 1985; Schatten et al., 1985; Batten et al., 1987; Noto et al., 1993; Calarco, 1995; Navara et al., 1995; Barnett et al., 1996; Hewitson et al., 1996; Luoh and Wu, 1996; Hewitson et al., 1997) of embryos from a number of species. Although vastly informative, such fixed specimen studies are limited: it is unsatisfactory to study a dynamic process, such as embryo development, with static images collected from different specimens at various time points. There are a number of problems with this approach including artefacts introduced by the fixation and processing but, more importantly, this approach limits both the amount of information gathered from a single embryo (you only get information about a single time point) as well as the reliability of the information (once the specimen is fixed you no longer have information about its developmental potential). Therefore, a certain amount of guesswork, albeit educated guesswork, is necessary to reconstruct the dynamic process of interest. Fixed tissue analysis is clearly not appropriate as a "non-invasive" technique for assessing embryo quality.

Therefore, live cell imaging is necessary not only for diagnostic but also for research purposes in order to understand what is happening morphologically and cell biologically. Those morphological and cellular events can then be correlated with developmental potential. Video microscopy has been the mainstay of live imaging in mammalian embryology (Lehtonen, 1980; Bavister, 1988; Simerly et al., 1993; Gonzales et al., 1995; Gonzales et al., 1996; Suenaga et al., 1996). This technique provides important information, but can be limited in its resolution. The use of fluorescence microscopy of live mammalian embryos has been accomplished with some success (Albertini, 1987; Adenot et al., 1991; Ferreira and Carmo-Fonseca., 1997). These studies tend to be for short imaging times and the imaging itself may be damaging to living cells (Terasaki and Dailey, 1995).

Thus, it is necessary to pursue technologies that permit high-resolution subcellular embryo imaging while maintaining embryo health and developmental potential. This is obviously important for any kind of diagnostic imaging in human, non-human primate, or domestic animal fields. It is also important in research where confidence in the results obtained dictates that the imaging technique does not interfere with the biology of interest. One such imaging technology is multiphoton laser scanning microscopy (MPLSM) and its application to the analysis of mammalian embryo development is just beginning to be explored.

2. Multiphoton Microscopy Basics

Although the concept of multiphoton excitation was put forth in the 1930's (reviewed in Denk and Svoboda, 1997) it wasn't until 1990 that its practical application to imaging was achieved (Denk et al., 1990). Since that time there have been a number of both general (Denk et al., 1995; Potter, 1996; Wokosin et al., 1996; Denk and Svoboda, 1997; Denk, 2000) and technical (Potter, 1996; Xu et al., 1996; Nakamura, 1999; Konig, 2000) papers describing and discussing the technique of multiphoton microscopy. Readers are referred to those papers for a more in depth explanation of the technology. However, in order to illustrate the potential benefits of multiphoton microscopy, a simplified explanation of this method is necessary. Fluorophores are excited when an electron encounters the energy of a photon of light and the electron is elevated into a higher energy state (Figure 1).

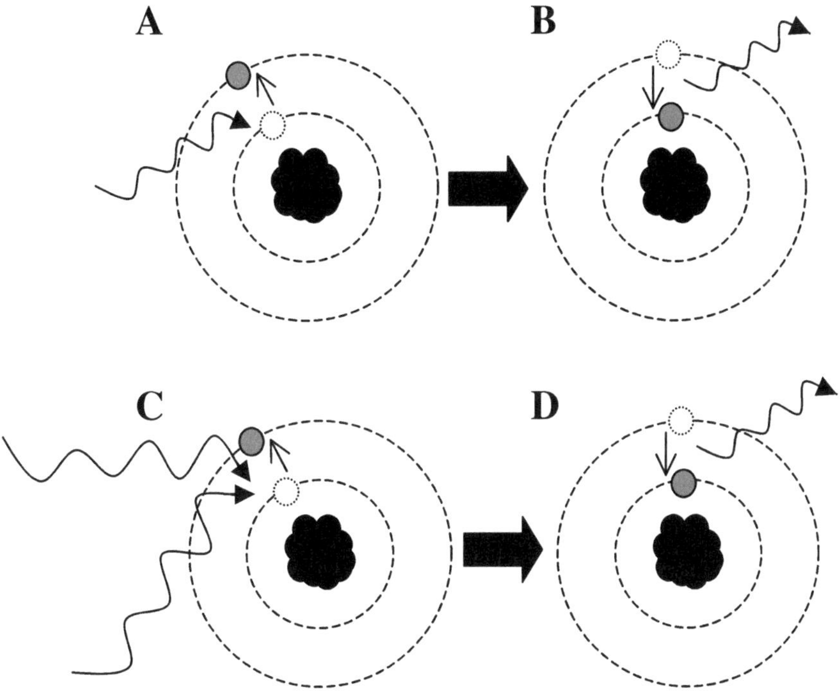

Figure 1: *Schematic diagram of one photon and two photon excitation. A single photon interacts with the electron and the energy of the photon raises the electron to a higher energy state (A). The electron eventually falls from this unstable state back to its more stable lower energy state. When this occurs, a photon is emitted (B). A similar transfer of energy occurs with two photon excitation, except that two photons (of lower individual energy than the single photon depicted in A) simultaneously interact with the electron, elevating it to its higher energy state (C). Again, when the electron returns to its more stable state, a photon is released (D). Note that this emitted photon is the same regardless of whether the excitation was single photon or two photon. This process can be extended to three photon excitation.*

The energy required for that excitation event is fluorophore-specific (excitation wavelength). This is an unstable situation and the electron eventually falls back to its more stable, lower energy state and a photon is released (emission wavelength). Multiphoton microscopy takes advantage of the fact that the electron can be excited either by a single high energy photon or by two (or more) low energy photons, as long as the photons reach the fluorophore simultaneously. Because this is a probabilistic event, a two or multi-photon event only occurs with when the density of photons is high, such as when a laser beam is focused through a microscope lens. Thus, multiphoton excitation occurs only at the plane of focus. There is no out of focus emission because fluorophores above and below the plane of focus are not excited by the lower energy wavelengths of the single photons. For practical purposes, the effective excitation wavelength for two photon excitation is half the source wavelength (e.g. a 1047 nm laser would have an effective excitation of about 523 nm) and for three photon excitation the effective excitation wavelength would be a third (e.g. ~350 nm effective excitation for a 1047 nm laser).

The use of lower energy wavelengths, combined with the fact that fluorophore is only excited in the plane of focus, provides numerous advantages for biological imaging. (1) This technique provides a method for optical sectioning since no out of focus illumination is generated. Furthermore, all of the emitted light is generated in the plane of focus and can be collected without the use of a pinhole aperture, used with confocal excitation, while still maintaining good resolution. (2) Although fluorophore excitation occurs at the plane of focus, the majority of the specimen encounters only the lower energy photons. This has positive implications for specimen viability, as will be discussed in detail. (3) The longer wavelengths should be scattered less by cellular components, such as lipids, than shorter, higher energy wavelengths. As discussed below, this aspect improves the imaging of highly light scattering specimens, such as lipid-rich bovine oocytes and embryos.

Of course, as with any technique, there are both theoretical and practical disadvantages. The optical section thickness is a product of the wavelength used and is therefore fixed for a given wavelength. This is a fluorescence excitation technique so the use of either a fluorophore or endogenous autofluorescence is necessary. Some of the wavelengths used for multiphoton excitation can be absorbed by water or pigments, causing heating of the specimen (Williams et al., 1994; Konig et al., 1996). The practical disadvantages include the cost of the system, which are commercially available (>$250,000 depending on what equipment you already have). Furthermore, because the lasers used are relatively sophisticated, such systems benefit from designated personnel. However, the commercial systems are becoming easier to use.

Commercial multiphoton microscope systems are available from Biorad, Leica and Zeiss. Price ranges are similar among the companies but the configuration of the microscope system varies. When considering the purchase of a multiphoton system, researchers should investigate all of the options of the available systems to make sure that the system chosen meets their needs. In some cases, an upgrade of an existing microscope system is possible. The wavelength range of the available lasers can differ. A review by Wise (Wise, 2000) describes the lasers which can be used for multiphoton microscopy. The most common is a titanium:sapphire laser, which is usually tuneable from about 775 nm to 910 nm. This range can vary slightly depending on the particular specification of the laser chosen. This wavelength range is suitable for green fluorescent protein (GFP) imaging. Although it appears to be appropriate for many types of live cell imaging (Mohler, 1999; Periasamy et al., 1999; Raich et al., 1999) its application to long term mammalian embryo imaging has yet to be investigated. A fixed wavelength 1047 nm yttrium-lithium-fluoride (YLF) laser is available on some systems and can be used for exciting fluorophores such as rhodamine or yellow fluorescent protein (YFP). The lasers can be femtosecond or picosecond pulsed. Although for some live cell imaging there does not seem a difference in specimen viability between these two pulse frequencies (Koester et al., 1999; Konig, 2000) the effect of a picosecond pulsed laser on mammalian embryo viability remains to be conclusively determined. Because of the high cost of these instruments, researchers should be aware of all of the options available and determine which are necessary for their particular applications. At this stage of the technology, a femtosecond 1047 nm laser seems to be the most suitable for long term imaging of mammalian embryos. This is not to say other options are not appropriate, but rather that this is the option that has been shown to be effective for this application.

3. Embryo Culture For Imaging

The importance of determining the proper conditions for culturing your embryos on the microscope stage can not be emphasized enough, and this applies to all types of live cell imaging, multiphoton or otherwise. High technology microscopy will not compensate for inadequate culture conditions or poor culture technique. The live imaging microscopist has to deal with the fact that there is essentially no way to make the microscope stage as stable as a water jacketed incubator, so by default the embryos are going to be cultured in a suboptimal environment. Therefore, controlling all of the parameters that can be controlled becomes crucial. The culture medium must be the best one available for the species of interest and must be meticulously made. A mini-chamber system has been previously described in detail (Bavister, 1988) for culturing mammalian embryos on a microscope stage and this method worked well for multiphoton imaging. Basically, a large Plexiglas chamber surrounds the entire microscope. This chamber contains a heater and thermostat for temperature control. A small Plexiglas chamber (~9.5 x 9.5 x 2 cm) is used to cover the culture dish. This mini-chamber is supplied with the appropriate mixed gases for the species of embryo imaged. Before reaching the culture chamber, the mixed gases are bubbled through culture quality water in an Erlenmeyer flask within the heated large Plexiglas chamber in order to humidify the gasses. Before and after each experiment the gas mixture should be checked to confirm that the mix is correct.

One of the most critical aspects of culturing embryos on the stage is temperature. It is insufficient to simply measure the temperature of the chamber. Before conducting any experiments, determine the temperature of the large and small chambers as well as the temperature within a culture drop under the conditions used for imaging. There are frequently temperature gradients under such conditions and the microscope itself can act as a heat sink. We found up to a 3° C differential between the temperature of the incoming gases and the culture drop. There are stage and lens heaters available that can be used, but these too should be tested for accuracy and stability. Fluctuations in room temperature can affect the temperature on the microscope stage, so it is important that you know how your system works under a variety of conditions. Periodic testing using a thermocouple in a culture drop is important for maintaining the reproducibility of your microscope stage culture environment.

There are a number of miscellaneous things to consider when preparing to culture embryos on a microscope stage. Because mammalian embryos are sensitive it is important to be aware of any potential sources of contamination: the quality of the gases, the tubing for delivering the gas to the mini-chamber, airborne toxins, or the material used to make the stopper in the Erlenmeyer flask (black rubber seems detrimental, Teflon seems better). One of the challenges for imaging embryos is what dish to culture the embryos in since the dish has to meet the demanding standard of both the embryos as well as the imaging. We used a Falcon dish (60 mm) with a half-inch hole drilled in the centre of the bottom half of the dish. The hole is covered with a siliconized cover glass. The cover glass is glued in place with epoxy (SuperGlue 30 minute epoxy: Mohler and Squirrell, 2000). Again, it is important to determine your embryos' sensitivity to the type of glue used prior to imaging. Once you have all of the parts in place - the dishes, the temperature and gases, the fluorescent label of choice, the lens - make sure you culture the fluorescently labelled embryos on the stage without imaging. Having confidence in your ability to culture embryos on the stage will not only save time and reduce frustration, it will also improve the reproducibility and reliability of your data.

4. Viability (Laboratory Species)

There are three main requirements for studying subcellular rearrangements in living mammalian embryos: (1) labelling the component of interest (2) optical sectioning capabilities and (3) maintenance of embryo viability during and following long periods of imaging. Because we were interested in mitochondrial re-arrangements, to satisfy first requirement we utilized the vital dye, Mitotracker Rosamine (Molecular Probes), a fluorescent label that accumulates in mitochondria (Whitaker et al., 1991), to label 2-cell hamster embryos. For the second requirement, we had the choice between confocal laser scanning microscopy (CLSM) and multiphoton laser scanning microscopy (MPLSM). To determine which of these methods was more appropriate for our application, we used a system that was built at the Integrated Microscopy Resource at the University of Wisconsin-Madison. This system has been previously described (Wokosin and White, 1996). Basically, it utilized a 1047 nm YLF femtosecond pulsed laser (Microlase / Coherent), a modified Biorad scanhead, and a Nikon microscope. The YLF laser could be replaced with other lasers so that the system could be converted to a confocal configuration. Thus, the efficacy of CSLM and MPLSM for long-term imaging of mammalian embryos could be compared on the same microscope system with the same culture set up.

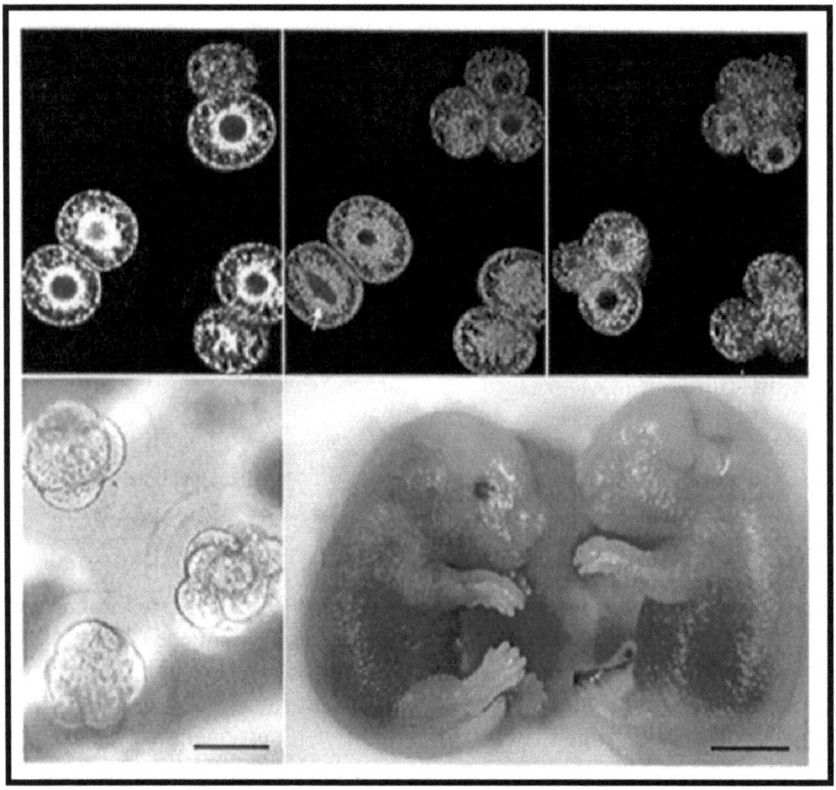

Figure 2: *Foetal development following long-term imaging.*
Two photon images of mitochondria-labelled embryos at (A) the initiation of imaging, (B) after 8.5 hr of imaging, showing a mitotic spindle (arrow) and (C) the completion of a 24 hr imaging sequence. After imaging embryos were cultured in the incubator until 82 hr post-egg activation (PEA) (Nomarski - D) at which point they were transferred to a recipient female. E. A black-eyed foetus that developed from one of these imaged embryos is shown next to an albino uterine-mate. Scale bar for A-D = 45 microns. Scale bar for E = 4.75 mm.

To compare CLSM and MPLSM imaging (Squirrell et al., 1999b) we used the same imaging regime (5 optical sections, 5 microns apart, collected every 15 minutes) for both techniques. Hamster embryos with labelled mitochondria developed from the 2-cell stage to the morula and blastocyst stages, at level comparable to non-imaged controls cultured in the same dish, when imaged for 24 hrs with MPLSM. In contrast, when imaged for as little as 8 hrs with CLSM, embryos never reached the morula stage, and many arrested at the 2- or 4-cell stage, even in the absence of a fluorophore. The MPLSM-imaged embryos, when transferred at the morula / blastocyst stage to pseudopregnant surrogate mothers, were capable of foetal development (Figure 2). Morula and blastocyst development was not impaired by MPLSM even when the imaging frequency was increased to every 2.5 minutes. Indeed, longer imaging may be possible, since embryos imaged for 40 hrs were able to develop on the microscope stage to the morula and blastocyst stage (Figure 3), such that blastocoel formation could be examined. A comparison between these two microscopy methods of the various parameters used in this viability study is shown in Table 1.

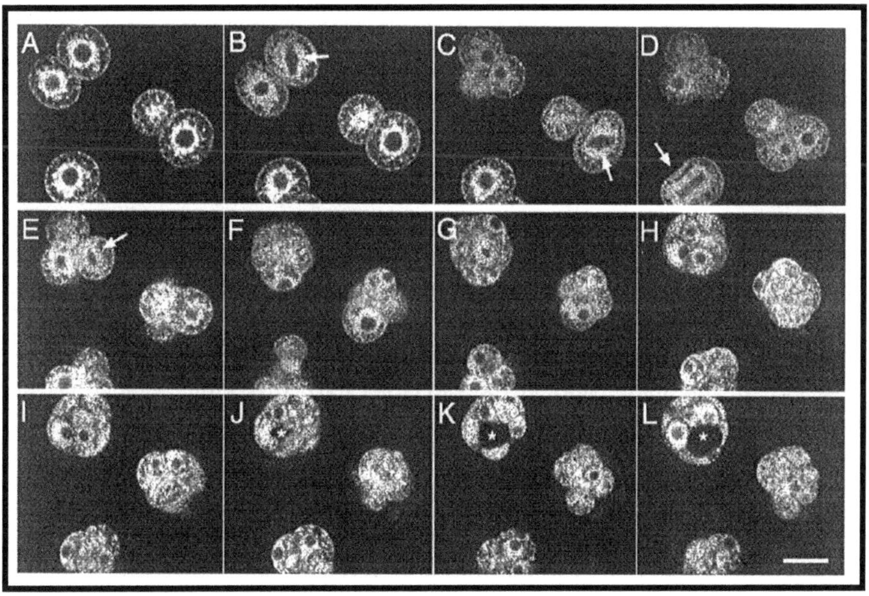

Figure 3: Compaction and blastocoel cavity formation in living hamster embryos imaged for 40 hr with MPLSM.
Development of embryos with fluorescently labelled mitochondria imaged with 1047 nm MPSLM for over 36 hrs (images collected every 15 min) from the 2-cell stage (A) through the blastocyst stage (K, L). Note mitotic spindles and dividing cells (arrows) and formation of blastocoel cavity (*). Images are single optical sections. At each time point 5 optical sections, 5 microns apart were collected. t = time after initiation of imaging. Scale bar = 50 microns.

Table 1: Comparison of various parameters of confocal laser scanning microscopy and multiphoton microscopy used in the described viability studies for hamster embryos.

CLSM	**MPLSM**
15 min imaging interval	2.5 min imaging interval
8 hr total imaging	24 hr total imaging
10 - 30 μW	13 mW
Unstained	Stained
160 images	>2800 images
Arrested development	Live offspring

For any imaging or analysis protocol of living specimens, the researcher must take the time and effort to confirm that the protocol itself does not impinge on the biological process of interest. Although it may be impossible to rule out all possible subtle effects, it is still important to develop criteria to assess the effects of the technologies utilized. In our studies (Squirrell et al., 1999b) the embryos were not only able to continue cell division, but were able to differentiate and even develop into offspring indicating the detrimental effects of 1047 nm MPLSM imaging, if any, were within an acceptable range for research.

This comparison of the effect of imaging techniques establishes the efficacy of multiphoton microscopy for long term imaging of mammalian embryos and opens the door to a multitude of applications for studying subcellular responses to a variety of conditions. In particular, we have used long term imaging to analyse the effects of various culture conditions on mitochondrial organisation and subsequent development (Squirrell et al., 2001). This method permits not only the analysis of changes over time but also the direct correlation of subcellular changes and developmental potential.

5. Imaging Depth (Domestic Species)

In addition to issues of viability, researchers working with embryos from domestic species have the problem of imaging through a large, lipid-rich cytoplasm. The large size of the oocyte/embryo contributes to attenuation of both the excitation and emission signal with increased imaging depth. The lipids scatter light, making the photons either unavailable to excite the fluorophore of interest or deflecting the emitted photons so that they are unable to pass through the pinhole aperture (used with CLSM) and thus go undetected. The wavelengths used for MPLSM permit imaging deeper into light scattering specimens (Denk and Svoboda, 1997; Wokosin et al., 1997; Centonze and White, 1998) because longer wavelengths of light are less prone to scattering. Furthermore, only the fluorophores in the plane of focus are excited, thus any emitted photons are of interest. The pinhole aperture can be eliminated, and all of the light emitted from the fluorophore can be collected (direct detection) (Wokosin et al., 1997; Centonze and White, 1998). A more complete description of the advantages of 1047 nm MPLSM for deep imaging are presented in Centonze and White (1998). In this paper, the authors show that MPLSM provides better deep imaging for bovine embryos than confocal microscopy (Figure 4).

The ability to gather good quality images from regions throughout the embryo with MPLSM, along with maintenance of embryo viability, provide the perfect combination for performing studies in which the embryos (or oocytes) are subjected to particular conditions. The effects on both the cytoplasmic organisation as well as developmental competence can be assessed *in the same specimens*. Krisher et al. (submitted) have conducted such an experiment. Bovine oocytes were matured in 3 different media. Following maturation, the oocytes were labelled with Mitotracker and imaged with 1047 nm MPLSM. The oocytes were then inseminated and their development in culture assessed. In this manner, the relationship between the maturation medium, mitochondrial organisation and embryo developmental potential was determined.

Figure 4: *Demonstration of increased imaging penetration depth with multi-photon microscopy in a living embryo.*

A bovine embryo stained with Mitotracker rosamine was imaged by confocal microscopy (left:3.5 μW of 532 nm light) and by multiphoton microscopy (direct detection, right:10 mW of 1047 nm light). Confocal microscopy gives the erroneous impression that mitochondria are localized to the region adjacent to the cell cortex. However, the multi-photon image indicates that the mitochondrion distribution is fairly uniform throughout the cytoplasm. Images were collected at a pixel resolution of 0.27 μm with a Kalman 3 collection filter. Scale bar = 20μm.

The results of this experiment suggest there are dramatic differences in both mitochondrial distribution and developmental potential in embryos cultured from oocytes matured in different media. However, the relationship between subtle differences in mitochondrial organisation and development of embryos from oocytes matured in the same medium is minimal.

6. Maximizing Information from Rare Specimens (Non-human primate)

Because of obvious ethical constraints, research progress in the understanding of early human embryo physiology must be accomplished using non-human primate embryos. Two of the disadvantages of using such embryos are that they are both difficult and expensive to acquire in significant numbers and there is great variability between embryos, even from the same female. This makes the fixed tissue approach to studying subcellular organisation of these embryos far from ideal. Information from a given oocyte/embryo can be gathered only at a single time point, the time at which that embryo was fixed. Heterogeneity among the embryos makes it difficult to form correlations between embryos fixed at an early time point and the developmental potential of cohorts. Live imaging, such as that which is possible with MPLSM solves both of these problems. Information can be collected at various time points from the same embryos, maximizing the amount of data gathered from each embryo. Furthermore, the developmental potential of individual embryos that have been imaged can be assessed, making direct correlations between early events and subsequent development possible.

Preliminary experiments using rhesus monkey embryos illustrate these advantages. In vivo matured oocytes were labelled with Mitotracker and images of each oocyte were collected. The oocytes were inseminated and images of the fertilized ovum were gathered at various time points and subsequent development was assessed. Some fertilized ova exhibited a transient disk-shaped accumulation of mitochondria between the pronuclei (Figure 5). A similar observation has been made in fertilized human oocytes (Noto et al., 1993; Van Blerkom et al., 2000). Although the significance of this accumulation of mitochondria is unclear, preliminary studies using MPLSM imaging suggest a positive correlation between the presence of this pronuclear accumulation and development to the blastocyst stage (data not shown). Additional confirmation is necessary before strong conclusions can be reached. Regardless, this type of live imaging does promise to permit great strides in the study of a variety of embryos, particularly rare mammalian embryos.

Figure 5: Relocation of mitochondria in a primate ovum undergoing fertilization. In vivo matured rhesus monkey oocyte undergoing IVF at 16 h post-insemination, just prior to syngamy, showing a dense, biconcave accumulation of mitochondria between the two pronuclei. Scale bar = 50 μM. Presumptive zygotes were imaged live with MPLSM.

7. Potential Future Directions

7.1. Embryo Research

It is clear that MPLSM has vast potential for advancing the study of cell biological processes in living mammalian early embryos. Several vital dyes are available for studying a variety of subcellular components and the number and quality of these probes will probably increase as more and more research is conducted on living cells. The optical sections collected from living embryos can be used to generate 4D reconstructions (Mohler and White, 1998; Mohler, 1999). Studies on the changes in subcellular organisation will not only improve our understanding of the importance of structural changes in embryo physiology at different points in development but also, as briefly described in this chapter, to show how that organisation is impacted by changes in the embryo's environment, particularly as it pertains to in vitro culture. The more that is understood about how the culture environment perturbs all aspects of embryo physiology, the better the culture conditions can be improved to reduce those perturbations and thereby improve embryo development. The ability to determine how in vitro manipulation alters embryo structure may be particularly important for understanding the effects of mechanically invasive procedures, such as intracytoplasmic sperm injection (Van Steirteghem, 1997; Sakkas et al., 1998; Hewitson et al., 2000) or nuclear transfer (McGrath and Solter, 1985; Howlett et al., 1987; Modlinski and Smorag, 1991; Dominko et al., 1999). It is similarly important to determine, and thereby perhaps correct, the effects of assisted reproductive procedures that could cause structural damage to the embryo, such as oocyte or embryo freezing (Bernart et al., 1987; Noto et al., 1993; Breed et al., 1994).

Additionally, MPLSM could be used for imaging the temporal and spatial patterns of fluorescently tagged proteins (Chalfie, 1995), such as GFP (using 900nm) or YFP (using 1047nm) in transgenic animals. Although this has not yet been published for mammalian embryos, MPLSM has been used for live imaging of GFP-tagged proteins in the embryos of other organisms (Mohler, 1999; Periasamy et al., 1999; Raich et al., 1999). Similarly, MPLSM could be used as part of a screening of presumptive transgenic blastocysts, using a GFP or YFP reporter, for embryo transfer in order to establish transgenic mouse strains. In this manner, only embryos in which the injection procedure was successful need be transferred to surrogate mothers, increasing the likelihood of getting transgenic animals from a given transfer. This has been previously proposed using the luciferase gene as a reporter (Menck et al., 1998). The MPLSM method has the advantage that it is the presence or absence of fluorescence that is of significance and an enzyme assay is not required.

7.2. Diagnostic

Although the application of MPLSM to mammalian embryo research is obvious and diverse, what role it will play as a diagnostic tool, particularly in a clinical situation, is less evident and probably impractical, at least in the near future. The creation of fluorescently labelled markers for gene expression, such as those used for fluorescent in situ hybridization (the use of multiphoton imaging and FISH is reviewed in Konig, 2000), could be used to screen for embryos which do, or do not, carry a gene of interest, such as a gender-specific genetic marker (which might be useful in domestic animal production) or a disease-related gene. However, much preliminary work would need to be done to develop non-toxic protocols for using these fluorescent probes with living embryos. At this point in time, it is not clear that this type of whole embryo screening would have substantial benefits over current blastomere PCR screening (Kunieda et al., 1992; Morsy et al., 1992; Cousens et al., 1994; Chang et al., 1996; Chen et al., 1999; Pauken and Capco, 2000).

MPLSM is being used in the laboratory to assess the relationship between cytoplasmic organisation and embryo development. This information could then be applied to the selection of embryos that have a greater developmental potential as suggested by their cytoplasmic organisation. Although this might be of interest to the research community and perhaps useful for domestic or endangered species production, the additional handling required for applying the fluorescent probe of interest or the necessity of having a fluorescent transgene make this impractical in the clinical realm. One way to avoid the use of fluorescent probes is to take advantage of the fact that MPLSM can be used to assess changes in metabolic status of a specimen by imaging autofluorescence, specifically the fluorescence of NADPH. This type of imaging has been successfully performed in other tissues (Piston et al., 1995; Masters et al., 1999; Piston and Knobel, 1999). How well it would work in mammalian embryos remains to be determined.

Although it has been shown that there is no obvious damage to mammalian embryo development following long term imaging with MPLSM (Squirrell et al., 1999b), which may be acceptable within the research community, it may not be acceptable to the general public for use in conjunction with human assisted reproductive technologies. Although the embryos are imaged with laser light that has low per photon energy, the overall power of the laser light hitting the specimen is still quite large (Table 1) and this may have subtle consequences that would not be acceptable in a clinical situation.

8. Conclusion

MPLSM will probably play only a limited role in the assessment of human embryos. However, particularly with the increased commercial availability and improvements in the ease of use of MPLSM, it is clear that this instrument can have a major impact on improving our understanding of what is happening inside the cells of developing mammalian embryos and how in vitro manipulations alter those processes. The investigations presented here are merely the tip of the iceberg of possible applications. The number and variety of projects that could be addressed with this type of imaging is limited only by the probes available and the imagination of the researchers.

References

Adenot PG, Szöllösi MS, Geze M, Renard JP, Debey P. 1991. Dynamics of paternal chromatin changes in live one-cell mouse embryo after natural fertilization. Mol Reprod Dev 28:23-34.

Albertini DF. 1987. Cytoplasmic reorganization during the resumption of meiosis in cultured preovulatory rat oocytes. Dev Biol 120:121-131.

Barnett DK, Kimura J, Bavister BD. 1996. Translocation of active mitochondria during hamster preimplantation embryo development studied by confocal laser scanning microscopy. Dev Dyn 205:64-72.

Batten BE, Albertini DF, Ducibella T. 1987. Patterns of organelle distribution in mouse embryos during preimplantation development. Am J Anat 178:204-213.

Bavister BD. 1988. A minichamber device for maintaining a constant carbon dioxide in air atmosphere during prolonged culture of cells on the stage of an inverted microscope. In Vitro Cell Dev Biol 24:759-763.

Bavister BD, Squirrell JM. 2000. Mitochondrial distribution and function in oocytes and early embryos. Hum Reprod 15 Suppl 2:189-198.

Bernart W, Rabe T, Schleiermacher E, Anton-Lamprecht I, Runnebaum B. 1987. Freezing-thawing behaviour and cell membrane ultrastructure of mouse embryos pre-cultured in B2-Menezo medium before cryopreservation. Hum Reprod 2:679-687.

Breed WG, Taggart DA, Bradtke V, Leigh CM, Gameau L, Carroll J. 1994. Effect of cryopreservation on development and ultrastructure of preimplantation embryos from the dasyurid marsupial Sminthopsis crassicaudata. J Reprod Fertil 100:429-438.

Calarco PG. 1995. Polarization of mitochondria in the unfertilized mouse oocyte. Dev Genet 16:36-43.

Calarco PG, McLaren A. 1976. Ultrastructural observations of preimplantation stages of the sheep. J. Embryol. exp. Morph. 36:609-622.

Centonze VE, White JG. 1998. Multiphoton excitation provides optical sections from deeper within scattering specimens than confocal imaging. Biophys J 75:2015-2024.

Chalfie M. 1995. Green fluorescent protein. Photochem Photobiol 62:651-656.

Chang MY, Soong YK, Wang ML. 1996. Preimplantation diagnosis of alpha-thalassemia by blastomere aspiration and polymerase chain reaction: preliminary experience. J Formos Med Assoc 95:203-208.

Chen CM, Hu CL, Wang CH, Hung CM, Wu HK, Choo KB, Cheng WT. 1999. Gender determination in single bovine blastomeres by polymerase chain reaction amplification of sex-specific polymorphic fragments in the amelogenin gene. Mol Reprod Dev 54:209-214.

Cousens C, Carver AS, Wilmut I, Colman A, Garner I, O'Neill GT. 1994. Use of PCR-based methods for selection of integrated transgenes in preimplantation embryos. Mol Reprod Dev 39:384-391.

Cran DG. 1985. Qualitative and quantitative structural changes during pig oocyte maturation. J Reprod Fertil 74:237-245.

Cran DG, Moor RM, Hay MF. 1980. Fine structure of the sheep oocyte during antral follicle development. J Reprod Fertil 59:125-132.

Denk W. 2000. Principles of multiphoton-excitation fluorescence microscopy. In: R. Yuste, K. Lanni, A. Konnerth, (Eds.), "Imaging neurons: a laboratory manual", Cold Spring Harbor Laboratory Press, pp 17.1-17.8.

Denk W, Piston D, Webb W. 1995. Two-photon molecular excitation in laser-scanning microscopy. In J. Pawley, (Ed.), "Handbook of Biological Confocal Microscopy", Plenum Press, pp 445-458.

Denk W, Strickler JH, Webb WW. 1990. Two-photon laser scanning fluorescence microscopy. Science 248:73-76.

Denk W, Svoboda K. 1997. Photon upmanship - why multiphoton imaging is more than a gimmick. Neuron 18:351-357.

Dominko T, Mitalipova M, Haley B, Beyhan Z, Memili E, McKusick B, First NL. 1999. Bovine oocyte cytoplasm supports development of embryos produced by nuclear transfer of somatic cell nuclei from various mammalian species. Biol Reprod 60:1496-1502.

Ferreira J, Carmo-Fonseca M. 1997. Genome replication in early mouse embryos follows a defined temporal and spatial order. J Cell Sci 110:889-897.

Gallicano GI, McGaughey RW, Capco DG. 1992. Cytoskeletal sheets appear as universal components of mammalian eggs. J Exp Zool 263:194-203.

Gonzales DS, Jones JM, Pinyopummintr T, Carnevale EM, Ginther OJ, Shapiro SS, Bavister BD. 1996. Trophectoderm projections: a potential means for locomotion, attachment and implantation of bovine, equine and human blastocysts. Hum Reprod 11:2739-2745.

Gonzales DS, Pinheiro JC, Bavister BD. 1995. Prediction of the developmental potential of hamster embryos in vitro by precise timing of the third cell cycle. J Reprod Fertil 105:1-8.

Hewitson L, Haavisto A, Simerly C, Jones J, Schatten G. 1997. Microtubule organization and chromatin configurations in hamster oocytes during fertilization and parthenogenetic activation, and after insemination with human sperm. Biol Reprod 57:967-975.

Hewitson L, Simerly C, Dominko T, Schatten G. 2000. Cellular and molecular events after in vitro fertilization and intracytoplasmic sperm injection. Theriogenology 53:95-104.

Hewitson LC, Simerly CR, Tengowski MW, Sutovsky P, Navara CS, Haavisto AJ, Schatten G. 1996. Microtubule and chromatin configurations during rhesus intracytoplasmic sperm injection: successes and failures. Biol Reprod 55:271-280.

Hillman N, Tasca R. 1983. Ultrastructural and autoradiographic studies of mouse cleavage stages. Am J Anat 126:151-174.

Howlett SK, Barton SC, Surani MA. 1987. Nuclear cytoplasmic interactions following nuclear transplantation in mouse embryos. Development 101:915-923.

Hyttel P, Niemann H. 1990. Ultrastructure of porcine embryos following development in vitro versus in vivo. Mol Reprod Dev 27:136-144.

Koester HJ, Baur D, Uhl R, Hell SW. 1999. Ca2+ fluorescence imaging with pico- and femtosecond two-photon excitation: signal and photodamage. Biophys J 77:2226-2236.

Konig K. 2000. Multiphoton microscopy in life sciences. J Microsc 200:83-104.

Konig K, Simon U, Halbhuber KJ. 1996. 3D resolved two photon fluorescence microscopy of living cells using a modified confocal laser scanning microscope. Cellular & Molecular Biology 42:1181-1194.

Kunieda T, Xian M, Kobayashi E, Imamichi T, Moriwaki K, Toyoda Y. 1992. Sexing of mouse preimplantation embryos by detection of Y chromosome-specific sequences using polymerase chain reaction. Biol Reprod 46:692-697.

Lehtonen E. 1980. Changes in cell dimensions and intercellular contacts during cleavage- stage cell cycles in mouse embryonic cells. J Embryol Exp Morphol 58:231-249.

Longo FJ. 1973. An ultrastructural analysis of spontaneous activation of hamster eggs aged in vivo. Anat Rec 79:27-56.

Luoh Y, Wu M. 1996. Mitochondrial distribution of swine ova and embryos analysed by rhodamine 123 fluorescent assay. J Chin Soc Anim Sci 25:53-66.

Maro B, Johnson JH, Pickering SJ. 1985. Changes in the distribution of membranous organelles during mouse early development. J Embryol exp Morph 90:287-309.

Maro B, Johnson MH, Pickering SJ, Flach G. 1984. Changes in actin distribution during fertilization of the mouse egg. J Embryol Exp Morphol 81:211-237.

Masters BR, So PT, Kim KH, Buehler C, Gratton E. 1999. Multiphoton excitation microscopy, confocal microscopy, and spectroscopy of living cells and tissues; functional metabolic imaging of human skin in vivo. Methods Enzymol 307:513-536.

McGrath J, Solter D. 1985. Nuclear and cytoplasmic transfer in mammalian embryos. Dev Biol N Y 4:37-55.

Menck M, Mercier Y, Campion E, Lobo RB, Heyman Y, Renard JP, Thompson EM. 1998. Prediction of transgene integration by noninvasive bioluminescent screening of microinjected bovine embryos. Transgenic Res 7:331-341.

Modlinski JA, Smorag Z. 1991. Preimplantation development of rabbit embryos after transfer of embryonic nuclei into different cytoplasmic environment. Mol Reprod Dev 28:361-372.

Mohler W, Squirrell JM. 2000. Imaging embryo development with multiphoton microscopy. In R. Yuste, K. Lanni, A. Konnerth, (Eds.), "Imaging Neurons: A Laboratory Manual", Cold Spring Harbor Press.

Mohler WA. 1999. Visual reality: using computer reconstruction and animation to magnify the microscopist's perception. Mol Biol Cell 10:3061-3065.

Mohler WA, White JG. 1998. Stereo-4-D reconstruction and animation from living fluorescent specimens. Biotechniques 24:1006-10, 1012.

Morsy M, Takeuchi K, Kaufmann R, Veeck L, Hodgen GD, Beebe SJ. 1992. Preclinical models for human pre-embryo biopsy and genetic diagnosis. II. Polymerase chain reaction amplification of deoxyribonucleic acid from single lymphoblasts and blastomeres with mutation detection. Fertil Steril 57:431-438.

Nakamura O. 1999. Fundamental of two-photon microscopy. Microsc Res Tech 47:165-171.

Navara CS, Wu GJ, Simerly C, Schatten G. 1995. Mammalian model systems for exploring cytoskeletal dynamics during fertilization. Curr Top Dev Biol 31:321-342.

Noto V, Campo R, Roziers P, Swinnen K, Vercruyssen M, Gordts S. 1993. Mitochondrial distribution after fast embryo freezing. Hum Reprod 8:2115-2118.

Oura C, Toshimori K. 1990. Ultrastructural studies on the fertilization of mammalian gametes. Int Rev Cytol 122:105-151.

Panigel M, Kraemer DC, Kalter SS, Smith GC, Heberling RL. 1975. Ultrastructure of cleavage stages and preimplantation embryos of the baboon. Anat Embr 147:45-62.

Pauken CM, Capco DG. 2000. The expression and stage-specific localization of protein kinase C isotypes during mouse preimplantation development. Dev Biol 223:411-421.

Periasamy A, Skoglund P, Noakes C, Keller R. 1999. An evaluation of two-photon excitation versus confocal and digital deconvolution fluorescence microscopy imaging in Xenopus morphogenesis. Microsc Res Tech 47:172-181.

Piston DW, Knobel SM. 1999. Quantitative imaging of metabolism by two-photon excitation microscopy. Methods Enzymol 307:351-368.

Piston DW, Masters BR, Webb WW. 1995. Three-dimensionally resolved NAD(P)H cellular metabolic redox imaging of the in situ cornea with two-photon excitation laser scanning microscopy. J Microsc 178:20-27.

Plante L, King WA. 1994. Light and electron microscopic analysis of bovine embryos derived by in vitro and in vivo fertilization. J Assist Reprod Genet 11:515-529.

Potter SM. 1996. Vital imaging: two photons are better than one. Curr Biol 6:1595-1598.

Raich WB, Agbunag C, Hardin J. 1999. Rapid epithelial-sheet sealing in the Caenorhabditis elegans embryo requires cadherin-dependent filopodial priming. Curr Biol 9:1139-1146.

Sakkas D, Urner F, Bizzaro D, Manicardi G, Bianchi PG, Shoukir Y, Campana A. 1998. Sperm nuclear DNA damage and altered chromatin structure: effect on fertilization and embryo development. Hum Reprod 13 Suppl 4:11-19.

Sathananthan AH. 1994. Ultrastructural changes during meiotic maturation in mammalian oocytes: unique aspects of the human oocyte. Microsc Res Tech 27:145-164.

Schatten G, Simerly C, Schatten H. 1985. Microtubule configurations during fertilization, mitosis, and early development in the mouse and the requirement for egg microtubule- mediated motility during mammalian fertilization. Proc Natl Acad Sci U S A 82:4152-4156.

Simerly CR, Hecht NB, Goldberg E, Schatten G. 1993. Tracing the incorporation of the sperm tail in the mouse zygote and early embryo using an anti-testicular alpha-tubulin antibody. Dev Biol 158:536-548.

Squirrell JM, Wokosin DL, Bavister BD, White JG. 1999a. Imaging subcellular changes in living mammalian embryos using 1047nm two photon excitation fluorescence microscopy. Microsc Microan 5:1061-1062.

Squirrell JM, Wokosin DL, White JG, Bavister BD. 1999b. Long-term two-photon fluorescence imaging of mammalian embryos without compromising viability. Nat Biotechnol 17:763-767.

Squirrell JM, Lane M, Bavister BD. 2001. Altering intracellular pH disrupts development and cellular organisation in preimplantation hamster embryos. Biol. Reprod. 64:1845-1854.

Stern S, Biggers JD, Anderson E. 1971. Mitochondria and early development of the mouse. J Exp Zool 176:179-191.

Suenaga A, Tachi C, Tojo H, Tanaka S, Tsutsumi O, Taketani Y. 1996. Quantitative analysis of the spreading of the mouse trophoblast in vitro: a model for early invasion. Placenta 17:583-590.

Terasaki M, Dailey ME. 1995). Confocal microscopy of living cells. In J. B. Pawley, (Ed.), "Confocal microscopy of living cells", Plenum Press, pp 327-346.

Van Blerkom J, Bell H, Weipz D. 1990. Cellular and Developmental Biological aspects of bovine meiotic maturation, fertilization, and preimplantation embryogenesis in vitro. J Elect Micro Tech 16:298-323.

Van Blerkom J, Davis P, Alexander S. 2000. Differential mitochondrial distribution in human pronuclear embryos leads to disproportionate inheritance between blastomeres: relationship to microtubular organization, ATP content and competence. Hum Reprod 15:2621-2633.

Van Blerkom J, Runner MN. 1984. Mitochondrial reorganization during resumption of arrested meiosis in the mouse oocyte. Am J Anat 171:335-355.

Van Steirteghem A. 1997. Intracytoplasmic sperm injection. Baillieres Clin Obstet Gynaecol 11:725-738.

Whitaker JE, Moore PL, Haugland RP. 1991. Dihydrotetramethylrosamine: a long wavelength, fluorogenic peroxidase substrate evaluated in vitro and in a model phagocyte. Biochem Biophys Res Commun 175:387-393.

Williams RM, Piston DW, Webb WW. 1994. Two-photon molecular excitation provides intrinsic 3-dimensional resolution for laser-based microscopy and microphotochemistry. Faseb J 8:804-813.

Wise FW. 2000. Lasers for multiphoton microscopy. In R. Yuste, K. Lanni, A. Konnerth, (Eds.), "Imaging Neurons: A Laboratory Manual", Cold Spring Harbor Laboratory Press, pp 18.1-18.9.

Wokosin DL, Centonze VE, White JG, Hird SN, Sepsenwol S, Malcolm GPA, Maker GT, Ferguson AL. 1996. Multiple-photon excitation imaging with an all-solid-state laser. SPIE 2678:38-49.

Wokosin DL, White JG. 1997. Optimization of the design of a multiple-photon excitation laser scanning fluoresence imaging system. In C. J. Cogswell, J.-A. Conchello, T. Wilson, (Eds.), "Three-Dimensional Microscopy: Image Acquisition and Processing" Vol. 2984, pp 24-29.

Xu C, Zipfel W, Shear JB, Williams RM, Webb WW. 1996. Multiphoton fluorescence excitation - new spectral windows for biological nonlinear microscopy. PNAS, USA 93:10763-10768.

Chapter 10

IS THE FREQUENCY OF CHROMOSOME ABNORMALITIES INFLUENCED BY IN VITRO TECHNIQUES?

Dorthe Viuff[1], Torben Greve[1] and Preben Dybdahl Thomsen[2]

[1] *Department of Clinical Studies, Reproduction, The Royal Veterinary and Agricultural University, Dyrlægevej 68, 1870 Frederiksberg C, Copenhagen, Denmark.*

[2] *Department of Anatomy and Physiology, The Royal Veterinary and Agricultural University, Groennegaardsvej 7, 1870 Frederiksberg C, Copenhagen, Denmark*

1. Introduction

The very specialized process of meiosis and the fact that the zygote is the stem cell of all cells in a new individual make the period of early embryonic development vulnerable to errors of chromosome segregation. If the zygote is aneuploid, i.e. not carrying the normal diploid chromosome set, it is most likely lethal to the embryo. However, major alterations of the genome are more harmful to the embryo than minor alterations. In other words, an additional copy of a chromosome (trisomy) will be less harmful than the presence of an entire extra chromosome set (triploidy), and it is expected that trisomic embryos will develop further than triploid embryos. For example, there are several reports of autosomal and sex chromosome trisomies in live born, although malformed, calves (see Popescu, 1990 for review; Herzog and Hoehn, 1991; Agerholm and Christensen, 1993; Christensen and Juul, 1999; Ducos et al., 2000) but there are no published reports of live born polyploid calves. This is at first surprising since the reports of live born, but severely malformed, triploid or even tetraploid children suggest that polyploid calves might survive to term. The lack of reports of such calves probably reflects the fact that malformed individuals are karyotyped with a higher frequency in humans compared to domestic animals and that malformed calves are of little commercial value.

The chromosome abnormalities mentioned above are all formed *de novo* and will not be passed on to the next generation. However, the inheritance of more subtle chromosome changes is well documented in all domestic animal species. The 1/29 Robertsonian translocation in cattle is a prominent example which has become widespread and of some economic consequence due to its high frequency in some cattle breeds. It is a centric fusion event that leads to a reduction of 7% in the 56-day non-return rate in carrier bulls as a result of the formation of unbalanced gametes (Dyrendahl and Gustavsson, 1979). However, carriers of chromosome translocations are rare and hereditary chromosome abnormalities are normally of minor importance for embryonic mortality. The rest of this chapter is therefore concerned with chromosome abnormalities that are spontaneously formed during early embryo development.

If faulty chromosome segregation occurs later than the zygote stage it will only affect some of the cells of an embryo and the consequences will be less severe. Polyploid or haploid cells have been found in embryos of the mammalian species investigated as early as the 2-cell stage (see review by King, 1990; Viuff et al., 1999, 2000, 2001; Slimane et al., 2000) and they are a normal rather than an exceptional finding, in the placenta and in the adult animal. Some organs, such as the liver, contain 35-40 % tetraploid cells but they also occur in other organs such as bladder epithelium, heart and ovary at a lower frequency (Keighren and West, 1993). However, little is known about the biological significance of the polyploid cells.

It is now more than 10 years since W. Allan King published his extensive review on chromosome abnormalities in the embryos of domestic animals (King, 1990). Since then an important technical improvement has emerged and it is therefore timely to summarize recent results and to evaluate whether there is any evidence suggesting that the culture steps of modern in vitro techniques influence the frequency of chromosome abnormalities. This chapter is based on data from cattle simply because most investigations are carried out on cattle embryos, but data from other species are included where relevant.

2. Methodology

The study of metaphase chromosomes (karyotyping) has until recently been the preferred method for analysing the frequency and nature of chromosomal abnormalities in the cells of domestic animal embryos. A method, which allows chromosome analysis of most if not all cells of an embryo is fluorescence in situ hybridization (FISH) with chromosome-specific DNA probes. In principle, FISH permits analysis of both numerical and structural aberrations in metaphase chromosomes as well as in interphase nuclei. However, the number of suitable probes for domestic animals has so far restricted the use of FISH for the detection of polyploid and haploid cells in embryos

2.1. Karyotyping

Karyotyping has, until recently, been the preferred method for analysing the frequency and nature of chromosomal abnormalities in the cells of domestic animal embryos. Embryonic cells are normally incubated overnight with colchicine to arrest dividing cells at metaphase, and are then fixed in methanol-acetic acid and prepared by the air-drying method to obtain metaphase spreads. Most metaphase spreads permit the counting of the chromosomes, and thereby the detection of any missing or supernumerary chromosomes. However, inherent difficulties in karyotyping and chromosome banding of the cattle chromosomes has made it difficult to identify the missing or supernumerary chromosomes on the often poorly spread and condensed chromosomes from embryonic cells. Although the situation has improved now that there are chromosome markers available for FISH-based identification of cattle chromosomes, we are not aware of any studies where such an approach has been used in studies of embryos from other domestic animals.

The major advantage of karyotyping is that all types of aneuploidies will be revealed and may be identified. The major drawback, however, is that the procedure is rather inefficient in obtaining metaphase preparations and therefore normally only a minor fraction of the embryonic cells can be analysed. The frequencies of polyploidy and mixoploidy frequencies that are estimated using this technique should therefore be used in the understanding that frequency of polyploid embryos will be overestimated because some mixoploid embryos will be diagnosed as polyploid. In addition, the mixoploidy frequency is most likely underestimated because some truly mixoploid embryos will go undetected if the frequency of polyploid cells is low in the embryos. New methods allowing to obtain reliable estimates of the mixoploidy and the polyploidy frequencies are therefore very interesting.

2.2. Interphase cytogenetic analysis

The major advantage of interphase cytogenetic analysis compared to karyotyping is that most if not all cells are included in the analysis. The major drawback of the method is its inability to reveal all chromosome abnormalities. The use of differentially labelled probes that hybridize to two different chromosomes is crucial to ensure that aneuploidy can be distinguished from polyploidy or haploidy and to ensure that haploidy can be distinguished from hybridization failure. However, most trisomies will go undetected and monosomies and haploidies may disappear in the nuclei categorized as hybridization errors.

2.2.1. Interphase cytogenetics on embryonic cells

Our laboratory has tried various methods for pre-treatment and hybridization of embryonic cells. It was found that these cells are more difficult to hybridize than the lymphocytes or fibroblast nuclei prepared by standard methods for cytogenetic analysis. It seems that the zona pellucida and cytoplasm provide a coating of nuclei that reduces the efficiency of hybridization. A similar observation has been made by Slimane et al. (2000). There are, however, pre-treatment protocols that produce a better FISH signal than others (Xu et al., 1998). We have adapted a method of nuclear extraction used on human pre-implantation cell biopsies by Dr. Lee F Rickords (Oklahoma State University College of Medicine, Tulsa, Oklahoma). In brief, whole embryos or cell biopsies are pre-treated in a solution containing 0.01 N HCl/0.1% Tween-20 to dissolve the zona pellucida and disrupt the plasma membrane, resulting in clean nuclei with good morphology. Furthermore, we find pepsin digestion prior to hybridization to be an efficient way of removing remnants of cytoplasm and making the DNA accessible to hybridization by the probes. Occasional nuclei that remain covered with cytoplasm fail to hybridize with one or both probes, thus demonstrating the importance of the pre-treatment steps.

Next to the pre-treatment, the most important steps to ensure a strong hybridization signal from both probes are an efficient labelling, a well-defined length of the probe and stringent hybridization. All of these factors influence the degree of background after FISH. Embryonic nuclei are hybridized simultaneously with two differentially labelled probes (two-colour-FISH), one probe is labelled with biotin and the other with digoxigenin by nick-translation. The nick-translation reaction has the advantage that the probe size can be controlled. We aim at using the DNAse activity that produces DNA fragments of 200-300 base pairs at subsequent agarose gel electrophoresis. This is the size that consistently gives the highest signal-to-noise ratio. Larger probe fragments generally produce high levels of fluorescent clumps and smaller probe fragments produce weaker signals. The hybridization conditions must reduce non-specific hybridization by increasing hybridization temperature to a stringency level where cross-hybridization is minimized and probe signal is maximized. In addition, cross hybridization is reduced by including unlabelled genomic DNA in the hybridization solution. Salmon sperm DNA is used to suppress unspecific binding of probe DNA to the slide and genomic cattle DNA to reduce the background due to interspersed sequences included in the probe. In addition, cattle Cot1-DNA (which is enriched for repeated sequences) has been used for blocking high background levels in some cases.

The hybridization sites of the probes are visualized by antibody binding. This introduces another possibility of increased background or a weak signal, since not all antibodies react equally strongly and since antibodies may disintegrate on storage. We therefore prepare new dilutions of antibodies for each batch of slides.

The many gene mapping projects of the nineties have produced a number potential probes for interphase cytogenetic chromosome analysis in domestic animal species, but only a few have been used for chromosome analysis of embryonic cells. The first report on in situ hybridization for the sexing of bovine embryonic cells made use of a tritiated Y probe (Thomsen et al., 1991). Later, Kobayashi et al. (1998) used bovine Y probes to sex bovine embryos by FISH. In our recent studies of chromosome abnormalities in cattle embryos, we have used chromosome-specific, repeated sequences from the pericentromere region of chromosome 6 and 7 for two-colour-FISH (Figure 1) (Viuff et al., 1999; 2000; 2001). We have tested a number of other genomic cattle clones and it is our impression that repeated probes, which produce a strong signal in a small domain of the interphase nucleus, should be preferred for this type of analysis. This is in good agreement with the preference of the repeated alpha satellites for preimplantation diagnosis of human embryos by FISH and with the report of Slimane et al. (2000) who compared unique sequences from cattle chromosome 1 and a cosmid clone containing a repeated motive present in the pericentromere region of cattle chromosomes 14, 20 and 25 for FISH based chromosome analysis of two-cell bovine embryos.

2.2.2. The evaluation of FISH results

Figure 1: *Fluorescence in situ hybridization (FISH) with chromosome 6- (red) and chromosome 7-specific (green) DNA probes on a metaphase from a lymphocyte culture A) and from extracted interphase nuclei from bovine embryos B) +C). The nuclei are counter stained with DAPI (blue). A) A normal diploid metaphase with two signals from each chromosome. B) A normal diploid interphase nucleus. C) A triploid interphase nucleus.*

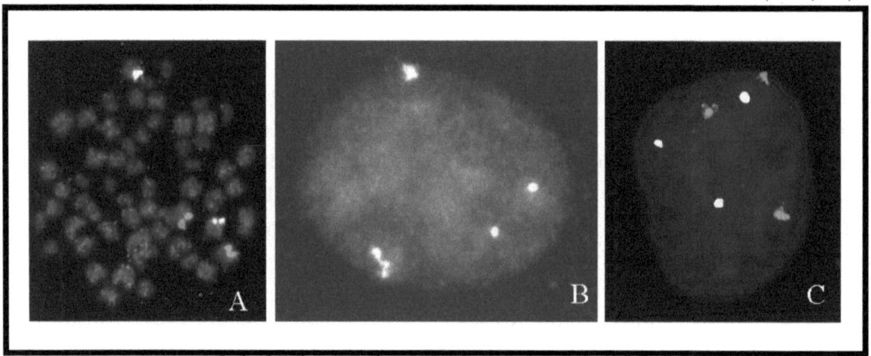

The evaluation of FISH results in embryonic nuclei requires carefully chosen scoring criteria since the signal depends on the type of probes, the types of cells and their fixation. As suggested by Munné et al. (1998) it is important that analytical criteria are re-evaluated for each new combination of probes and cell types. In our studies, nuclei are only scored if they were intact and non-overlapping and if the signals were of similar size, shape and intensity. The effect of the so-called "split-spots" phenomenon, which are separated, but closely linked signals, was minimized by only accepting 2 spots as 2 signals if they were more than the diameter of a single signal apart. For two-colour FISH experiments, a normal cell is 2+2 (2 green and 2 red signals as illustrated in Figure 1A and Figure 1B), but 2+1 or 2+0 signals were also scored as diploid, except if all cells of the embryo showed the same pattern. Thus, nuclei with potential monosomy of either chromosome 6 or 7 were disregarded because of the risk of including hybridization errors as genuine chromosome errors. Chromosomal abnormal cells were scored as triploid if they were 3+3 as illustrated in Figure 1C, or if they were 3+2, 3+1 or 3+0.

Tetraploid or higher ploidies were classified accordingly. Embryos, where all cells were 1+1, were scored as haploid and haploid mosaics if they had more than two 1+1 nuclei and none of the false negative types. Nuclei lacking signals such as 0+0, 0+1, 1+diffuse, 0+diffuse together with 1+1 signals were counted as false negative; only embryos having less than 20% false negative nuclei were included in the analysis. Embryos where all cells could not be analysed need special consideration, and we found it most sensible to be conservative in these cases. We realize that the frequency of false negative hybridization will be limiting to the accuracy of the method and we therefore stress the importance of monitoring the percentage of nuclei showing false negative results as a quality control of pre-treatment steps and the importance of including proper controls to make sure that probes, hybridizations as well as detection steps are optimal.

3. Chromosome abnormalities in embryos

3.1. Incidence of chromosome abnormalities in embryos developed in vivo

Most of the data has been obtained by karyotyping. The 20 published studies summarized by King (1990) include a total of 683 analysed cattle embryos, 606 sheep embryos, 421 pig embryos and 92 horse embryos. There was a mean frequency of chromosome abnormalities of 10% in cattle, 7% in sheep, 5 % in pigs and 7% in horses. About half of the abnormal embryos were mixoploid, the other half were either aneuploid or ploidy errors in about equal proportions. The mixoploidy frequency is close to the 42 % reported by Hare et al. (1980) based on a total of 159 day 12-18 cattle embryos. The high frequency of mixoploidy was surprising at the time that Hare et al. (1980) published their data, since previous studies had revealed only few diploid-tetraploid mosaics in pig, cattle, sheep, rabbit and mouse embryos. It was in fact only Nechiporenko and Oprishko (1977) that had reported similar frequencies of anomalies. These authors investigated 361 pig embryos and found a mixoploidy rate of 67 %. Later investigations have, however, clearly documented that mixoploid embryos are a frequent finding in mammalian embryos. Furthermore, there have been karyotyping studies showing that the mixoploidy frequency increases during development. In sheep, the cytogenetic analysis of 103 day 13-14 blastocysts and 116 day 24-32 foetuses showed an increase in the mixoploidy frequency from 51% to 69% (Murray et al., 1986). The same tendency was found in pigs: Long and Williams (1982) and van der Hoeven et al. (1985) analysed 86 day 3-4 and 31 day 10 pig embryos, respectively. The frequency of mixoploidy increased in pig embryos from 7% at day 3-4 to 65% at day 10.

The interesting question is of course to what extent the mixoploid embryos will develop to living offspring. Hare et al. (1980) point out that bovine embryos with a low level of mixoploidy are most likely developmentally competent. Other results suggest that morphologically abnormal embryos have a higher frequency of chromosome abnormalities than morphologically normal bovine embryos. King et al. (1987) found 21 % chromosomal abnormal embryos in 75 embryos classified as morphologically abnormal whereas no abnormalities were seen in 23 morphological normal day 7 embryos. Moreover Rubes et al. (1988) investigated 48 excellent quality bovine embryos (group A), 76 embryos with moderate degenerative changes (group B) and 31 embryos with pronounced degeneration (group C) and reported a frequency of mixoploidy of 3%, 17% and 56%, respectively, in these day 7 embryos. It is known that the pregnancy rate following transfer of group B and C embryos is lower than that of group A, and it is thus tempting to speculate that embryos with a low frequency of polyploid cells have better chances of surviving. However, karyotyping has severe limitations as a method to resolve this question because estimates of polyploid cells are based only on a few cells - cells that may not even be representative of the population. The FISH-based estimates of the fraction of polyploid cells in normal embryos are particularly interesting in this respect.

The first estimates of chromosome abnormality in cattle embryos that were based on FISH analysis of interphase nuclei revealed that 25 % of cattle embryos were mixoploid and that the mixoploid embryos contained few (generally less than 10%) polyploid cells (Viuff et al., 1999). This study did, however, only include 28 embryos at day 7/8, but it formed a starting point for further studies that were designed to elucidate at what stage of development the polyploid cells would be formed and what their fate was later in development. We have not yet finished these studies, but results of our analysis of 256 cattle embryos isolated on days 2, 3, 4 and 5 post ovulation (p.o.) revealed an increase in the frequency of mixoploidy from day 2 to day 5.

Figure 2: *Increase in mixoploidy frequency during initial development of bovine embryos* in vivo. *Data marked ■ are FISH-based and obtained by interphase cytogenetic analysis (Viuff et al., 1999; 2001) whereas the data marked ▲ are derived by karyotyping (Hare et al., 1980) and should be considered minimum estimates.*

As illustrated in Figure 2, the percentage of mixoploidy on days 2, 3, 4 and 5 was 5%, 13%, 16% and 31%, respectively. The corresponding percentages

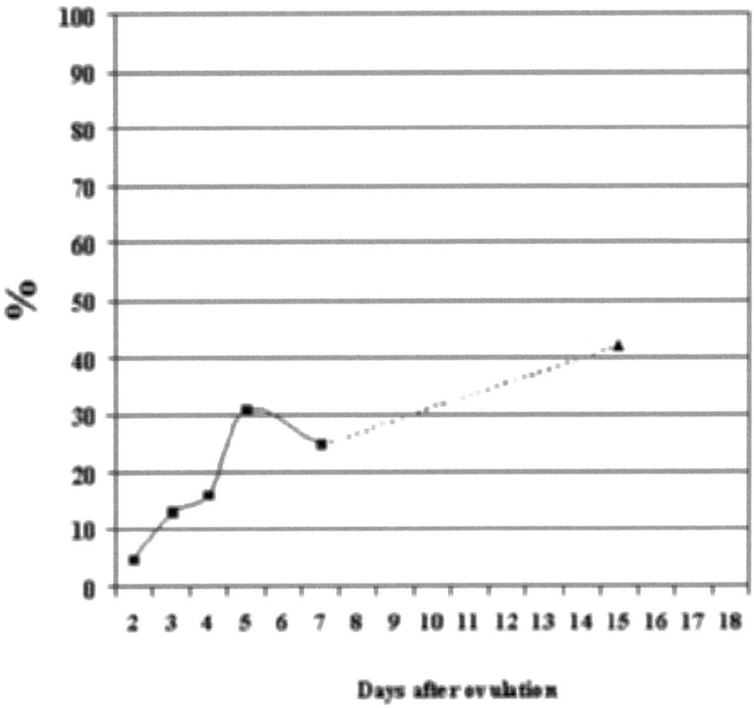

of polyploidy were 2%, 2%, 0% and 0%. Thus, mixoploidy was significantly more frequent than polyploidy. Another striking finding was that the fraction of polyploid cells was low in the mixoploid embryos: Around 90% of the mixoploid embryos contained less than 10 % polyploid cells.

Table 1: Summary of chromosomal analysis of cattle embryos developed in vivo and produced in vitro using karyotyping or FISH.

Post insemination	Number	In vivo	In vitro	Method		Chromosome Abnormalities (%)	References
				Karyotype	FISH		
Day 2-16	683	+		+		0-36	King et al., 1990
Day 7	85	+		+		25	Rubes et al., 1988
Day 2-7	284	+			+	7-31	Viuff et al.,1999; 2001
Day 1	41		+	+		12	Iwasaki et al., 1992
Day 1-2	1005		+	+		14	Iwasaki et al., 1989.
Day 1-2	324		+		+	34	Slimane et al., 2000
Day 2	81		+	+		24	Iwasaki et al., 1992
Day 2	204		+	+		36	Kawarsky et al., 1996
Day 2	112		+		+	27	Viuff et al., 2000
Day 2-3	202		+	+		23	Lechniak et al., 1996
Day 3	22		+	+		36	Iwasaki et al., 1992
Day 3	20		+	+		80	Yoshizawa et al., 1999
Day 3	113		+		+	28	Viuff et al., 2000
Day 4	102		+		+	19	Viuff et al., 2000
Day 5	14		+	+		7	Iwasaki et al., 1992
Day 5	158		+	+		39	Kawarsky et al., 1996
Day 5	99		+		+	42	Viuff et al., 2000
Day 7/8	72		+	+		38	Iwasaki and Nakahara, 1990
Day 7/8	151		+		+	72	Viuff et al., 1999

In the 40 mixoploid embryos, diploidy/triploidy was the most frequent finding (50%), whereas diploid/haploid, diploid / tetraploid and diploid / triploid / haploid + diploid / triploid / tetraploid mosaics were observed in 12.5, 25 and 12.5% of embryos, respectively. Diploid / triploid mosaics were present throughout the early developmental stages whereas diploid / tetraploid mosaics first appeared on days 4 and 5 (with one exception on day 3). The three polyploid embryos of this study were triploid.

The data obtained on the frequency of chromosome abnormalities of cattle embryos developed in vivo are summarized in Table 1.

3.2. Incidence of chromosome abnormalities in embryos produced in vitro (IVP-embryos)

There have been a number of studies investigating the frequency of chromosome abnormalities in cattle embryos produced in vitro. All but three have used karyotyping of metaphase spreads and these results are variable (Table 1). The differences may to some extent be related to the different culture systems used by the different authors but it may also reflect the inherent problem in obtaining reliable estimates of mixoploidy frequency by karyotyping. It is therefore relevant to compare the results of the recently published FISH-based chromosome analyses of in vitro produced cattle embryos with those of karyotyping. The first data (Viuff et al., 1999), which included 151 IVP blastocysts isolated at day 7-8, showed that 72 % were mixoploid. The frequency of polyploid cells was low in the mixoploid embryos:83% contained less than 10% polyploid cells, 13% contained from 11-25% polyploid cells and only 4% contained more than 25% polyploid cells. In a larger study of earlier developmental stages (Viuff et al., 2000) it was possible to document an increase of mixoploidy frequency around day 5: Among 426 IVP embryos, we found that 22 % were mixoploid at day 2, 15% at day 3, 16% at day 4 and 42% at day 5. The corresponding percentages of polyploidy were 5%, 13%, 3% and 0%. In the 99 mixoploid embryos, diploidy-triploidy was the most frequent abnormality (65%), whereas diploid-tetraploid and diploid-triploid-tetraploid mosaics were observed in 11% and 24% of embryos, respectively. Among the polyploid embryos, triploidy was the most common aberration. In addition to the lack of polyploid embryos at day 5 it was striking that the speed of development of polyploid embryos was significantly slower than that of the diploid and the mixoploid embryos and that few if any polyploid embryos progressed beyond the third cell cycle, whereas the mixoploid embryos seem to continue development unaffected beyond this stage.

These studies have been complemented by another FISH-based investigation of the chromosome abnormalities of bovine embryos produced in vitro. Slimane et al. (2000) reported on 111 2-cell embryos isolated 24-40 hours after fertilization and found that 29-45% of the embryos contained chromosome abnormalities, depending on the probes used for the analysis.

The abnormalities were 14-30 % haploid embryos, 5-10 % mixoploid and 5-7 % polyploid embryos. The haploidy frequency obtained by Slimane et al. (2000) is not comparable to that from our study because of differences in scoring criteria and experimental design, a fact which underlines the importance of standardisation of scoring criteria. It is however interesting that the frequencies of polyploidy are quite similar in the two studies, and that the mixoploidy frequency that we find at day 2 (22 %) is much higher than the 5-10 % that was found by Slimane et al. (2000).

4. Is the incidence of chromosomal abnormalities increased by in vitro culture?

The data obtained in cattle by interphase molecular cytogenetics make it possible to compare cattle embryos produced in vitro with in vivo developed embryos. This comparison shows that in vitro produced embryos exhibit a significantly higher frequency of mixoploidy and polyploidy than their in vivo developed counterparts. The polyploid embryos showed retarded growth and were not found after the 8 cell stage suggesting that they are eliminated early in development. Comparison of the differences of mixoploidy at different stages of development showed that mixoploidy increases around day 4 in both in vivo developed and in vitro produced embryos but that the increase was more prominent in those embryos that were produced in vitro compared to their in vivo developed counterparts.

Which factors might increase the incidence of chromosome abnormalities during culture? From the previous paragraph it is clear that the conditions of in vitro production increase the frequency of polyploidy in cattle embryos. It is our hypothesis that the increased polyploidy frequency is caused by an inability of the in vitro maturation system to properly mature the oocytes, with the consequence that polyspermy or lack of polar body extrusion may occur. Preliminary studies (Hendriksen et al., 2001) have shown that day 3 cattle embryos, derived from in vivo matured oocytes (aspirated from pre-LH or 24h-after-LH follicles from superovulated cows) followed by in vitro fertilization and -culture, had a lower polyploidy frequency than the control IVP group. The molecular mechanism is far from clarified, but Pellestor et al. (1994) suggested that insufficient maturation may lead to an incomplete cortical reaction with consequent polyspermy, whereas overmaturity may result in failure to extrude the second polar body. This view is supported by data from the chicken which clearly show the triploidy rate to be influenced by the dam (Fechheimer, 1990) and by ultrastructural studies which have demonstrated disturbances in the peripheral migration of the cortical granules during in vitro maturation of cattle oocytes (Hyttel et al., 1986).

The most logical consequence of polyspermy or deficiencies of polar body extrusion is a triploid embryo, but mosaic embryos may also form. It has been shown that tri-pronuclear human zygotes do not all develop into triploid embryos but that some of them form mosaics (Kola et al., 1987; Ma et al., 1995; Handyside and Delhanty, 1997) and some poly-pronuclear pig zygotes have even been shown to develop to term (Han et al., 1999). Thus, our finding that most of the mixoploid cattle embryos in the initial cleavage stages are diploid/triploid rather than diploid/tetraploid could be explained by a higher incidence of polyspermy by two rather than by three spermatozoa. It does however not explain why the frequency of diploid/tetraploid shows an increase on day 4 and 5 in embryos developed in vivo and why the mixoploidy frequency increases from day 2 to day 5 in both in vivo developed and in vitro produced embryos. This could in turn be related to the fact that cell cycle regulation changes from a maternal dominance to an embryonic dominance during the 3rd cell cycle if we assume that cell-cycle checkpoints may not operate effectively during the cleavage divisions that occur before activation of the embryonic genome.

The generally higher level of mixoploidy in IVP embryos than in embryos developed in vivo does, however, suggest that there are also factors related to the culture system, which influence the generation of polyploid cells. That we do not find the same increase in the proportion of tetraploid cells in mixoploid IVP embryos as we have demonstrated in the in vivo produced material supports this notion. In this context it is interesting to note that even temperature differences of 1°C can disturb normal spindle structure (Pickering et al., 1988) leading to a disorganized movement of chromosomes and formation of mosaics (Kola et al., 1987). Such small temperature differences could easily occur during handling of oocytes, zygotes and embryos.

It would be valuable to clarify the origin of the increased incidence of embryonic polyploidy and mixoploidy after in vitro culture because it could assist in optimising in vitro production systems. Moreover, it would also be interesting to know the fate of embryos containing these polyploid cells. It is known that 100% polyploid embryos are eliminated, but what is the consequence for an embryo, when 50% or 25% of the cells are polyploid? One of the few relevant observations in this respect was made by Hare et al. (1980), who found a mixoploidy level of 41.5% in normal in vivo developed cattle embryos at day 12-18. These authors suggested that up to at least 25 %

polyploid cells in an embryo might be compatible with pregnancy. Another line of relevant evidence is that tetraploid cells in mouse embryos are preferentially eliminated or allocated to the extra-embryonic membranes (Nagy et al., 1993; James et al., 1995, Everett et al., 1998). That this mechanism may also be active in cattle was indicated by preliminary data from our laboratory, which showed an increased frequency of polyploid cells in the trophoblast versus the inner cell mass at day 7 and day 12 and a decreased frequency of chromosomal abnormal cells when comparing day 7 and day 12 cattle embryos. It is still unknown, however, what level of tetraploid cells can be tolerated in an embryo or whether an eventual negative effect may be cumulative.

5. Conclusion

The ultimate proof of the value of any embryo production system is the birth of living, healthy offspring. However, the high cost and the long period of time which is necessary to provide this proof necessitates the development of markers of embryo quality and viability during the in vitro phase. The frequency of ploidy errors may be one important reference point for monitoring the differences in development that may be induced by in vitro manipulations of cattle oocytes. Demonstrating significant differences between in vivo and IVP cattle embryos with respect to the frequency of polyploidy have illustrated this point. It is, however, also clear that there are other differences between in vivo and in vitro produced embryos that could be related to their developmental potential. Two such differences are the increased frequency of mixoploidy and the slower development rate of IVP embryos compared to that of embryos produced in vivo after superovulation (Viuff et al., 2001). We do not know yet whether these changes are interrelated and whether they are induced by the culture system. On the other hand, these cellular alterations may be entirely reversible and in this case, the biological significance of such changes would be none. The other possibility is that the cellular alterations do induce a biological disadvantage and that such embryos are preferentially lost. It is too early to tell which hypothesis is correct, but this is certainly an important question to answer. Meanwhile, the frequency of polyploidy, mixoploidy and the developmental rate can be used as markers of embryonic growth. More importantly, we have provided sufficient data from cattle embryos produced in vivo to establish a firm basis for comparing embryo development after different in vitro production protocols with embryo development within the normal uterine environment.

Acknowledgements

The authors acknowledge the skilled help of Ms. Jytte Nielsen for preparing the microscopical illustrations. Figs, 3, 4, 5, 6 and 7 are reproduced from Anatomy and Embryology with kind permission from Springer-Verlag, Figs. 8, 9, 10, 13, 14, 15, 16 and 17 from Journal of Reproduction and Fertility with kind permission from Journals of Reproduction and Fertility, Ltd., Figs. 20, 21, 22, 23, 28 and 30 from Biology of Reproduction with kind permission from Society of for the Study of Reproduction, Inc., and Figs. 24, 25, 26, 29, 31 and 33 from Molecular Reproduction and Development with kind permission from Wiley-Liss, Inc.

References

Agerholm JS, Christensen K. 1993. Trisomy 22 in a calf. Zentralbl Veterinärmed A 40:576-581.

Christensen K, Juul L. 1999. A case of trisomy 22 in a live Hereford calf. Acta Vet Scand 40:85-88.

Ducos A, Seguela A, Pinton A, Berland H, Brun-Baronnat C, Darre A, Manesse M, Darre R. 2000. Trisomy 26 mosaicism in a sterile Holstein-Friesian heifer. Vet Rec 146:163-164.

Dyrendahl I, Gustavsson I. 1979. Sexual functions, semen characteristics and fertility of bulls carrying the 1/29 translocations. Hereditas 90:281-289.

Everett CA, West JD. 1998. Evidence for selection against tetraploid cells in tetraploid/diploid mouse chimaeras before the late blastocyst stage. Genet Res 72:225-228.

Fechheimer NS. 1990. Chromosomes of chickens. In: McFeely RA (ed), Advances in Veterinary Science and Comparative Medicine. San Diego, CA: Academic Press 34:169-206.

Han YM, Wang WH, Abeydeera LR, Petersen AL, Kim JH, Murphy C, Day BN, Prather RS. 1999. Pronuclear location before the first cell division determines ploidy of polyspermic pig embryos. Biol Reprod 61:1340-1346.

Handyside AH, Delhanty JDA. 1997. Preimplantation genetic diagnosis: strategies and surprises. Trends in Genetics 13:270-275.

Hare WCD, Singh EL, Betteridge KJ, Eaglesome MD, Randall GCB, Mitchell D, Bilton RJ, Trounson AO. 1980. Chromosomal analysis of 159 bovine embryos collected 12 to 18 days after estrus. Can J Genet Cytol 22:615-626.

Hendriksen PJM, Viuff D, Greve T, Vos PLAM, Mullaart E, Bevers MM, Dieleman SJ. 2001. Effect of the origin of bovine oocytes on cell numbers and numerical chromosome aberrations in in vitro embryos collected at day 3 post-insemination. Theriogenology 55:475.

Herzog A, Hoehn H. 1991. Two additional cases of autosomal trisomy, 61,XY,+12 and 61,XX,+12, in cattle. Cytogenet Cell Genet 57:211-213.

Hyttel P, Xu KP, Smith S, Greve T. 1986. Ultrastructure of in vitro oocyte maturation in cattle. J Reprod Fertil 78:615-625.

Iwasaki S, Shioya Y, Masuda H, Hanada A, Nakahara T. 1989. Incidence of chromosomal anomalies in early bovine embryos derived from in vitro fertilization. Gamete Res 22:83-91.

Iwasaki S, Nakahara T. 1990. Incidence of embryos with chromosomal anomalies in the inner cell mass among bovine blastocysts fertilized in vitro. Theriogenology 34:683-690.

Iwasaki S, Hamano S, Kuwayama M, Yamashita M, Ushijima H, Nagaoka S, Nakahara T. 1992. Developmental changes in the incidence of chromosome anomalies of bovine embryos fertilized in vitro. J Experimental Zoology 261:79-85.

James RM, Klerkx AHEM, Keighren M, Flockhart JH, West JD. 1995. Restricted distribution of tetraploid cells in mouse tetraploid-diploid chimaeras. Dev Biol 167:213-226.

Kawarsky SJ, Basrur PK, Stubbings RB, Hansen PJ, King WA. 1996. Chromosomal abnormalities in bovine embryos and their influence on development. Biol Reprod 54:53-59.

Keighren M, West JD. 1993. Analysis of cell ploidy in histological sections of mouse tissues by DNA-DNA in situ hybridization with digoxigenin-labelled probes. Histochem J 25:30-44.

King WA, Guay P, Picard L. 1987. A cytogenetical study of 7-day-old bovine embryos of poor morphological quality. Genome 29:160-164.

King WA. 1990. Chromosome abnormalities and pregnancy failure in domestic animals. In: McFeely RA (ed), Advances in Veterinary Science and Comparative Medicine. San Diego, CA: Academic Press 34:229-250.

Kobayashi J, Sekimoto A, Uchida H, Wada T, Sasaki K, Sasada H, Umezu M, Sato E. 1998. Rapid detection of male-specific DNA sequence in bovine embryos using fluorescence in situ hybridization. Mol Reprod Dev 51:390-394.

Kola I, Trounson A, Dawson G, Rogers P. 1987. Tripronuclear human oocytes: altered cleavage patterns and subsequent karyotypic analysis of embryos. Biol Reprod 37:395-401.

Lechniak D, Switonski M, Sosnowski M. 1996. The incidence of bovine diploid oocytes matured in vitro. Theriogenology 46:267-277.

Long SE, Williams CV. 1982. A comparison of the chromosome complement of inner cell mass and trophoblast cells in day-10 pig embryos. J Reprod Fertil 66:645-648.

Ma S, Kalousek DK, Yuen BH, Moon YS. 1995. The chromosome pattern of embryos derived from tripronuclear zygotes studied by cytogenetic analysis and fluorescence in situ hybridization. Fertil Steril 63:1246-1250.

Munné S, Márquez C, Magli C, Morton P, Morrison L. 1998. Scoring criteria for preimplantation genetic diagnosis of numerical abnormalities for chromosomes X, Y, 13, 16, 18 and 21. Mol Hum Reprod 9:4:863-870.

Murray JD, Moran C, Boland MP, Nancarrow, Sutton R, Hoskinson RM, Scaramuzzi RJ. 1986. Polyploid cells in blastocysts and early fetuses from Australian Merino sheep. J Reprod Fertil 78:439-446.

Nagy A, Rossant J, Nagy R, Abramow-Newerly W, Roder JC. 1993. Derivation of completely cell culture-derived mice from early-passage embryonic stem cells. Proc Natl Acad Sci USA 90:8424-8428.

Nechiporenko V KH, Oprishko NN. 1977. Studies of chromosome in pig embryos at early stages of embryogenesis. Tsitolog. Genet 11:457-460 (in Russian).

Pellestor F, Girardet A, Andrèo B, Arnal F, Humeau C.1994. Relationship between morphology and chromosomal constitution in human preimplantation embryo. Mol Reprod Dev 39:141-146.

Pickering SJ, Johnson MH, Braude PR, Houliston E. 1988. Cytoskeletal organisation in fresh, aged and spontaneously activated human oocytes. Hum Reprod 3:978-989.

Popescu P. 1990. Chromosomes of the cow and bull. Advances in Veterinary Science and Comparative Medicine. San Diego, CA: Academic Press 34:41-71.

Rubes J, Zák M, Horinová Z, Machatková M. 1988. Cytogenetic examination of Bovine embryos of different quality. 8th European colloquium on cytogenetics of domestic animals 134-138.

Slimane W, Heyman Y, Lavergne Y, Humbolt P, Renard JP. 2000. Assessing chromosomal abnormalities in two-cell bovine in vitro-fertilized embryos by using fluorescent in situ hybridization with three different cloned probes. Biol Reprod 62:628-635.

Thomsen PD, Schwerin M, Poulsen PH, Avery B. 1991. Sex determination of bovine embryo biopsies using in situ hybridization. Reprod Dom Anim 26:66-69.

Van der Hoeven FA, Cuijpers MP, de Boer P. 1985. Karyotypes of 3- or 4-day-old pig embryos after short in vitro culture. J Reprod Fertil 75:593-597.

Viuff D, Rickords L, Offenberg H, Hyttel P, Avery B, Greve T, Olsaker I, Williams JL, Callesen H, Thomsen PD. 1999. A high proportion of bovine blastocysts produced in vitro are mixoploid. Biol Reprod 60:1273-1278.

Viuff D, Greve T, Avery B, Hyttel P, Brockhoff PB, Thomsen PD. 2000. Chromosome aberrations in in vitro-produced bovine embryos at days 2-5 post-insemination. Biol Reprod 63:1143-1148.

Viuff D, Hendriksen PJM, Vos PLAM, Dieleman SJ, Bibby BM, Greve T, Hyttel P, Thomsen PD. 2001. Chromosomal abnormalities and developmental kinetics in cattle embryos at days 2 to 5 after ovulation. Biol Reprod, 65:204-208.

Xu K, Huang T, Liu T, Shi Z, Rosenwaks Z. 1998. Improving the fixation method for preimplantation genetic diagnosis by fluorescent in situ hybridization. J Assisted Reprod Gen 15:9:570-574.

Yoshizawa M, Konno H, Zhu S, Kageyama S, Fukui E, Muramatsu S, Kim S, Araki Y. 1999. Chromosomal diagnosis in each individual blastomere of 5- to 10-cell bovine embryos derived from in vitro fertilization. Theriogenology 51:1239-1250.

Chapter 11

DIFFERENTIAL STAINING OF INNER AND OUTER CELLS TO ASSESS MAMMALIAN EMBRYO QUALITY

Ann Van Soom[1], Marleen Boerjan[2] and Kate Hardy[3]

[1] *Department of Reproduction, Obstetrics and Herd Health, Faculty of Veterinary Medicine, Salisburylaan 133, B-9820 Merelbeke, Belgium*

[2] *PasReform, Hatchery technologies, P.O. Box 2, NL-7038 ZG Zeddam, the Netherlands*

[3] *Department for Reproductive Science and Medicine, Institute of Reproductive and Developmental Biology, Imperial College, Hammersmith Hospital, Du Cane Road, London, W12 0NN, UK*

1. Introduction

The famous statement "A rose is a rose is a rose" by Gertrude Stein does not apply to the mammalian embryo. At first sight, the preimplantation eutherian or mammalian embryo shows many similarities between different species. The fertilized mammalian oocyte follows a common pattern of cleavage divisions, compaction and blastulation. Furthermore, the initial size of the embryo does not vary greatly between species. It is therefore difficult for an inexperienced observer to differentiate between laboratory animal embryos, domestic animal embryos and human embryos by routine stereomicroscopical evaluation. Upon more detailed examination, pre-implantation mammalian embryos from even closely related species differ with regard to their timing of development, their culture/metabolic requirements, the way they expand at the blastocyst stage, their behaviour at and after hatching and in the lipid content of their cells. These differences are also reflected in the various ways mammalian embryos resist cryopreservation and the efficiency of in vitro production procedures.

The temporal sequence of early developmental events and the study of cell potency during cleavage and blastocyst formation have been studied predominantly in the mouse (for reviews see Biggers, 1987 and Johnson, 1996). In recent years, more data have become available in a wider range of laboratory animals such as the rabbit and the rat, in domestic species such as cow, sheep and pig and in primates, including human. It is our objective to review what is known about cell fate and cell allocation during early preimplantation development in different mammalian species, and to assess whether this knowledge can provide us with useful information on mammalian embryo quality.

2. Embryonic development

2.1. Cleavage to compacted morula

Mammalian development is initiated by the fusion of one spermatozoon with the plasma membrane of the oocyte. Paternal and maternal pronuclei are formed within a few hours after sperm penetration. The newly formed individual, still protected by the zona pellucida, starts its development by cleavage divisions. The presence of the zona pellucida is of the uttermost importance at this stage of development. It serves as a physical barrier between the embryo and its environment. Removal of the zona pellucida leads to dissociation of the blastomeres and subsequent adherence of these blastomeres to oviductal epithelium. The zona pellucida also protects the embryo from invasion by pathogens (Vanroose et al., 1997, 1998, 2000).

The cells of the early cleavage stage embryo are characterized by their totipotency or their capacity to contribute to any cell lineage. This is demonstrated by the ability to disaggregate the embryo into individual cells, which subsequently can develop further. In some cases, these cells retain their capacity to develop into small blastocysts (Pedersen, 1986). Two-cell blastomeres appear to be developmentally totipotent in most mammalian species: when disaggregated blastomeres are injected into the blastocoel of another blastocyst, they can contribute to the different lineages of this embryo (Pedersen, 1986). Isolated 8-cell blastomeres from sheep embryos produced identical, non-chimeric lambs when combined with blastomeres from younger embryos and after transfer to a recipient animal (Fehilly and Willadsen, 1986).

Cleavage stage blastomeres are also characterized by their radial distribution of organelles and membrane structures. The first event which leads to differentiation between the embryonic cells is called compaction when individual blastomeres can no longer be distinguished because of the increased cell-cell interactions. The cleavage stage embryo with all blastomeres individually visible develops into a smoothly contoured morula (Ducibella and Anderson, 1975). Compaction can be defined as a progressive increase in the complexity of intercellular adhesion and maximization of cellular contact between the outer cells of the embryo (Johnson and Maro, 1986; Johnson, 1996). Compaction is thus merely a transitional developmental stage between the morula with individual, more or less identical, blastomeres and the blastocyst consisting of two cell lineages with different fates. The underlying cellular processes of compaction have mainly been studied in more detail for mouse embryos. In the mouse, compaction occurs during the fourth cell cycle. Cellular flattening is accompanied by the formation of polarized cells from the radially symmetric blastomeres. The cytoplasmic distribution of organelles becomes polarized and microvilli form apically. In the subsequent mitosis the polarblasts divide either in two polarblasts or in one polarblast and one pluriblast (inner cell) (Johnson and Selwood, 1996). The polarblasts differentiate to the epithelial trophoblast cells in the subsequent cell cycles. The pluriblasts constitute the inner cell mass and will eventually differentiate to hypoblast and pluripotent epiblast cells. In conclusion, the mouse 8-cell blastomere is competent to polarize if it experiences asymmetric cell-cell interactions *(*for review see Johnson and Maro, 1986*)*.
The process of cell flattening as the initial step of compaction occurs in all mammalian species studied so far. The timing varies and is not always accompanied by polarisation of the outer cells (Table 1).

The latter statement must be taken with caution because polarisation of embryonic cells has been studied superficially in most species except the mouse. In the pig, compaction occurs during the third cell cycle (Boerjan and te Kronnie, 1993; Reima et al., 1993) while polarisation of the outer cells is completed shortly before blastulation (Reima et al., 1993). In the bovine embryo polarity of the outer cells was not manifested until after the 16-cell stage with approximately 40% polar cells per embryo (Koyama et al., 1994) and compaction occurs between the 32-64-cell stage (Van Soom et al., 1997). In the rabbit embryo, polarized cells are observed not earlier than the 32-cell stage (Koyama et al., 1994)

Table 1: Overview of timing of compaction and polarisation of embryonic blastomeres in different species

Species	Cell flattening	Polarisation	Reference
Laboratory			
Mouse	8-cell	8-cell (actin)	Reeve and Ziomek, 1981
Rat	4-cell	8-cell	Schlafke and Enders 1967; Reeve, 1981
Hamster	8-cell	8-cell (microvilli)	Suzuki et al., 1999
Guinea pig			Hunter et al., 1969
Rabbit	32-cell	After 32-cell (microvilli)	Ziomek et al., 1990; Koyama et al., 1994
Domestic			
Bovine	16-32 cell	16-cell (microvilli)	Koyama et al., 1994; Van Soom et al., 1996
Sheep		ND	De Paz et al., 1994
Porcine	4 cell	32-64 (actin)	Boerjan and te Kronnie, 1993; Reima et al., 1993
Human	8-16-cell	10-18 cell (microvilli)	Hardy et al., 1996; Nikas et al., 1996

In the mouse embryo, the increased adhesion between the cells is caused by the cell-adhesive properties of E-cadherin (Takeichi, 1988; Fleming et al., 1992; Barcroft et al., 1998). E-cadherin (uvomorulin) forms the main component of the adherent junction and is typically expressed in epithelial tissues and in compacting embryos. A system of intracellular intermediate filaments develops between the outer cells and thus forms an impermeable seal of outer cells between the inside cells and the outside cells of the embryo (Becker et al., 1992). Furthermore gap junctions, zonular tight junctions and point desmosomes are formed between the outer cells (mouse: Fleming et al., 1992; pig, Reima et al., 1993*)*. In most embryos studied so far, the process of compaction is mediated by proteins involved in cytoskeleton formation and intercellular interaction. Timing and the level of involvement might differ between species. Gap junction mediated cell-to-cell coupling in the bovine embryo is limited to 18% of the compacted superovulated morulae and 25% of the expanded superovulated blastocysts.

In the mouse, cell-to cell coupling was observed in 95% of the morulae (Prather and First, 1993). Whether gap junction mediated cell-to-cell coupling is characteristic for viable embryos is not known. Interestingly, the absence of cellular coupling (Prather and First, 1993) as well as the absence of transcripts of the gap junction gene connexin43 in in vitro produced bovine embryos (Wrenzycki et al., 1996) indicates that culture conditions influence the final differentiation of the outer cells.

By the time of compaction two types of cells are thus recognized in the preimplantation mammalian embryo: (1) the flattened outer cells and (2) the pluripotent more spherical inner cells. The former develop to the trophectoderm and will finaly develop to extra embryonic tissue and placenta. The latter develop to the inner cell mass (ICM) in the blastocyst.

2.2. The blastocyst

The blastocyst stage is characteristic of the eutherian embryo (De Loof, 1992). In most eutherian mammals blastocoel formation occurs after a fixed number of nuclear divisions (Smith and McLaren, 1977). In the blastocyst the system of intracellular intermediate filaments develops and generates together with the increased intercellular contacts a seal separating the inside from the outside of the embryo. The seal is permeable and selective with respect to transport from the outside to the inside of the embryo: it regulates the vectorial transport of fluids to form and sustain the blastocoel (Biggers et al., 1988, Watson, 1992). The formation of the tight junctions between the outer cells is essential in the formation of this epithelium. The formation of the permeability seal is also the basis for the technique of differential staining.

Both cell types in the blastocyst form distinct cell lineages each with a specific fate and role in future development of the conceptus. Or in other words, from now on, the two cell lineages develop in different directions which is reflected in the different patterns of gene and protein expression. It also means that the allocation of cells to either the inner cell mass or the trophectoderm has finished and thus the number of cells present in both lineage is set.

Gene expression in the trophectoderm is directed to fluid transport (reviewed by Baltz, 1993) or embryo-maternal interaction (Wiley et al., 1995). As development proceeds the gene expression related to embryo-maternal interaction becomes more and more species specific.

In the morula some cells become internalized and remain apolar for a certain period of time. These inner cells are sequestered at one pole of the nascent blastocyst and thereby the dorsal-ventral axis of the mammalian embryo has been specified (Gardner, 1997). As the blastocoel expands, a subsequent round of differentiation occurs in the ICM. The cells adjacent to the blastocoel side differentiate to endoderm cells in the mouse recognized by a specific expression of the homeobox gene HEX (Rossant, 1986; Thomas et al., 1998) while the inner epiblast cells are pluripotent as characterized by OCT-4 expression (Schöler et al., 1990). As development proceeds, the epiblast differentiates to pregastrula embryonic ectoderm recognized by pregastrula gene expression like in the mouse. In the pig, expression of the mesoderm specific gene goosecoid has been determined in the day 9 embryo (Meijer et al., 2000).

3. Differential staining of the blastocyst

3.1. The ratio of inner and outer cells

As explained above the inner cells and outer cells of the blastocyst are the inner cell mass and the trophectoderm respectively. Therefore, the ratio of the number of inner and outer cells in the early blastocyst is an important factor for normal development. The technique of differential staining of the inner and outer cells is a tool to determine this ratio. If the pattern of allocation of inner and outer cells is disturbed it often leads to non-viable embryos as is discussed below. In in vivo grown embryos the ratio varies between 30-40% for the different species studied so far (Table 2). It has been shown that in different species allocation of inner cells can take place during a few consecutive cell cycles: in mouse inner cells are segregated predominantly during the fourth cleavage division whilst in porcine embryos inner cells are segregated during transition from the 4-to 8-cell stage and inner cells are added during two subsequent cell cycles (Boerjan and te Kronnie, 1993).

Table 2: Overview of appearance of inner cells and ICM/total cell ratio in different species

Species	**First appearance of inner cells**	**ICM / Total cell ratio in early to mid blastocysts**	**Reference**
Laboratory			
Mouse	13-16 cells	0.40	Handyside and Hunter, 1984; Boerjan and te Kronnie, 1993
Rat	11-23 cells	0.28-0.32	Pampfer et al., 1990
Rabbit	Blastocyst	0.21-0.30	Giles and Foote, 1995
Domestic			
Bovine	10-32 cells	0.36	Van Soom et al., 1996
Sheep		0.44	Handyside et al., 1987
Porcine	9-12 cells	0.22-0.25 0.25-0.36	Boerjan and te Kronnie, 1993; Rivera et al., 1996
Human	20 cells +*	0.25 - 0.33	Hardy et al., 1989

* morulae were not evaluated, only early blastocysts.

3.2. Applications of differential staining for the evaluation of mammalian blastocysts

3.2.1. Inner cell allocation and embryonic origin

Comparative studies of in vivo and in vitro development in different species indicate that cell allocation to the ICM is skewed under in vitro circumstances (Iwasaki et al., 1990; Erbach et al.,1994; Rath et al.,1995; Giles and Foote, 1995; Du et al., 1996; Van Soom et al., 1997a).

Table 3: Influence of embryonic origin (vivo vs. vitro) on inner cell allocation in expanded blastocysts

Species	ICM cell no.		Proportion ICM / Total cell no.		Reference
	Vivo	Vitro	Vivo	Vitro	
Bovine	42	36	0.27	0.30	Van Soom et al., 1996, 1997a, 1997b
	53	39	0.27	0.23	De La Fuente and King, 1997
	23	11	0.21	0.15	Iwasaki et al., 1990
Buffalo*	---	31	---	0.24	Narula et al., 1996
Porcine	19	---	0.21	---	De La Fuente and King, 1997
	13.5	5	0.23	0.32	Rath et al., 1995
	18.1	3.3	0.17	0.50	Macháty et al., 1998
Murine	20	---	0.27	---	De La Fuente and King, 1997; Devreker and Hardy, 1997 Hardy and Handyside, 1996
Rabbit	238	58	0.30	0.25	Giles and Foote, 1995
Rat	32	---	0.32	---	Pampfer et al., 1990

* *In this study, hatched blastocysts were used*

From the data compiled in table 3 it is clear that in vitro culture reduces the number of ICM-cells at the expanded blastocyst stage in all species, although the proportion of ICM to total cell number is not always lower. Moreover, the incidence of embryos with low ICM-cell numbers (less than 20) is much higher when embryos are cultured in vitro than for in vivo produced embryos, at least in the bovine (Table 4).

Table 4: Prevalence of low ICM cell numbers in bovine blastocysts generated in vivo or in vitro.

Embryo source	No of embryos evaluated	Percentage of blastocysts with ICM-cells			
		0	1-10	11-20	>20
VitroTCM199	89	-	7%	16%	78%
VitroB2	133	2%	6%	15%	77%
Vivo	26	-	-	-	100%

Certain aspects of in vivo embryo development have been studied by differential staining e.g. embryo differentiation in pig breeds with a different fertility, such as the Meishan and the Yorkshire pig (Rivera et al., 1996), as well as the influence of a severe reduction in food intake on embryo differentiation in cattle and mouse (Nolan et al., 1998; Fleming et al., 1992). Interestingly, in a comparative study of in vivo and in vitro development in pigs, it was found that in vitro culture of porcine embryos severely decreased the number of ICM-cells compared with their in vivo cultured counterparts (Macháty et al., 1998). However , in spite of these low cell numbers, 37 % of these embryos was developmentally competent after transfer (formation of day 40 conceptuses), while this rate was 54 % for in vivo developed embryos (p= 0.78) (Macháty et al., 1998). This raises again the following question: what is the minimal number of ICM-cells which must be present to produce normal foetal development?

Also the genetic make-up of an embryo might influence embryo differentiation. In mice, differential staining has been used to study the importance of the pre-implantation-embryo-development (Ped) gene on blastocyst cell differentiation (Roudebush, 1998). Qa-2(a) negative (Ped-) knock-out mouse blastocysts had significantly lower ICM-cell numbers compared with controls. With the recent evidence for epigenetic effects upon development (Boerjan et al., 2000), this area of research will surely expand in the near future.

Knowledge on the distribution of ICM- and TE-cells is of considerable value for the clinician if he wants to transfer the embryo, because it is thought that a minimal number of ICM-cells is required to obtain a pregnancy. Besides, excessive allocation of cells to the TE might be responsible for pregnancy anomalies which are encountered after transferring in vitro cultured bovine embryos such as increased incidence of hydroallantois (Kruip and den Daas, 1997; Hasler, 1998).

3.2.2. Inner cell allocation and embryo morphology

Since morphological embryo assessment is still the most valuable technique to select embryos for transfer, it might be interesting to link embryo morphology with inner cell allocation (Hardy et al., 1989). We found that some in vitro produced cow blastocysts displayed very low ICM-cell numbers (less than 10 ICM-cells in day 7 blastocysts (Table 4, Van Soom et al., 1997b). Initially we thought that this was due to poor quality of the embryo at the morula stage, because a sufficient number of internal cells has to be enclosed at this stage as a basis for the ICM. If a lot of the cells are fragmented or dead at the morula stage, it would be logical to presume that these embryos would display a lower number of cells at the blastocyst stage. However, this was not the case since all morphological grades produced comparable percentages of hatched blastocysts with lower ICM-cell numbers (Table 5) (Figure 1a and b), except for the embryo group with poor morphology but this might be biased due to a lower number of embryos in this group. This observation is probably an indication for the fact that not only the number of inner cells present at the morula stage is important for the final number of ICM-cells at the blastocyst stage, but that there might also be a substantial loss of ICM-cells during blastocyst formation, which is caused by apoptosis. Recent investigations on apoptosis in bovine and human embryos confirm this hypothesis since blastocyst formation is associated with a marked increase in incidence of apoptosis (Hardy et al., 1989; Byrne et al., 1999, Hardy et al., 2001). Another way to relate embryo morphology with allocation of ICM-cells is to estimate the size of the inner cell mass as it can be evaluated by means of routine stereomicroscopical evaluation. We did this in a small study using in vitro produced bovine blastocysts.

Table 5: Prevalence of low ICM-cell-numbers in bovine hatched blastocysts generated in vivo or in vitro

Embryo source	**Grade at morula stage**	**No. of embryos evaluated**	**Percentage of hatched blastocysts with ICM-cells**				
			0	1-10	11-20	21-30	>30
VitroB2	Excellent	56	-	7%	16%	29%	48%
	Good	44	-	4%	23%	25%	48%
	Fair	27	4%	7%	22%	37%	30%
	Poor	10	-	20%	50%	20%	10%
Vivo		8	-	-	-	13%	87%

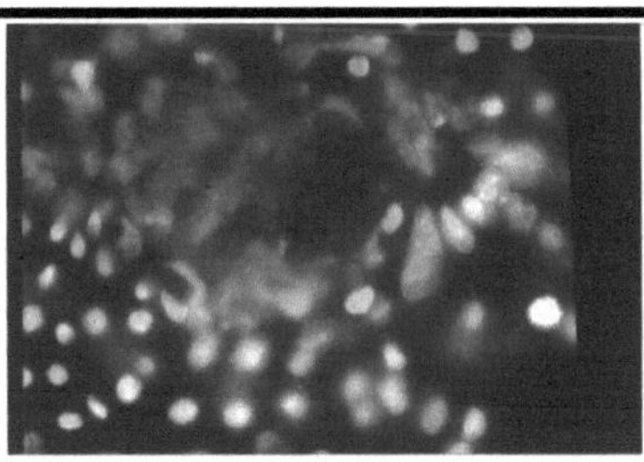

Figure 1a: In vitro produced hatched blastocyst with large number of ICM-cells

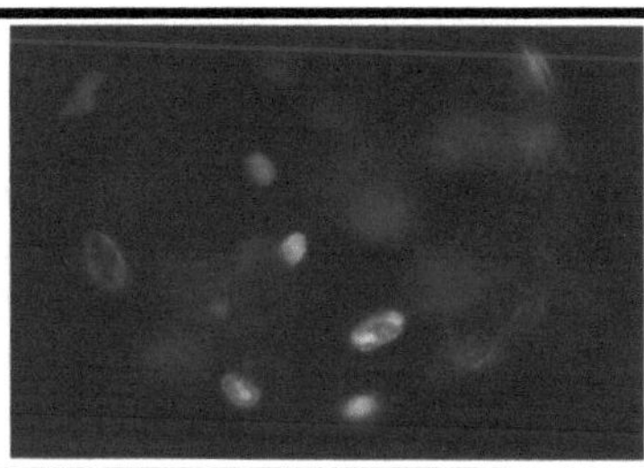

Figure 1b: In vitro produced hatched blastocyst without ICM (trophectodermic vesicle)

The stereomicroscopical evaluation of the inner cell mass is not very easy, because it is influenced by the level of expansion of the blastocyst, and moreover, fragmented cells inside the blastocoel can have an appearance which is very similar to the inner cell mass. Also the presence of apoptotic cells cannot be evaluated by means of stereomicroscopy. Despite these problems, we scored the size of the ICM by means of a stereo microscope in bovine blastocysts which were starting to expand at 168 hpi and which were produced in bovine oviduct epithelial cell (BOEC) coculture B2-medium. Blastocysts were divided into two groups: large and small according to the criteria presented in Figure 2 and then subjected to differential staining to control the accuracy of the stereomicroscopical evaluation of the ICM appearance.

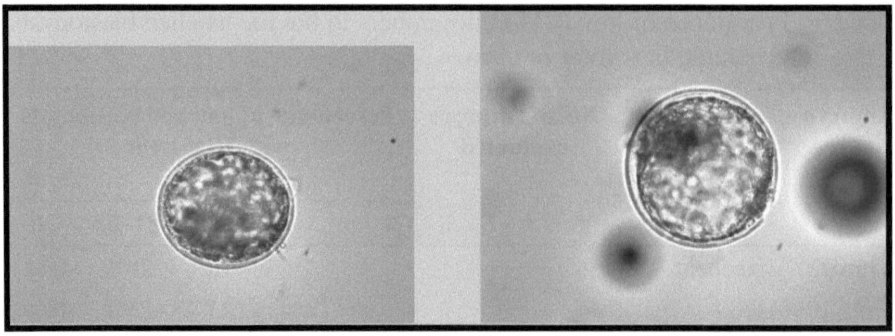

Figure 2a: *Bovine blastocyst with large ICM, which is present over about half the blastocoel of the embryo*

Figure 2b: *Bovine blastocyst with small ICM, which is more compact and is confined to less than half of the blastocoel*

A total of 3082 zygotes developed into 1220 8-cell embryos at 3 dpi (40 %) and 853 were scored for morula morphology at 6 dpi (28 %). Finally 245 blastocysts were succesfully stained differentially. Results of ICM-grading and cell counts are presented in table 6.

Table 6: Relationship between visual score of ICM and cell count of ICM after differential staining of in vitro produced bovine blastocysts

Visual score ICM	**No. blastocysts evaluated**	**No. ICM-cells (Mean±SD)**	**Total cell no. (Mean±SD)**	**Proportion ICM/total (Mean±SD)**
Large	154	41 ± 14 [a]	116 ± 32 [a]	0.36 ± 0.08 [a]
Small	91	32 ± 12 [b]	83 ± 27 [b]	0.38 ± 0.09 [b]

[a,b] Values with different superscript within the same column are significantly different (two sample t-test, P<0.01)

From the results it is clear that a visual estimation of the ICM can give useful information on the number of cells present in the ICM, since ICM-cell no. between embryos with large and small ICMs indeed differred significantly. Unexpectedly, the proportion of ICM-cells in the embryo was higher for embryos with a small ICM than for embryos with a large ICM. This means that a the presence of the ICM over about half the embryo (as in large ICM) is an indication that the embryo has more ICM-cells, but that these ICM-cells do not proportionally add to the total cell number. The large ICM as it can be evaluated microscopically, is probably caused by a stretching of the ICM-cells following the expansion of the blastocyst. The trophectoderm cells make a more important contribution to the increase of the total cell no. than the ICM-cells in this case. In conclusion, visual scoring of the size of the ICM can be useful to estimate ICM cell no. but differential staining remains necessary to evaluate proportion ICM/total cell number.

3.2.3. Inner cell allocation and timing of development

Although it is known that early cleaving embryos are more likely to develop into morula or blastocyst stages (Van Soom et al., 1992; Grisart et al., 1994, Van Soom et al., 1997b), no relationship between timing of early cleavage and number of inner cells at the blastocyst stage was detected in bovine embryos (Van Soom et al., 1997b). However, when timing of bovine blastocyst formation was assessed, it was clear that blastocysts produced at day 6 after fertilization had significantly higher numbers of inner cells than blastocysts which developed only at day 7 or 8 (Van Soom et al., 1997b).

3.2.4. Inner cell allocation and oocyte maturation or fertilization conditions

Little is known about the influence of oocyte origin or fertilization conditions on cell allocation at the blastocyst stage. Since embryos are depending on maternal stores for the first cleavage divisions but not anymore for inner cell allocation, this might be too farfetched especially since it is known that embryos from PCO women (Polycystic Ovarian Syndrome) develop normally (Hardy et al., 1995). Anyhow, calf oocytes are notorious for the low viability of the blastocysts they produce, which might be caused by an abnormal cell number of ICM and TE. However, no differences could be observed in inner cell allocation between blastocysts derived form calf and cow oocytes (Khatir et al., 1998; Majerus et al., 2000). Superovulation regimen affects blastocyst formation and cell number in the human (Hardy et al., 1995).

Abnormal fertilization leading to polyspermy, as frequently occurs during porcine and human in vitro fertilization, has a significant influence on the number of ICM-cells. Polyspermic pig embryos posses a much lower number of ICM-cells than normal embryos, while the growth of TE-Cells is the same in both groups (Han et al., 1999). This was explained by a growth retardation of the polyspermic embryos, which led to a reduced number of ICM-cells but not TE- or total cells

3.2.5. Inner cell allocation and embryo environment.

It has been known for quite some time that the environment, both in vivo and in vitro, can have a major effect upon embryo development. In vivo it is the oviduct or uterine environment, which can be changed by nutrition, disease, pharmaceutical or chemical treatment of the mother. It is known that various chemical insults present before implantation can disrupt embryonic and foetal development (Iannaccone et al., 1987). Diabetic rats for instance, produce lower numbers of blastocysts with lower ICM cell numbers, due to increased cell death in the ICM compared to controls (Pampfer et al., 1990). The increased cell death in the ICM could be reversed by insulin treatment in diabetic rats (De Hertogh et al., 1992). Moreover, the effect of diabetes on ICM-cells in rat blastocysts could be reproduced in vitro by increasing glucose levels in the medium which was inhibitory mainly for ICM development (De Hertogh et al., 1991). Lower ICM-cell numbers probably do not impair implantation, which is merely dependant upon a sufficient number of TE-Cells, but can affect post-implantation development in a negative way (Snow et al., 1976; Tam, 1988). Changes in nutrition can also affect embryo number and quality, at least in cattle (Nagrao et al., 1997): low energy intake prior to and during superovulation increases the number of follicles and leads to improved embryo quality, but this was not reflected by a change in the relative number of ICM-cells (Nolan et al., 1998). Feeding a low protein diet to pregnant Wistar rats resulted in embryos with significantly reduced ICM cell numbers in early blastocysts (Kwong et al., 2000). In later stages also TE cell numbers were affected. It was proposed that the mildly hyperglycemic maternal environment generated by the low protein diet could be responsible for the reduced ICM and TE cell numbers (Kwong et al., 2000).

Quite a large number of embryos are being cultured before they are transferred to a recipient, hence various in vitro culture techniques possibly influence cell allocation. Sometimes it might be wise to sacrifice some embryos to test the influence of media or incubation conditions on cell allocation. ICM-cell numbers in ruminant embryos are definitely influenced by the media which are being used (Van Soom et al., 1996). Porcine embryos cultured in medium with serum displayed higher cell numbers in the ICM, although the ratio of ICM / total cell number was not increased

(Eckert et al., 1997). NCSU23 medium increased ICM and TE numbers in porcine blastocysts compared with KSOM/AA medium but had again no effect upon the ratio (Macháty et al., 1998). In the same study, culturing porcine embryos in NCSU23 medium under low oxygen compared to high oxygen tension did not influence inner cell allocation. Rat blastocysts cultured in the presence of high levels of glucose or of high glucose and tumour necrosis factor-α showed increased nuclear fragmentation, occurring mainly in the inner cell mass (Pampfer et al., 1997). In mouse embryos, it has been shown that coculture conditions induced elevated ICM cell numbers compared with embryos grown in medium alone (Leppens et al., 1996). In the human, supplementation of culture medium with insulin-like growth factor-I (IGF-I) significantly increases the number of inner cell mass cells (Lighten et al., 1998), this increase mainly resulting from IGF-I acting as a survival factor and reducing apoptosis (Spanos et al., 2000).

3.2.6. Inner cell allocation in parthenotes

Oocytes and embryos can be manipulated to induce development without fertilization. Examples of such manipulations are parthenotes, which arise after artificial activation of the oocyte - they are haploid embryos. When the embryos are also exposed to cytochalasin D after activation, they become diploid parthenotes because the second polar body is not extruded (Mognetti and Sakkas, 1996). Parthenotes are characterized by poor development of trophectoderm derivatives and this can be attributed to reduced cell numbers in both TE and ICM in mouse (Hardy and Handyside, 1996; Mognetti and Sakkas, 1996) and cow (De la Fuente and King, 1998; Van de Velde et al., 1999) embryos. However, this could not be confirmed in buffalo embryos (Narula et al., 1996).

3.2.7. Inner cell allocation in cryopreserved embryos

Differential staining or immunosurgery has been used to evaluate specific resistance of ICM-cells to different freezing protocols (Iwasaki et al., 1994; Takagi et al., 1996; Kaidi et al., 1999). This can be valuable since the sensitivity of the cell types to cryoprotectants or freezing rates can be different and since freezing damage e.g in the ICM-cells can be present in morphologically normal cattle embryos. Moreover, differential staining can be used to examine the influence of the basic medium used for freezing or the influence of embryonic stage (Iwasaki et al., 1994). It has also been demonstrated that proliferation of ICM-cells in frozen-thawed in vitro produced cattle embryos tends to be lower than that of unfrozen embryos (Takagi et al., 1996). An adaptation of differential staining (without immunosurgery) has been used by Kaidi et al. (1999) to discriminate between different types of cellular injuries in bovine embryos after various steps of a vitrification protocol.

3.2.8. Inner cell allocation in biopsied embryos

The ratio inner/outer cells provides important data on the impact of bisecting embryos or when taking biopsies for preimplantation genetic diagnosis or sexing (Hardy et al., 1990; Hardy and Handyside, 1993; Tao et al., 1995; Eckert et al., 1997; Rho et al., 1998). Splitting bovine embryos into demi- or even quarter embryos does not seem to harm them a lot. Despite the fact that total cell numbers of the embryos were decreased, the ICM/total cell ratio was significantly increased in these embryos and moreover, after transferring quarter embryos, two out of six recipients were pregnant (Rho et al., 1998). These findings contrast other studies in pigs, in which a significant proportion of demi-embryos lacked an ICM (Tao et al., 1995). This can be explained by the fact that in the bovine study only top quality embryos were selected for bisection and maybe also that pig embryos are less suitable for culture without zona pellucida (Rho et al., 1998), since it has been found that culturing pig embryos without zona reduced the number of cells found in the ICM (Eckert et al., 1997). Bisecting mouse embryos led to an increase in TE cell number in half embryos which was significantly more than half the number of TE-Cells in control unmanipulated embryos. As the surface area of a half embryo is greater than half the intact control, it was thought that the TE-Cells were dividing faster to reach an optimal packing density similar to that in intact embryos (Hardy and Handyside, 1993). Removing one or two cells from 8-cell stage embryos decreased the numbers of cells in the ICM and TE of the resulting blastocysts. However, this decrease was only in proportion to the number of cells removed, and ratio of ICM:TE-Cells was not disturbed (Hardy et al., 1990).

3.2.9. Inner cell allocation in cloned embryos:

Circumventing fertilization also occurs in nuclear transfer embryos or clones. In rabbits differential staining has been used to investigate normality of cloned embryos derived by nuclear transfer of putative embryonic stem cells (Du et al., 1995). Although cloned embryos had the same numbers of ICM cells as controls, no transfers were performed, hence it was not possible to conclude anything about embryo viability.

A limited study performed on bovine nuclear transfer embryos showed a decrease in the proportion of ICM-cells compared with in vivo controls (Du et al., 1996).

3.3. Technical aspects: Methods to determine the number of the inner cell mass and trophectoderm cells

3.3.1. History

Inner and outer cells in embryos can be identified by analysing serially sectioned embryos as has been performed earlier in the mouse (Barlow et al., 1972) and in the horse embryo (Bruyas et al., 1993). Difficulties with ease of evaluation and interpretation of cell position have led to abandoning of these techniques.

As the outer cells of the blastocyst form a seal of zonular tight junction, a quick procedure to differentiate between ICM- and TE-cells was developed using an immunological approach (immunosurgery: Solter and Knowles, 1975; Handyside and Hunter, 1984).

For an elaborate review on technical aspects of differential staining, the reader is referred to the appendix and the review by Van Soom et al., (2001). In this section, we will briefly discuss four different approaches which can be used to differentiate between ICM- and TE-cells.

3.3.2. Techniques for differential staining

3.3.2.1. Zona removal

Although immunosurgery can be performed with success on zona intact mouse embryos (Hogan et al., 1994), it is essential to remove the zona in embryos of most other species since partial removal of the zona in porcine

(Papaioannou and Ebert, 1988), bovine (Van Soom, unpublished) and human embryos (Hardy, unpublished) results in incomplete lysis of outer cells. Chemical (acid Tyrode solution) or enzymatical (pronase) treatments can both be used to remove the zona. An alternative is tearing the zona by means of needles or micromanipulation. It is also possible to use hatched blastocysts. Techniques for zona removal in different species are listed in Van Soom et al., (2001).

3.3.2.2. Differential staining after lysis induced by species-specific antiserum followed by complement

Differential staining was first described by Handyside and Hunter in 1984 for mouse embryos: a heat-inactivated antiserum generated in rabbits against mouse spleen cells, followed by guinea pig complement was used for immunosurgery. Outer cells which had been exposed to antiserum were lysed by the complement treatment. Propidium iodide present in the complement solution entered the lysed TE-Cells and stained the nuclei red or pink. ICM-cells were protected against the complement action, since antibodies will not go through the single cell layer formed by the trophectoderm cells. ICM-cells remain intact and cannot be stained by propidium iodide (PI), but are stained by a second fluorochrome during final alcohol fixation (bisbenzimide or Hoechst 33342).

The advantage of this technique is that it is rapid. The disadvantages are that antiserum must be prepared and tested. Antiserum is mostly prepared in rabbits and differences between batches of antiserum are common. Moreover, the complement is thermolabile and must therefore be thawed on ice and we recommend using it within one hour of thawing. Since the antiserum is species-specific the technique is not applicable in different species.

3.3.2.3. Differential staining after artificial labelling of outer cells with trinitrobenzenesulphonic acid and lysis induced by antiserum against dinitrophenyl groups

To circumvent the production of the antiserum, embryonic cell surface proteins can be artificially labelled by trinitrobenzene sulphonic acid (TNBS) (Hardy et al., 1989). Labelled outer cells can be recognized by a non-species specific antiserum against dinitrophenyl groups, which crossreacts with TNBS (incubation at 37°C). Since both antiserum and

TNBS are commercially available, preparation of antiserum is not necessary and it can be used with any species, as illustrated by the differentially stained bovine and porcine blastocysts in Figure 3. It takes a little more time than the previous method.

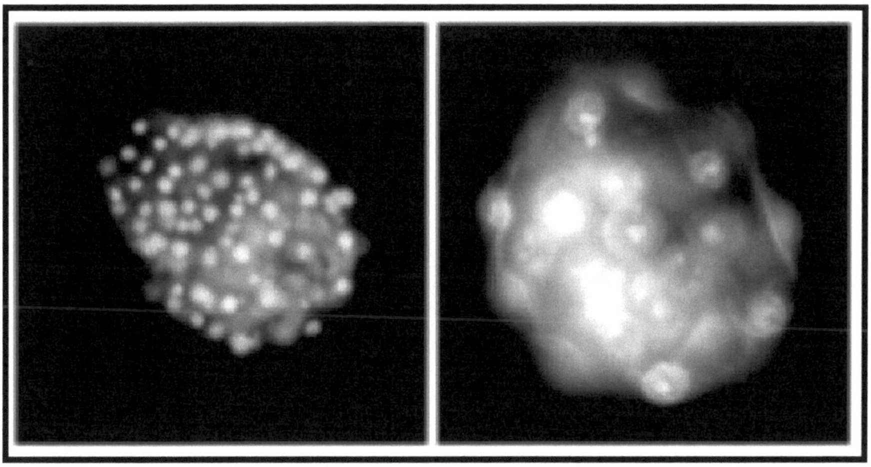

Figure 3a: Differentially stained day 7 bovine blastocyst

Figure 3b: Differentially stained day 6 porcine blastocyst

3.3.2.4. *Differential staining after lysis induced by chemical permeabilisation of trophectoderm*

Immunosurgery is not always necessary to obtain differential staining. Trophectoderm cells can be permeabilized by chemical treatment with Ca^{2+} ionophore A23187. This was first described by de la Fuente and King (1997) Differential staining of TE and ICM of porcine, murine and bovine embryos was successful using this method. Cells show an osmotic response to Ca^{2+} ionophore treatment, leading to cell lysis (Surani et al. 1978). This lysis is visualized by vesiculation of the cell membranes. Temperature can influence the $[Ca^{2+}]_i$-mediated cell injury, leading in some cases to overexposure and subsequent permeabilization of the ICM, hence careful observation of cellular responses is required. The conventional fluorochrome combination Hoechst 33342 or DAPI to stain intact ICM-cells, and PI for lysed TE-cells, cannot be used to evaluate with most laser scanning confocal microscopes. This problem was also solved by de la Fuente and King (1997) who found that SYTO 17, a red vital nucleic acid stain could be used in combination with SYTOX green nuclei acid stain (for permeated cells).

3.3.2.5. Differential staining by artificial labelling of outer cells with FITC-labelled wheat germ agglutinin (WGA)

Another adaptation of differential staining was performed in another study by de la Fuente and King (1998). They labelled the outer cells of zona-free bovine blastocysts with FITC labelled wheat germ agglutinin (WGA). WGA is a lectin which can be used as a cell surface probe to bind to saccharide residues on the plasma membrane of TE-cells. After fixing, permeabilizing and staining with PI, TE-Cells fluoresce with green cytoplasm and red nuclei while ICM-cells only show red nuclei.

3.3.3. Reliability of the method

Differential staining can be a rather tricky technique. Outcome can be biased by embryo quality, temperature, length of incubation, problems with zona removal, inadequate storage of complement, washing out of the dye and antiserum problems (Figure 4a-c). Following antiserum preparation, it is essential to test its (complement dependent) capacity to lyse cells of the species you intend to use it for. For instance, in the case of cow embryos, a suspension of bovine spleen cells is used to immunize a rabbit. To determine the working concentration of the antiserum, different dilutions of the antiserum should be added to, for example, a bovine oviduct monolayer, before addition of complement and evaluation of how many cells come off (Papaioannou and Ebert, 1988) or are stained by trypan blue.

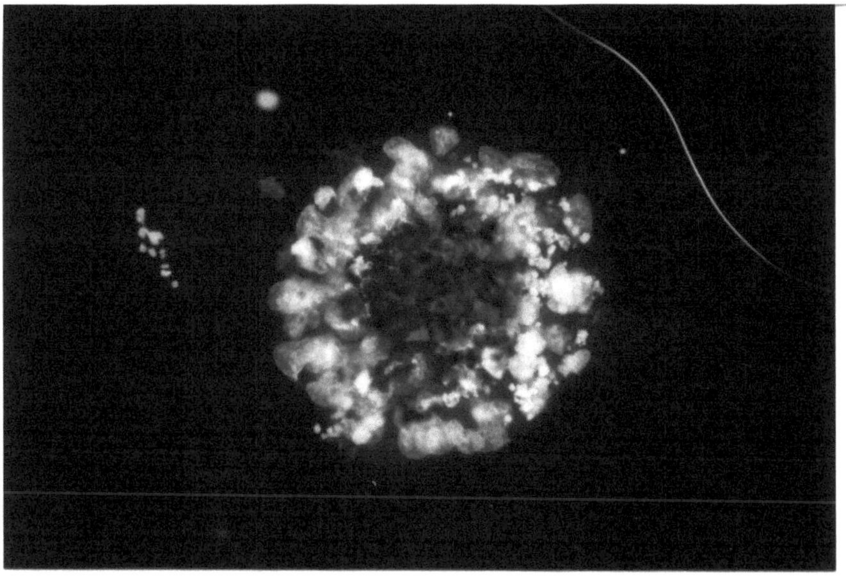

Figure 4 a: Bad staining due to inferior embryo quality

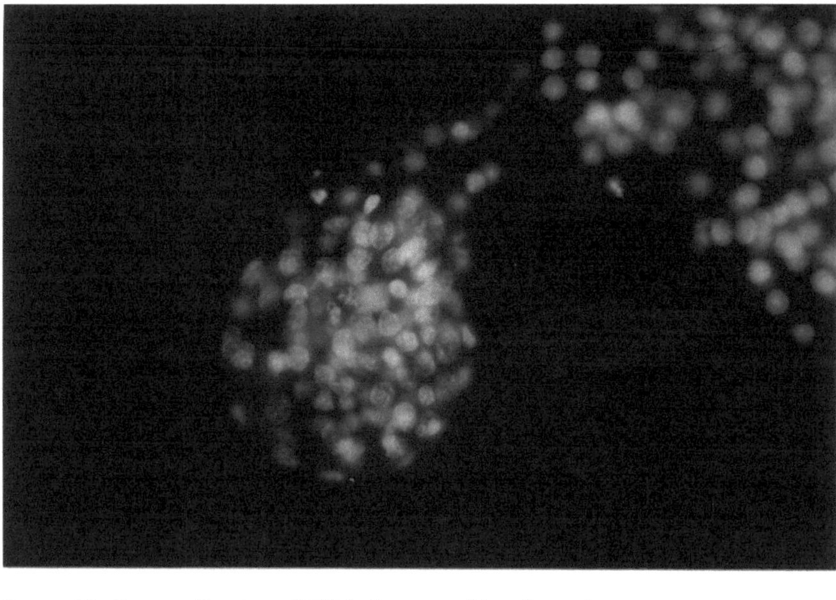

Figure 4 b: Externalisation of ICM after squashing the embryo

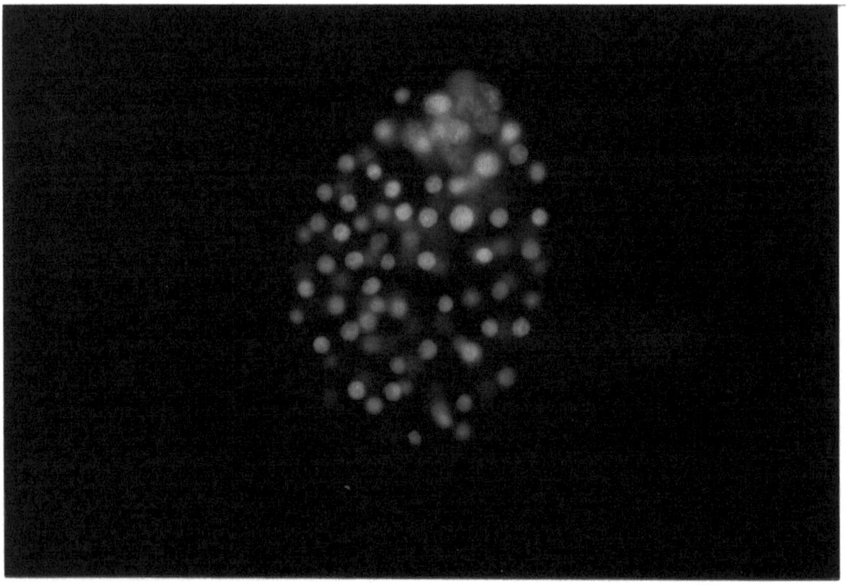

Figure 4 c: Wash-out of PI dye after prolonged storage of preparation

Commercial TNBS and anti DNP-BSA, should also be tested in a preliminary embryo test. To make sure that TNBS followed by rabbit anti DNP-BSA is uniformly labelling the outer blastomeres, embryos can be incubated in TNBS, washed and incubated in different anti DNP-BSA serum dilutions on ice for 30 min followed by fluorescein conjugated goat anti-rabbit serum (1:20 (Nordic)) for 30 min at 4°C. When evaluated by means of confocal scanning laser microscopy, absence of staining of the centre of the embryo can be confirmed. At the dilution which causes more spotted labelling on the outer blastomeres, insufficient antibody binding can be suspected (Van Soom et al., 1996). It is also advisable to include control embryos which are not treated with complement, to determine the number of cells which are lysed by the zona removal procedure or by the antiserum treatment. Normally these control embryos must stain blue, although a small percentage of the cells might stain red (Van Soom et al., 1996). If embryos which have undergone the complete procedure stain blue, this might indicate incomplete zona removal or loss of activity of antiserum or complement.

4. Conclusion

Some people complain that differential staining does not work properly. However, the use of the right controls and careful attention to technical aspects allow a better estimation of embryo quality and embryo differentiation. Because differential staining can be used in many species, it allows a good comparison of ICM development in different mammalian species under different culture conditions and to control for deviations in development due to micromanipulation, environmental influences or genetic make-up. Newer developments of the technique with new fluorochromic stains such as SYTO 17 and SYTOX allow the use of confocal scanning laser microscopy for assessing differentially labelled embryos (de la Fuente and King, 1997). Overall, reliable differential cell counts can be expected in approximately 85% of the embryos assessed. Also with the techniques which make use of immunosurgery, it is a common finding that not all stained embryos can be evaluated properly.

Probably more variations of the differential staining technique as described by de la Fuente and King (1998) are possible. The advantage is that this technique probably can be used in combination with other staining techniques, because no immunosurgery is involved. Definitely, in combination with confocal scanning laser microscopy is this last technique the most powerful. The listing of possible applications makes clear that differential staining has been applied during the last twenty years in a number of species to test a wide range of conditions. Despite its shortcomings, it is a valuable technique to study embryo differentiation and quality.

References

Baltz JM. 1993. Intracellular pH regulation in the early embryo. BioEssays 15:523-530.

Barcroft LC, Hay-Schmidt A, Caveney A, Gilfoyle E, Overstrom EW, Hyttel P, Watson AJ. 1998. Trophectoderm differentiation in the bovine embryo: characterization of a polarized epithelium. J Reprod Fertil 114:327-339.

Barlow P, Owen DAJ, Graham C. 1972. DNA-synthesis in the preimplantation mouse embryo. J Embryol Exp Morphol 27:431-445.

Becker DL, Leclerc-David C, Warner A. 1992. The relationship of gap junctions and compaction in the preimplantation mouse embryo. Development Suppl. 1992:113-118.

Biggers JD, Bell JE, Benos DJ. 1988. Mammalian blastocyst: transport functions in a developing epithelium. Am J Physiol 255:C419-C432.

Biggers JD. 1987. Pioneering mammalian embryo culture. In: The mammalian preimplantation embryo:regulation of growth and differentiation in vitro. Eds BD Bavister, Plenum Press, NY

Boerjan ML, Den Daas JHG, Dieleman SJ. 2000. Embryonic origins of health: long term effects of IVF on human and livestock. Theriogenology 53:537-547.

Boerjan ML, te Kronnie G. 1993. The segregation of inner and outer cells in porcine embryos follows a different pattern compared to the segregation in mouse embryos. Roux's Arch Dev Biol 203:113-116.

Bruyas J-F, Bézard J, Lagneaux D, Palmer E. 1993. Quantitative analysis of morphological modifications of day 6.5 horse embryos after cryopreservation: differential effects on inner cell mass cells and trophoblast cells. J Reprod Fertil 99:15-23.

Byrne AT, Southgate J, Brison DR, Leese HJ. 1999. Analysis of apoptosis in the preimplantation bovine embryo using TUNEL. J Reprod Fertil 117:97-105.

De Hertogh R, Vanderheyden I, Pampfer S, Robin D, Delcourt J. 1992. Maternal insulin treatment improves pre-implantation embryo development in diabetic rats. Diabetologia 35:406-408.

De Hertogh R, Vanderheyden I, Pampfer S, Robin D, Dufrasne E, Delcourt J. 1991. Stimulatory and inhibitory effect of glucose and insulin on rat blastocyst development in vitro. Diabetes 40:641-647.

De Hertogh R, Vanderheyden I, Pampfer S, Robin D, Dufrasne E, Delcourt J. 1991. Stimulatory and inhibitory effects of glucose and insulin on rat blastocyst development in vitro. Diabetes 40:641-647.

De la Fuente R, King WA. 1997. Use of a chemically defined system for the direct comparison of inner cell mass and trophectoderm distribution in murine, porcine and bovine embryos. Zygote 5:309-320.

De la Fuente R, King WA. 1998. Developmental consequences of karyokinesis without cytokinesis during the first mitotic cell cycle of bovine parthenotes. Biol Reprod 58:952-962.

De Loof A. 1992. All animals develop from a blastula: consequences of an undervalued definition for thinking on development. BioEssays 14:573-575.

De Paz P, Sanchez AJ, Fernandez JG, Garcia C, Chamorro CA, Anel L. 1994. Ultrastructural localization of lectin receptors in the preimplantation ovine embryo. Anat Rec 240:537-544.

Devreker F, Hardy K. 1997. Effects of glutamine and taurine on preimplantation development and cleavage of mouse embryos in vitro. Biol Reprod 57:921-928

Du F, Giles JR, Foote RH, Graves KH, Yang X, Moreadith RW. 1995. Nuclear transfer of putative rabbit embryonic stem cells leads to normal blastocyst development. J Reprod Fertil 104:219-223.

Du F, Looney CR, Yang X. 1996. Evaluation of bovine embryos produced in vitro vs in vivo by differential staining of inner cell mass and trophectoderm cells. Theriogenology 45:211.

Ducibella T, Anderson E. 1975. Cell shape and membrane changes in the eight cell embryo; prerequisites for morphogenesis of the blastocyst. Dev Biol 47:45-58.

Eckert J, Tao T, Niemann H. 1997. Ratio of inner cell mass and trophoblastic cells in blastocysts derived from porcine 4- and 8-cell embryos and isolated blastomeres cultured in vitro in the presence or absence of protein and human leukemia inhibitory factor. Biol Reprod 57:552-560.

Fehilly CB, Willadsen SM. 1986. Embryo manipulation in farm animals. In: Oxford reviews of Reproductive Biology, vol 8. Ed. J.R. Clarke, Clarendon Press, Oxford, pp 379-413.

Fleming TP, Javed Q, Hay M. 1992. Epithelial differentiation and intercellular junction formation in the mouse early embryo. Development Suppl. 1992:105-112

Gardner RL. 1997. The early blastocyst is bilaterally symmetrical and its axis of symmetry is aligned with the animal-vegetal axis of the zygote in the mouse. Development 12:289-301.

Giles JR, Foote RH. 1995. Rabbit blastocyst: allocation of cells to the inner cell mass and trophectoderm. Mol Reprod Dev 41:204-211.

Grisart B, Massip A, Dessy F. 1994. Cinematographic analysis of bovine embryo development in serum-free oviduct-conditioned medium. J Reprod Fertil 101:257-264.

Han YM, Abeydeera LR, Kim JH, Moon HB, Cabot RA, Day BN, Prather RS. 1999. Growth retardation of inner cell mass cells in polyspermic porcine embryos produced in vitro. Biol Reprod 60:1110-1113.

Handyside AH, Hooper ML, Kaufman MH, Wilmut I. 1987. Towards the isolation of embryonal stem cell lines from the sheep. Roux's Archiv Dev Biol 196:185-190.

Handyside AH, Hunter S. 1984. A rapid procedure for visualising the inner cell mass and trophectoderm nuclei of mouse blastocysts in situ using polynucleotide-specific fluorochromes. J Exp Zool 231:429-434.

Hardy K, Handyside AH, Winston RH. 1989. The human blastocyst: cell number, death and allocation during late preimplantation development in vitro. Development 107:597-604.

Hardy K, Handyside AH. 1993. Cell allocation in twin half mouse embryos bisected at the 8-cell stage: implications for preimplantation diagnosis. Mol Reprod Dev 36:16-22.

Hardy K, Handyside AH. 1996. Metabolism and cell allocation during parthenogenetic preimplantation mouse development. Mol Reprod Dev 43:313-322.

Hardy K, Martin KL, Leese HJ, Winston RM, Handyside AH. 1990. Human preimplantation development in vitro is not adversely affected by biopsy at the 8-cell stage. Hum Reprod 5:708-714.

Hardy K, Robinson FM, Paraschos T, Wicks R, Franks S, Winston RM. 1995. Normal development and metabolic activity of preimplantation embryos in vitro from patients with polycystic ovaries. Hum Reprod 19910:2125-2135.

Hardy K, Spanos S, Becker D, Iannelli P, Winston RM, Stark J. 2001. From cell death to embryo arrest: mathematical models of human preimplantation embryo development. Proc Natl Acad Sci U S A. 298:1655-1660.

Hardy K, Warner A, Winston RM, Becker DL. 1996. Expression of intercellular junctions during preimplantation development of the human embryo. Mol Hum Reprod 2:621-632.

Hasler JF. 1998. The current status of oocyte recovery, in vitro embryo production , and embryo transfer in domestic animals, with an emphasis on the bovine. J Anim Sci 76 Suppl 3:52-74.

Hogan B, Beddington R, Costantini F, Lacy E. 1994. In: Manipulating the mouse embryo: A laboratory Manual (2nd Edition) Cold Spring Harbor Laboratory Press, p 195.

Hunter RHF, Hunt DM, Chang MC. 1969. Temporal and cytological aspects of fertilization and early development in the guinea pig Cavia Procellus. Anat Rec 155:411-430.

Iannaccone PM, Bossert NL, Connelly CS. 1987. Disruption of embryonic and fetal development due to preimplantation chemical insults: a critical review. Am J Obstet Gynecol 157:476-484.

Iwasaki S, Yoshibe N, Ushijima H, Watanabe S, Nakahara T. 1990. Morphology and proportion of inner cell mass of bovine blastocysts fertilized in vitro and in vivo. J Reprod Fertil 90:279-284.

Iwasaki S, Yoshikane Y , Watanabe S, Nakahara T 1994. Effects of freezing of bovine preimplantation embryos derived from oocytes fertilized in vitro on survival of their inner cell mass cells. Mol Reprod Dev 37:272-275.

Johnson MH, Maro B. 1986. Time and space in the mouse early embryo: a cell biological approach into cell diversification. In: Experimental approaches to mammalian development. Eds. Rossant J and Pedersen E, Cambridge University Press, NY: pp 35-66.

Johnson MH, Selwood L. 1996. Nomenclature of early development in mammals in marsupial gametes and embryos. Reprod Fertil Dev 8:759-764.

Johnson MH. 1996. Origins of pluriblast and trophoblast in the eutherian conceptus. Reprod Fertil Dev 8:699-709.

Kaidi S, Van Langendonckt A, Massip A, Dessy F, Donnay I. 1999. Cellular alteration after dilution of cryoprotective solutions used for the vitrification of in vitro-produced bovine embryos. Theriogenology 52:515-525.

Khatir H, Lonergan P, Touzé JL, Mermillod P. 1998. The characterization of bovine embryos obtained from prepuberal calf oocytes and their viability after non surgical embryo transfer. Theriogenology 50:1201-1210.

Koyama H, Suzuki H, Yang X, Jiang S, Foote RH. 1994. Analysis of polarity of bovine and rabbit embryos by scanning electron microscopy. Biol Reprod 50:163-170.

Kruip ThAM, Den Daas HG. 1997. In vitro produced and cloned embryos: Effects on pregnancy, parturition and offspring. Theriogenology 42:675-684.

Kwong WY, Wild AE, Roberts P, Willis AC, Fleming TP. 2000. Maternal undernutrition during the preimplantation period of rat development causes blastocyst abnormalities and programming of postnatal hypertension. Development 127:4195-4202.

Leppens G, Gardner DK, Sakkas D. 1996. Co-culture of 1-cell outbred mouse embryos on bovine kidney epithelial cels: effect on development, glycolytic activity, inner cell mass:trophectoderm ratios and viability. Hum Reprod 11:598-603.

Lighten AD, Moore GE, Winston RM, Hardy K. 1998. Routine addition of human insulin-like growth factor-I ligand could benefit clinical in vitro fertilization culture. Hum Reprod 13:3144-3150.

Macháty Z, Day BN, Prather RS. 1998. Development of early porcine embryos in vitro and in vivo. Biol Reprod 59:451-455.

Majerus V, Lequarre AS, Ferguson EM, Kaidi S, Massip A, Dessy F, Donnay I. 2000. Characterization of embryos derived from calf oocytes: kinetics of cleavage, cell allocation to inner cell mass, and trophectoderm and lipid metabolism. Mol Reprod Dev 57:346-352.

Meijer HA, Van De Pavert SA, Stroband HW, Boerjan ML. 2000. Expression of the organizer specific homeobox gene goosecoid (gsc) in porcine embryos. Mol Reprod Dev 55:1-7.

Mognetti B, Sakkas D. 1996. Defects in the allocation to the inner cell mass and trophectoderm of parthenogenetic mouse blastocysts. Reprod Fertil Dev 8:1193-1197.

Nagrao S, Nibart M, Humblot P. 1997. Negative effects of overfeeding on superovulatory response and embryo production in dairy heifers. Proc 13th Sci Meet AETE Lyon, abstr.

Narula A, Taneja M, Totey S. 1996. Morphological development, cell number and allocation of cells to trophectoderm and inner cell mass of in vitro fertilized and parthenogenetically developed buffalo embryos: the effect of IGF-I. Mol Reprod Dev 44:343-351.

Nikas G, Ao A; Winston RML, Handyside AH. 1996. Compaction and surface polarity in the human embryo in vitro. Biol Reprod 55:32-37.

Nolan R, O'Callaghan D, Duby RT, Lonergan P, Boland M. 1998. The influence of short-term nutrient changes on follicle growth and embryo production following superovulation in beef heifers. Theriogenology 50:1263- 1274.

Pampfer S, Vanderheyden I, McCracken JE, Vesela J, De Hertogh R. 1997. Increased cell death in rat blastocysts exposed to maternal diabetes in utero and to high glucose or tumor necrosis factor-alpha in vitro. Development 124:4827-4836.

Pampfer S, Vanderheyden I, Michiels B, De Hertogh R. 1990. Cell allocation to the inner cell mass and the trophectoderm in rat embryos during in vivo preimplantation development. Roux's Arch Dev Biol 198:257-263.

Papaioannou VE, Ebert KM. 1988. The preimplantation embryo: cell number and allocation to trophectoderm and inner cell mass of the blastocyst in vivo and in vitro. Development 102:793-803.

Pedersen RA. 1986. Potency, lineage, and allocation in preimplantation mouse embryos. In: Experimental approaches to mammalian embryonic development. Eds. J Rossant and RA Pedersen. Cambridge University Press, NY, pp. 3-34.

Prather RS, First NL. 1993. Cell-to-cell coupling in early-stage bovine embryos: a preliminary report. Theriogenology 39:561-567.

Rath D, Niemann H, Tao T, Boerjan ML. 1995. Ratio and number of inner cell mass and trophoblast cells of in vitro and in vivo produced porcine embryos. Theriogenology 43:304.

Reeve WJD, Ziomek CA. 1981. Distribution of mictovilli on dissociated blastomeres from mouse embryos: evidence for surface polarization at compaction. J Embryol Exp Morphol 62:339-350.

Reeve WJD. 1981. Cytoplasmic polarity develops at compaction in rat and mouse embryos. J Embryol Exp Morphol 62:351-367.

Reima I, Lehtonen E, Virtanen I, Fléchon J-E. 1993. The cytoskeleton and associated proteins during cleavage, compaction and blastocyst differentiation in the pig. Differentiation 54:35-45.

Rho GJ, Johnson WH, Betteridge KJ. 1998. Cellular composition and viability of demi- and quarter embryos made from bisected bovine morulae and blastocysts produced in vitro. Theriogenology 50:885-895.

Rivera RM, Youngs CR, Ford SP. 1996. A comparison of the number of inner cell mass and trophectoderm cells of preimplantation Meishan and Yorkshire pig embryos at similar developmental stages. J Reprod Fertil 106:111-116.

Rossant J. 1986. Development of extraembryonic cell lineages in het mouse embryo. In: Experimental approaches to mammalian embryonic development. Ed. J Rossant and RA Pedersen. Cambridge University Press, NY, pp. 97-120.

Roudebush WE. 1998. Influence of the preimplantation-embryo-development (Ped) gene on mouse blastocyst differentiation. Zoological Science 15:359-362.

Schlafke S, Enders AC. 1967. Cytological changes during cleavage and blastocyst formation in the rat. J Anat 102:13-32.

Schöler H, Ruppert S, Suzuki N, Chowdhury K, Gruss P. 1990. New type of POU domain in germ line-specific protein Oct-4. Nature 344:435-439.

Smith R, McLaren A. 1977. Factors affecting the time of formation of the mouse blastocoel. J Embryol Exp Morphol 41:79-92.

Snow MHL, Aitken J, Ansell JD. 1976. Role of inner cell mass in controlling implantation in the mouse. J Reprod Fertil 48:403-404.

Solter D, Knowles BB. 1975. Immunosurgery of mouse blastocyst. Proc. Natl Acad Sci USA 72:5099-5102.

Spanos S, Becker DL, Winston RM, Hardy K. 2000. Anti-apoptotic action of insulin-like growth factor-I during human preimplantation embryo development. Biol Reprod 63:1413-1420.

Surani MAH, Torchianan D, Barton SC. 1978. Isolation and development of the inner cell mass after exposure of mouse embryos to calcium ionophore A23187. J Embryol Exp Morphol 45:237-247.

Suzuki H, Azuma T, Koyama H, Yang XZ. 1999. Development of cellular polarity of hamster embryos during compaction. Biol Reprod 61:521-526.

Takagi M, Sakonju I, Suzuki T. 1996. Effect of cryopreservation on DNA-synthesis in the inner cell mass of in vitro matured/in vitro fertilized bovine embryos frozen in various cryoprotectants. J Vet Med Sci 58:1237-1238.

Takeichi M. 1988. The cadherins: cell-cell adhesion molecules controlling animal morphogenesis. Development 102:639-655.

Tam PPL. 1988. Postimplantation development of mytomycin-C treated mouse blastocysts. Teratology 37:205-212.

Tao T, Reichelt B, Niemann H. 1995 Ratio of inner cell mass and trophoblastic cells in demi- and intact pig embryos. J Reprod Fertil 104:251-258.

Thomas PO, Brown A, Beddington RSP. 1998. Hex: a homeobox gene revealing peri-implantation asymmetry in the mouse embryo and an early transient marker of endothelial cell precursors. Development 125:85-94.

Van De Velde A, Liu L, Bols PE, Ysebaert MT, Yang X. 1999. Cell allocation and chromosomal complement of parthenogenetic and IVF bovine embryos. Mol Reprod Dev 54:57-62.

Van Soom A, Boerjan ML, Bols PEJ, Vanroose G, Lein A, Coryn M, de Kruif A., 1997a. Timing of compaction and inner cell allocation in bovine embryos produced in vivo after superovulation. Biol Reprod 57:1041-1049.

Van Soom A, Boerjan ML, Ysebaert MT, de Kruif A. 1996. Cell allocation to the inner cell mass and the trophectoderm in bovine embryos cultured in two different media. Mol Reprod Dev 45:171-182.

Van Soom A, Van Vlaenderen I, Mahmoudzadeh AR, de Kruif A. 1992. Compaction rate of in vitro fertilized bovine embryos related to the interval from insemination to first cleavage. Theriogenology 38:905-919.

Van Soom A, Vanroose G, de Kruif A. 2001. Blastocyst evaluation by means of differential staining: a practical approach. Reprod Dom Anim 36:29-35.

Van Soom A, Ysebaert MT, de Kruif A. 1997b. Relationship between timing of development, morula morphology and cell allocation to inner cell mass and trophectoderm in in vitro produced bovine embryos. Mol Reprod Dev 47:47-56.

Vanroose G, Nauwynck H, Van Soom A, Vanopdenbosch E, de Kruif A. 1998. Replication of cytopathic and noncytopathic bovine viral diarrhea virus in zona-free and zona-intact in vitro produced bovine embryos and the effect on embryo quality. Biol Reprod 58:857-866.

Vanroose G, Nauwynck H, Van Soom A, Vanopdenbosch E, de Kruif A. 1997. Susceptibility of zona-intact and zona-free in vitro produced bovine embryos at different stages of development to an infection with bovine herpesvirus-I. Theriogenology 47:1389-1402.

Vanroose G, Nauwynck H, Van Soom A, Ysebaert MT, Charlier G, Van Oostveldt P, de Kruif A. 2000. Structural aspects of the zona pellucida of in vitro produced bovine embryos: a scanning electron and confocal laser scanning microscopic study. Biol Reprod 62:463-469.

Watson AJ. 1992. The cell biology of blastocyst development. Mol Reprod Dev 33:492-504.

Wiley LM, Adamson ED, Tsark EC. 1995. Epidermal growth factor receptor function in early mammalian development. BioEssays 17:839-846.

Wrenzycki C, Herrmann D, Carnwath JW, Niemann H. 1996. Expression of the gap junction gene connexin43 (Cx43) in preimplantation bovine embryos derived in vitro or in vivo. J Reprod Fertil 108:17-24.

Ziomek CA, Chatot CL, Manes C. 1990. Polarization of blastomeres in the cleaving rabbit embryo. J Exp Zool 256:84-91.

Chapter 12

APOPTOSIS IN MAMMALIAN EMBRYOS

Kate Hardy and Sophie Spanos
Department of Reproductive Science and Medicine,
Institute of Reproductive and Developmental Biology, Faculty of Medicine, Imperial College, Hammersmith Hospital, Du Cane Road, London W12 0NN, UK.

1. Introduction

The preimplantation development of eutherian mammals is remarkably similar, spanning cleavage of the fertilized ovum; compaction and morula formation; and finally cavitation, with the formation of a blastocyst. Blastocyst formation marks the differentiation of the epithelial troph-ectoderm (TE), which is specialized for implantation, and the establishment of the inner cell mass (ICM), from which the foetus is derived.

In addition to cell division and differentiation within the preimplantation embryo, it is becoming apparent that cell death by apoptosis is common, even in vivo (Handyside and Hunter, 1986), with the majority of mouse blastocysts having one or more dying cells. Within a blastocyst, the percentage of dying cells is sufficiently high (10-20%) to suggest that apoptosis is playing a developmental role. The onset of apoptosis in the mammalian blastocyst coincides with activation of the embryonic genome (Jurisicova et al., 1998a; Byrne et al., 1999; Hardy et al., 2001).

Apoptotic nuclei have been seen in blastocysts from many mammalian species, including mouse (El-Shershaby and Hinchliffe, 1974; Copp, 1978; Handyside and Hunter, 1986), rat (Pampfer et al., 1990), cow (Mohr and Trounson, 1982; Byrne et al., 1999; Matwee et al., 2000), baboon (Enders et al., 1990), rhesus monkey (Hurst et al., 1978; Enders and Schlafke, 1981;

Enders et al., 1982) and human (Mohr and Trounson, 1982; Lopata et al., 1982; Hardy et al., 1989; Jurisicova et al., 1996; Spanos et al., 2000; Hardy et al., 2001). These dying cells are characterized by blebbing of the nuclear membrane, chromatin condensation, cytoplasmic vacuoles and the formation of both nuclear and cytoplasmic fragments.

Apoptosis is an essential feature of many normal processes and pathological conditions and serves a variety of purposes, including tissue homeostasis and remodelling, and the removal of unwanted cells (Wyllie et al., 1980). It may result from activation of an endogenous developmental programme, or be induced by a stimulus, such as ligand-receptor interactions, growth factor deprivation, disruption of cell-cell or cell-matrix interactions, or DNA damage. Examples of developmental events that require extensive apoptosis include the sculpting of digits in the developing limb bud, and the removal of the tail of the tadpole during amphibian metamorphosis. During early rodent development, apoptosis plays a crucial role in amniotic cavity formation, soon after implantation (Coucouvanis and Martin, 1995). There is now increasing evidence that apoptosis is important before implantation, as soon as the totipotent cells start to differentiate (for review, see (Hardy, 1997).

The role of apoptosis at preimplantation stages remains unknown. The presence of apoptotic cells in vivo suggests a developmental role, which has not yet been elucidated. It is possible that the higher levels of apoptosis seen in the ICM of the mouse blastocyst are to remove ICM-cells which retain the potential to form TE (Pierce et al., 1989), which could lead to ectopic expression of TE-Cells in the germ layers during later development. In addition, levels of apoptosis can be modulated by environmental factors. Apoptosis in preimplantation embryos is increased in suboptimal culture media (Devreker and Hardy, 1997), in the presence of high concentrations of glucose (Pampfer et al., 1997; Moley et al., 1998a), and in the absence of specific growth factors (Brison and Schultz, 1997; O'Neill, 1998; Herrler et al., 1998; Spanos et al., 2000). Furthermore, apoptosis has been associated with increased levels of reactive oxygen species thought to result from high oxygen tension during culture (Yang et al., 1998), exposure to high concentrations of sperm during human IVF and increased maternal age (Jurisicova et al., 1998b). Finally, nuclear (Hardy et al., 1993) and chromosomal (Jamieson et al., 1994; Munné et al., 1995) abnormalities are common in human embryos and it has been suggested that one role of apoptosis may be to remove such defective cells (Hardy, 1997).

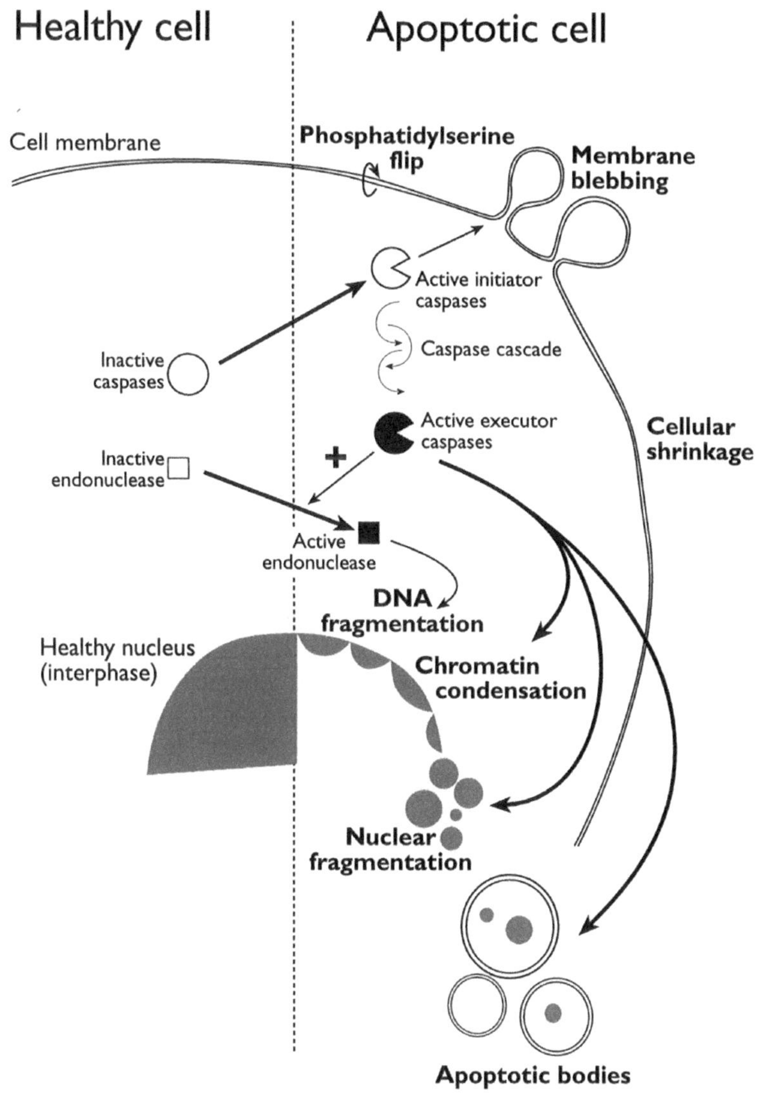

Figure 1: Schematic diagram of cellular changes during apoptosis

There are two forms of cell death, necrosis and apoptosis, which are morphologically distinct.

2. Definitions of cell death

2.1. Necrosis

Necrosis results from injury and affects large groups of cells, causing cellular swelling and membrane rupture (Wyllie et al., 1980) which invoke an inflammatory response in adjacent healthy tissue. Necrosis is induced by physical, chemical or biological events such as ischaemia or a physical trauma.

2.2. Apoptosis

Apoptosis is the mechanism by which single cells in isolation are disposed of, with no associated inflammation. Apoptosis is characterized by a series of consecutive, morphologically distinct phases (Figure 1). Initially, chromatin condenses into large compact granular clumps on the inner nuclear membrane; the cytoplasm condenses and the nuclear and cytoplasmic membranes become grossly indented. The nucleus fragments, and the whole cell blebs and fragments into membrane bound apoptotic bodies, which may contain nuclear fragments. The production of membrane bound apoptotic bodies prevents the release of intracellular contents from the dying cell which would result in damage to adjacent healthy cells and inflammation. These bodies are either dispersed in the intercellular tissue spaces and extruded from the tissue or phagocytosed by neighbouring tissue cells. Apoptotic cells that are not phagocytosed, for example those shed from an epithelium into a duct lumen, undergo secondary necrosis, characterized by an increase in membrane permeability and distension of cytosolic membrane structures. This classical sequence of events has been seen at the ultrastructural level in a wide variety of different cell types (Wyllie et al., 1980; Wyllie, 1997; Harmon et al., 1998) and is under physiological control.

In addition to the classic morphological features described here, which distinguish apoptotic cells from those undergoing necrosis, biochemical changes in cell surface morphology and DNA structure have been associated with apoptosis (Figure 1). Apoptosis is heralded by cell rounding, indicating that changes in cell adhesion occur (Wyllie et al., 1980; Collins et al., 1997). Activation of initiator caspases (proteases) such as caspase 8 leads to cleavage of cytoskeletal proteins, which may result in blebbing of the cell surface (Huppertz et al., 1999). Further early cell surface changes include redistribution of membrane phospholipids such as phosphatidylserine within the lipid bilayer of the cell membrane (Martin et al., 1995). Exposure of phosphatidylserine on the cell surface during apoptosis may act as the signal, promoting phagocytosis by neighbouring healthy cells. Initiator caspases cleave and activate a second subpopulation of caspases known as the execution caspases. These are regulated by the bcl-2 family of mitochondrial proteins. Various members of this family promote (bad, bax, bak, bcl-x_S, bik, hrk) and inhibit (A1, bcl-2, bcl-w, bcl-x_L, bfl-1, mcl-1 etc) cleavage and activation of execution caspases. These caspases in turn cleave a range of proteins critical for cell survival, including those involved in the cytoskeleton, DNA repair, nuclear envelope integrity and cell cycle control (Wyllie, 1997). Execution caspases also activate endonucleases, resulting in the cleavage of DNA between nucleosomes into oligonucleosomal fragments (Wyllie et al., 1980; Arends and Wyllie, 1991). Agarose gel electrophoresis of such DNA shows a characteristic 'ladder' pattern, with bands at 180 - 200 bp intervals. In contrast, DNA from necrotic cells is cleaved randomly and forms a 'smear' on agarose gels. These final apoptotic events result in the characteristic alterations in nuclear and cellular shape (Figure 1).

3. Characteristic features of apoptosis in mammalian embryos: incidence and detection

Investigations of apoptosis in preimplantation embryos are constrained by limited material, especially in humans, and to a lesser extent also for laboratory and domestic animal embryos. Even following superovulation, relatively few embryos are available, and each embryo contains between one and a hundred or so cells. Furthermore, it is difficult to estimate the full extent of cell death in the mammalian embryo as it is not known how long dead cells persist for, therefore current techniques only provide a 'snapshot' of the embryo at a specific time.

3.1. Membrane Changes

Early apoptotic events involve changes in the cell membrane, which are controlled by initiator caspases. These include modifications of the plasma membrane itself, and alterations in the cytoskeleton leading to aberrant membrane behaviour such as blebbing (reviewed in (Huppertz et al., 1999).

3.1.1. Redistribution Of Phosphatidylserine

An early event of apoptosis is a redistribution of the phospholipid phosphatidylserine in the cell membrane. In healthy cells, phosphatidylserine is normally confined to the inner cytoplasmic leaflet of the membrane. An early feature of apoptosis is the loss of this asymmetry, with exposure of phosphatidylserine on the outer leaflet (Martin et al., 1995). It has been suggested that externalisation of phosphatidylserine acts as a signal promoting phagocytosis of the dying cell by macrophages or other cells.

Annexin V (an anticoagulant protein) has a specific and high affinity for phosphatidylserine. Healthy cells incubated in fluorescein isothiocyanate (FITC)-conjugated annexin V remain unlabelled, because annexin V cannot access the phosphatidylserine in the inner leaflet of the plasma membrane. In contrast, apoptotic cells show marked annexin V binding, because annexin V can access the phosphatidylserine which has been translocated to the outer leaflet (Martin et al., 1995). Annexin V is a marker for the early stages of apoptosis, as externalisation of phosphatidylserine precedes other morphological changes such as nuclear condensation and cell shrinkage.

Annexin V labelling has been examined in early human embryos and oocytes (Van Blerkom and Davis, 1998; Levy et al., 1998; Antczak and Van Blerkom, 1999), with conflicting results. Levy et al. (1998) found extensive labelling in arrested and fragmented human embryos. However, the nuclei in these embryos also labelled with propidium iodide, indicating that the plasma membrane was permeable. Therefore it was probable that these arrested embryos were in fact undergoing secondary necrosis. In contrast, Antczak and Van Blerkom (1999) found no annexin V or propidium iodide labelling in cleavage stage human embryos with varying degrees of fragmentation, strongly suggesting that there is no correlation between fragmentation and apoptosis (Table 1).

3.1.2. Membrane Integrity

Membrane integrity is maintained until the late stages of apoptosis (Wyllie et al., 1980; Martin et al., 1995; Collins et al., 1997). It is important to use a marker of membrane permeability in addition to annexin V labelling to differentiate between cells undergoing apoptosis and those undergoing necrosis. One such marker is the polynucleotide-specific fluorochrome propidium iodide, which can only enter cells with 'leaky' membranes, labelling the DNA in the nucleus. While cells in the early stages of apoptosis label with annexin V but not with propidium iodide (confirming that these cells still have intact membranes), necrotic cells with permeable membranes allow both propidium iodide and annexin V to enter the cell (Martin et al., 1995) and label the nucleus and the phosphatidylserine in the inner leaflet, respectively.

3.1.3. Membrane Blebbing

Video time-lapse cinematography is a useful tool for observing the kinetics of apoptosis. Using this technique, violent membrane blebbing in trophectoderm and inner cell mass cells over a time period of up to an hour, followed by rupture of the cell and its contents into the blastocoel cavity have been observed (S. Spanos and K. Hardy, unpublished). Similar cytoplasmic blebbing has been seen in other cell types undergoing apoptosis using time-lapse cinematography (Collins et al., 1997; Wyllie, 1997).

3.2. Cytoplasmic Vacuoles (Phagolysosomes)

Translucent vacuoles are frequently associated with cytoplasmic condensation typical of cells undergoing apoptosis (Wyllie et al., 1980) (Table 1). Vacuoles are a common feature of human preimplantation embryos and have been seen by light (Hardy et al., 1996), electron (El-Shershaby and Hinchliffe, 1974; Sathananthan et al., 1982; Enders et al., 1990) and laser scanning confocal (Hardy et al., 1996) microscopy.

3.3. Changes In Expression Of Genes Involved In Regulation And Execution Of Apoptosis

The Bcl-2 family of proteins promotes or inhibits the activation of execution caspases by initiator caspases. Activation of execution caspases results in the classic morphological changes associated with apoptosis. Targets of caspases include cytoplasmic and nuclear proteins such as poly ADP ribose polymerase (PARP) and lamins. The precise mechanisms by which this activation is controlled are under debate, and probably differ depending on the cell type and the nature of the apoptotic trigger. Certainly the detail is beyond the scope of this review, and readers are directed towards a number of recent excellent reviews (Thornberry and Lazebnik, 1998; Datta et al., 1999; Earnshaw et al., 1999; Tsujimoto and Shimizu, 2000).

Currently, expression of mRNA and protein for members of the caspase and Bcl-2 families of genes in preimplantation embryos have been studied using reverse transcription polymerase chain reaction (RT-PCR) and immuno-histochemistry.

3.3.1. RT-PCR

Expression of mRNA for genes associated with the apoptosis cascade (e.g. Bcl-2 family, caspase family), can be examined in single embryos using RT-PCR. This provides information about the timing of gene expression. In addition, it is possible to use semi- or even fully-quantitative RT-PCR which allows accurate detection of alterations in gene expression during preimplantation development.

mRNA from lysed single embryos can be reverse transcribed to complementary DNA (cDNA). The synthesized cDNA is used as a template for PCR amplification. Specific primers for the mRNA of interest are designed to include an intron in the amplicon to allow discrimination of products of cDNA from genomic DNA. Following PCR, the products can be loaded on agarose gels for size determination. The identities of RT-PCR products can be verified by restriction digest analysis and DNA sequencing (Jurisicova et al., 1998a; Exley et al., 1999).

3.3.2. Immunohistochemistry

Expression of proteins, such as members of the Bcl-2 family, can be examined using immunohistochemistry. Embryos can be fixed in paraformaldehyde, embedded in agarose for ease of handling during dehydration, and blocked in paraffin wax. Serial sections can then be processed for immunohistochemistry for a number of proteins (Hardy, 1999). Alternatively, embryos can be fixed, permeabilized as described in Hardy et al., 1996 processed for immunohistochemistry and mounted in whole mount in an antifade mountant, before examination by laser scanning confocal microscopy. Appropriate combinations of rabbit and mouse primary antibodies with FITC-labelled anti-rabbit or CY5-labelled anti-mouse secondary antibodies, together with a polynucleotide-specific fluorochrome will allow concurrent visualisation of nuclear morphology and the detection of two proteins.

The presence of active caspases within cells indicates that the execution phase of the apoptotic cascade is underway. There are various commercial assays using either broad spectrum or specific caspase inhibitors that bind only to active caspases (e.g. CaspaTag™ Fluorescein Caspase Activity kit, Invitrogen). Zona-free embryos can be incubated in such a fluorescently-tagged caspase inhibitor. The inhibitor enters the cells and binds irreversibly to activated caspases. Specimens can then be fixed in paraformaldehyde. Positive controls are incubated in staurosporine (an inducer of apoptosis) prior to caspase activity detection and fixation. Following labelling, fixed embryos can be mounted in an antifade mountant such as Vectashield (Vector Laboratories, Peterborough, UK) containing DAPI before examining with two channel laser scanning confocal microscopy. The cytoplasm of cells which contain active caspases, and are undergoing apoptosis, are fluorescently labelled. Healthy cells remain unlabelled.

3.3.3. Expression Of Cell Death Genes In Embryos

The temporal expression pattern of cell death suppressor, promoter and executor genes have been examined in mouse preimplantation embryos, providing evidence that preimplantation oocytes and embryos of all stages express the relevant molecular machinery to undergo this type of cell death (Jurisicova et al., 1998a). In addition to members of the Bcl-2 family, caspase expression has been detected in mouse embryos throughout preimplantation development (Exley et al., 1999). Quantitative RT-PCR has been used to detect increased expression of Bax mRNA in mouse embryos exposed to high glucose concentrations in the culture medium (Moley et al., 1998a). mRNA for BAX and BCL-2 has also been detected by RT-PCR in mouse and human preimplantation embryos (Warner et al., 1998) (Table 1).

BAX, BCL-2, BCL-x and BCL-w protein has been detected in human embryos using immunohistochemistry (Hardy, 1999; Antczak and Van Blerkom, 1999; Brison, 2000) (Table 1).

3.4. Chromatin Condensation

The earliest morphological change indicative of apoptosis, which in fact is a late event within the apoptosis cascade, is the condensation of chromatin on the inner nuclear membrane. Initially, this was detected by transmission electron microscopy (El-Shershaby and Hinchliffe, 1974; Jurisicova et al., 1996), later by labelling DNA with polynucleotide specific fluorochromes such as propidium iodide, Hoechst or 4',6-diamidino-2-phenylindole (DAPI) (Table 1). Nuclei with clumped chromatin have been seen in human blastocysts using propidium iodide or Hoechst staining and standard fluorescence microscopy (Hardy et al., 1989). More recently it has been observed in cleavage stage embryos (Hardy et al., 1996; Levy et al., 1998) and blastocysts (Hardy, 1999) using laser scanning confocal microscopy.

With the same approach of labelling DNA, condensed and misshapen nuclei have been seen in fragmented arrested (Jurisicova et al., 1996; Levy et al., 1998) and developing (Spanos et al., 2000) human embryos. These may be precursors of fragmented nuclei (see below).

3.5. Nuclear Fragmentation

Nuclear fragmentation, which is perhaps the most classic morphological feature of apoptosis, can be observed using a variety of techniques, including transmission electron microscopy, examination of histological sections by light microscopy, differential staining (Chapter 11) and labelling of nuclei with polynucleotide-specific fluorochromes combined with fluorescence or laser scanning confocal microscopy (Table 1).

Fragmented nuclei have been visualized using transmission electron microscopy in embryos from a variety of species including mouse (El-Shershaby and Hinchliffe, 1974) and human (Yang et al., 1998). Importantly, pyknotic and fragmented nuclei have been described in paraffin embedded sections of human embryos in vivo (Hertig et al., 1954). This is unique evidence that apoptosis also occurs in vivo during human preimplantation development.

An alternative technique is differential staining of the TE and ICM nuclei of blastocysts with polynucleotide-specific fluorochromes (Handyside and Hunter, 1984) (Chapter 11). This technique allows the rapid identification, localisation and quantification of fragmenting nuclei, characteristic of apoptosis. It involves the specific permeabilization of TE-Cells, allowing nuclear labelling by propidium iodide, followed by alcohol fixation of the whole embryo in the presence of Hoechst. Since the emission spectra of propidium iodide and Hoechst are different, TE and ICM nuclei can be distinguished by the colour of their fluorescence. Using fluorescence microscopy, dead cells in either lineage can be identified as discrete clusters of fluorochrome-labelled nuclear fragments, which are smaller than intact healthy nuclei. Differential staining has been successfully used in a variety of species including mouse (Handyside and Hunter, 1986; Hardy and Handyside, 1996; Devreker and Hardy, 1997; Brison and Schultz, 1997), human (Hardy et al., 1989), rat (Pampfer et al., 1990; Lea et al., 1996; Pampfer et al., 1997), rabbit (Giles and Foote, 1995), cow (Iwasaki et al., 1990; Van Soom et al., 1996) and pig (Papaioannou and Ebert, 1988).

If the aim of an experiment is quantification of apoptosis, rather than localisation, an extremely rapid method of identifying fragmenting nuclei is to simply fix zona-free embryos in alcohol or methanol and mount in a commercial anti-fade glycerol based mountant such as Vectashield (Vector Laboratories, Peterborough, UK), containing a polynucleotide-specific fluorochrome. The embryo can then be disaggregated by gentle pressure on the coverslip and the individual nuclei examined using a fluorescence microscope, or the embryo can be kept intact and optical sections examined using laser scanning confocal microscopy.

3.5.1. Fragmenting Nuclei In Embryos

Fragmenting nuclei are common in mouse blastocysts in vivo, with the majority of blastocysts having one or more dead cells, predominantly in the ICM, as detected by differential labelling (Handyside and Hunter, 1986).

Fragmenting nuclei are also seen in mouse, rat, cow and human blastocysts in vitro (Hardy et al., 1989; Pampfer et al., 1990; Moley et al., 1998a; Byrne et al., 1999; Matwee et al., 2000) (Table 1). In the mouse, fragmented nuclei are only seen at the blastocyst stage (Handyside and Hunter, 1986; Brison and Schultz, 1997). In the human, during normal development in vitro, fragmented nuclei first appear at the 8-cell stage and become more common following compaction and cavitation (Hardy et al., 2001). In the human blastocyst, cell death is present equally in the trophectoderm (TE) and the inner cell mass (ICM), and the incidence appears to be correlated with embryo quality (Hardy et al., 1989).

3.6. DNA Fragmentation

Another classic feature of apoptosis is the degradation of DNA into oligonucleosomal fragments. Owing to the small numbers of cells in preimplantation embryos, it is impossible to use electrophoretic techniques to look for DNA laddering typical of apoptotic nuclei. However, the development of terminal deoxynucleotidyl transferase-mediated dUTP nick-end labelling (TUNEL) (Gavrieli et al., 1992) allows the assessment of nuclear DNA fragmentation in situ. This technique can be used on specimens with only a few cells, and is based on the fluorescent labelling of the 3' end of oligonucleosome fragments. This technique has the additional advantages of allowing both the localisation and the quantification of the percentage of nuclei with DNA fragmentation (Gavrieli et al., 1992).

The enzyme terminal deoxynucleotidyl transferase (TdT) binds to exposed 3'-OH ends of DNA single strand breaks and catalyzes the addition of deoxynucleotides that have been labelled with a marker. This can be biotin, which following specific binding by streptavidin can be visualized with DAB, resulting in an insoluble coloured substrate visible with light microscopy. In this case nuclei with fragmented DNA will appear brown, and this approach is suitable for paraffin-embedded tissue sections. Alternatively, the deoxynucleotides can be labelled with a fluorescent label such as FITC, in which case TUNEL positive nuclei will be clearly seen using fluorescent microscopy or laser scanning confocal microscopy. This latter approach allows the processing of a whole embryo through the different steps. Zona-free embryos can be fixed in 4% paraformaldehyde, before incubating in fluorescein-conjugated deoxynucleotides (dUTP) and TdT for one hour in the dark. Following TUNEL, washed embryos are mounted in Vectashield with DAPI (Vector Laboratories, Peterborough, UK) before examining with laser scanning confocal microscopy. Positive controls are embryos treated with DNase prior to TUNEL labelling. Negative controls omit the TdT, and are simply exposed to labelled dUTP. This approach has been used by a number of investigators, e.g. Jurisicova et al., 1996; Brison and Schultz, 1997; Spanos et al., 2000; Hardy et al., 2001.

3.6.1. DNA Fragmentation In Embryos

Positive labelling of fragmented nuclei in mouse (Jurisicova et al., 1995; Weil et al., 1996; Brison and Schultz, 1997; Warner et al., 1998; O'Neill, 1998; Moley et al., 1998a; Moley et al., 1998b; Wuu et al., 1999; Exley et al., 1999; Rinkenberger et al., 2000), human (Jurisicova et al., 1996; Levy et al., 1998; Warner et al., 1998; Yang et al., 1998; Antczak and Van Blerkom, 1999; Spanos et al., 2000; Hardy et al., 2001), rat (Pampfer et al., 1997), rabbit (Herrler et al., 1998), cow (Byrne et al., 1999; Watson et al., 2000; Matwee et al., 2000) and pig (Long et al., 1998) embryos has been demonstrated, providing evidence that DNA is being degraded into oligonucleosomal fragments (Table 1).

DNA fragmentation is only seen at the blastocyst stage in mice (Brison and Schultz, 1997), after compaction in human embryos (Hardy et al., 2001) and after the 8-cell stage in cow embryos (Byrne et al., 1999).

TUNEL labelling has been used to assess the anti- and pro-apoptotic effects of a variety of growth factors including IGF (Herrler et al., 1998; Spanos et al., 2000), TGF-α (Brison and Schultz, 1997) and TNF-α (Wuu et al., 1999).

3.6.2. Cautionary Note

A number of workers advocate caution in the interpretation of TUNEL results. It has been proposed that DNA strand breaks may occasionally occur as artefacts of fixation and specimen preparation (Collins et al., 1997). In addition to labelling apoptotic nuclei, positive TUNEL labelling has also been seen in cells undergoing necrotic cell death (Grasl-Kraupp et al., 1995; Charriaut-Marlangue and Ben-Ari, 1995; Bicknell and Cohen, 1995), so it is important to confirm apoptosis using other morphological markers, such as nuclear morphology. In necrotic cells, the nature of the TUNEL labelling is different: the signal is not confined to the nuclei but is seen as diffuse staining in the cytoplasm as well (Charriaut-Marlangue and Ben-Ari, 1995;

Jurisicova et al., 1996). Furthermore, morphological features of apoptosis are not always associated with DNA fragmentation (Cohen et al., 1992). Fragmented nuclei without a positive TUNEL signal occur in human blastocysts (Spanos et al., 2000). Conversely, TUNEL-positive nuclei can sometimes be seen in the absence of morphological features of apoptosis (Sanders and Wride, 1996), indicating that DNA fragmentation is an early event in apoptosis. Where TUNEL labelling has been seen in arrested embryos, it is possible that DNA degradation is occurring in cells undergoing the early stages of secondary necrosis, rather than apoptosis.

3.7. Morphological Changes

3.7.1. Excluded Cells

The earliest events in apoptosis involve the cell membrane, with the apoptotic cell rounding up and separating from neighbouring cells (Wyllie, 1997). At the time of compaction and blastocyst formation, it is not uncommon to see excluded cells which lie between the developing embryo and the zona pellucida. Immunocytochemical examination of these embryos shows that excluded cells have poor gap junction communication with the embryo (Hardy et al., 1996). Transmission electron microscopy and laser scanning confocal microscopy have often revealed extruded cells into the perivitelline space or blastocyst cavity (Lopata et al., 1982; Matwee et al., 2000).

3.7.2. Cytoplasmic Fragmentation

Over 75% of human embryos generated following IVF have various degrees of cytoplasmic fragmentation. These fragments resemble apoptotic bodies seen in other cell types. Ultrastructural examination of fragmented embryos showed that these membrane bound fragments contain organelles (Jurisicova et al., 1996; Yang et al., 1998). In the light of these observations, Jurisicova et al. (1996) proposed that these fragments represented apoptotic bodies. However, failure of the majority of fragmented embryos to show either TUNEL labelling (indicating DNA fragmentation) or annexin V labelling (indicating phosphatidylserine redistribution in the plasma membrane) led Antczak and Van Blerkom (1999) to postulate that fragmentation was not part of, or a consequence of, apoptosis. Fragmentation has also been seen in human embryos in vivo (Hertig et al., 1954; Pereda and Croxatto, 1978; Buster et al., 1985), indicating that fragmentation is not an in vitro artefact.

3.7.3. Phagocytosis

Further degeneration results in the formation of extracellular debris (Wyllie et al., 1980), with the presence of mitochondria and nuclear remnants confirming its cellular origin, for example, in arrested human embryos (Jurisicova et al., 1996). Confocal microscopy of human and cow blastocysts has shown cells being engulfed by neighbouring cells (Plante and King, 1994; Hardy et al., 1996). Autophagic vacuoles have been seen at the ultrastructural level using transmission electron microscopy in human (Mohr and Trounson, 1982), cow (Plante and King, 1994), rhesus (Enders and Schlafke, 1981) and mouse blastocysts (El-Shershaby and Hinchliffe, 1974). However, while it appears that many dead cells are cleared by phagocytosis within the blastocyst, other apparently arrested cells simply persist. In other cell types, apoptotic cells formed in single-layered epithelia may be extruded into lumina, and thus escape phagocytosis (Wyllie et al., 1980; Harmon et al., 1998). This process appears to occur in human blastocysts, with isolated cells being observed in the blastocoel cavity or between the trophectoderm and the zona pellucida (Lopata et al., 1982; Hardy, 1999). These cells and fragments are excluded from normal development. These isolated cells in the blastocoel cavity have been shown to be TUNEL labelled (Hardy, 1999), as have similar cells in mouse blastocysts (Brison and Schultz, 1997). In some cases, these cells are large, and this, coupled with the lack of intercellular junctions, poorly differentiated mitochondria and paucity of rough endoplasmic reticulum, suggests that these cells originated during early cleavage. It is possible that such cells lack cell surface markers which would promote their ingestion by neighbouring cells, causing them to persist throughout preimplantation development.

Table 1: Summary of morphological, molecular and biochemical features of apoptosis seen in preimplantation mammalian embryos

	TECHNIQUE*	SPECIES AND REFERENCES
MORPHOLOGICAL		
Chromatin condensation	*TEM(*)*	***Rhesus*** *Hurst '78; Enders, '82;* ***Human*** *Jurisicova '96;* ***Mouse*** *El-Shershaby & Hinchlcliffe '74;* ***Cow*** *Plante & King '94*
	Light microscopy	***Mouse*** *Copp '78*
	DNA fluorochrome/ confocal	***Human*** *Hardy '96; Jurisicova '96; Levy '98; Spanos '00; Hardy '01;* ***Mouse*** *Jurisicova '95,*

		'98; Brison & Schultz '97; ***Cow*** *Matwee, '00*
	Differential labelling	***Human*** *Hardy '89*
Nuclear fragmentation	*Light microscopy*	***Human*** *Hertig '54;* ***Mouse*** *Copp '78*
	TEM	***Rhesus*** *Enders '82* ***Mouse*** *El-Shershaby & Hinchlcliffe '74*
	DNA fluorochrome/ confocal	***Human*** *Jurisicova '96; Spanos '00; Levy '98; Hardy '96, '99, '01;* ***Cow*** *Matwee '00*
	Hoechst 33258/FM(*)	***Human*** *Hardy '93;* ***Mouse*** *O'Neill '98*
	Differential labelling	***Human*** *Hardy '89;* ***Mouse*** *Handyside & Hunter '86; Pampfer '94; Hardy & Handyside '96; Hardy '97; Devreker & Hardy '97; Brison & Schultz '97; Wuu '99;* ***Rat*** *Pampfer '90, '97;* ***Cow*** *Byrne '99*
Vacuole formation	*Light microscopy*	***Human*** *Hardy '96*
	Confocal	***Human*** *Hardy '96, '99*
	TEM	***Human*** *Sathananthan '82;* ***Rhesus*** *Hurst '78; Enders '82;* ***Mouse*** *El-Shershaby & Hinchlcliffe '74;* ***Baboon*** *Enders '90*
Excluded cells	*Light microscopy*	***Human*** *Hardy '96*
	Confocal	***Human*** *Hardy '96, '99; Jurisicova '96* ***Mouse*** *Brison & Schultz '97;* ***Cow*** *Byrne '99; Matwee '00*
	TEM	***Mouse*** *El-Shershaby & Hinchlcliffe '74* ***Rhesus*** *Enders '82;* ***Baboon*** *Enders '90*
Cytoplasmic condensation	*TEM*	***Human*** *Jurisicova '96; Yang, '98* ***Mouse*** *El-Shershaby & Hincliffe '74*

* TEM: Transmission electron microscopy; SEM: Scanning electron microscopy;

* FM: Fluorescence microscopy

Table 1 (continued): Summary of morphological, molecular and biochemical features of apoptosis seen in preimplantation mammalian embryos

	TECHNIQUE	SPECIES AND REFERENCES
Cytoplasmic fragmentation	*Light microscopy*	**Human** *Hardy '96, '99, Yang '98; Warner '98; Antzak & Van Blerkom '99; Spanos '00;* **Mouse** *Warner '98*
	SEM	**Human** *Jurisicova '96, '97*
	TEM	**Human** *Jurisicova '96;* **Rhesus** *Hurst '78; Enders, '82;* **Cow** *Plante & King '94;*
	Confocal	**Mouse** *Jurisicova '98*
Phagocytosis	*Confocal*	**Human** *Hardy '96, '99;* **Mouse** *Brison & Schultz, '97*
	TEM	**Mouse** *El-Shershaby & Hinchlcliffe '74* **Cow** *Plante & King '94;* **Rhesus** *Enders & Schlafke, '81, '82*
Expression of caspases/Bcl2 family	*RT-PCR*	**Mouse** *Exley '99; Jurisicova '98*
Expression of other apoptotic related proteins	*RT-PCR*	**Mouse** *Jurisicova '98 (MA-3, p53, Ttg)*
	Immunohisto-chemistry	**Human** *Galan '00 (Fas-L);* **Cow** *Matwee '00 (p53)*
	In situ hybridization	**Rat** *Pampfer '97 (Clusterin);* **Mouse** *Rickenberker '00 (Mcl-1)*
BIOCHEMICAL		
DNA fragmentation	*TUNEL labelling*	**Human** *Jurisicova '96; Yang '98; Levy '98; Warner '98a,b; Antzak & Van Blerkom '99; Spanos '00; Hardy '99, '01;* **Cow** *Byrne '99; Matwee '00; Watson '00;* **Mouse** *Weil '96; Brison & Schultz '97; '98; Jurisicova '95, '98; O'Neill '98; Warner '98; Moley '98; Exley '99; Wuu '99; Rinkenberger '00;* **Rat** *Pampfer '97;* **Rabbit** *Herrler '98;* **Pig** *Long '98*
Caspase activity	*Active caspase labelling*	**Mouse** *Exley '99;* **Human** *Spanos (unpublished)*
Phosphatidyl-serine translocation	*Annexin V labelling*	**Human** *Levy '98; Antzak & Van Blerkom '99*
Cytoplasmic activity	*Live/dead assays (Calcein-AM and Ethidium Homodimer)*	**Mouse** *O'Neill '98; Weil '96*
Loss of gap junctions	*Immunohistochemistry/ Confocal*	**Human** *Hardy '96*

4. Examples of application

It is now becoming clear that healthy preimplantation development involves a delicate balance between cell division and cell death. Considerable work is needed to elucidate the role of apoptosis in embryonic loss, ageing and the removal of unwanted cells.

The evolution and optimisation of new culture media currently uses a number of measures of preimplantation developmental potential, including blastocyst formation and cell number. Quantification of apoptosis augments these approaches. Several factors have been shown to influence the extent of apoptosis compared to that seen in vivo. These include the basic composition of the medium (Devreker and Hardy, 1997; Moley et al., 1998a) and supplementation with growth factors (Brison and Schultz, 1997; Herrler et al., 1998; Spanos et al., 2000). Furthermore, the physical aspects of embryo culture have been shown to affect apoptosis, including embryo density (Brison and Schultz, 1997; O'Neill, 1998) and oxygen concentration (Yang et al., 1998).

Assessment of apoptosis is also useful in evaluating the impact of a variety of pathological conditions and toxic insults on preimplantation development. Such studies include examining the effect of diabetes on preimplantation development in rat (Lea et al., 1996; Pampfer et al., 1997) and mouse (Moley et al., 1998b), and looking at growth factor modulation of UV irradiation damage in rabbit embryos (Herrler et al., 1998).

Finally, assessment of apoptosis has proved useful in examining the phenotype of preimplantation embryos from transgenic and knockout animals such as TGF-α (Brison and Schultz, 1998), Mcl-1 (Rinkenberger et al., 2000) and Bax (Moley et al., 1998b) knockouts.

Table 2: Advantages and disadvantages of different techniques to examine apoptotic morphology in preimplantation embryos

Technique	Advantages	Disadvantages
Transmission electron microscopy	*Accurate identification of morphological features of apoptosis*	*Labour intensive* *Time-consuming*
Serial paraffin-embedded sections	*Can visualize indented nuclear membrane, chromatin condensation, nuclear and cytoplasmic fragmentation* *Can localize apoptotic nuclei within embryo* *Can store specimens indefinitely*	*Technically challenging to embed, serially section, mount and stain single embryos* *Need to reconstruct serial sections, therefore time-consuming to analyse more than a few embryos*
Polynucleotide-specific fluorochromes	*Technique to visualize nuclear morphology* *Very quick, simply the time it takes to mount embryos in mountant on slides* *Can be carried out with a standard fluorescence microscope or with a laser scanning confocal microscope* *Easy to count number of interphase, metaphase and apoptotic nuclei in each embryo*	*Specimens fade, cannot be stored for more than a week or so*
Differential labelling	*Technique to visualize and localize fragmented nuclei in TE and ICM* *Quick, takes 1-2 hours to process a large number of embryos* *Can be carried out with a standard fluorescence microscope* *Easy to count number of interphase, metaphase and apoptotic nuclei in each lineage*	*Specimens fade, cannot be stored for more than a week or so* *Only suitable for blastocysts*

5. Advantages and disadvantages of different techniques

The best technique to choose for examining apoptosis depends on the endpoint of the study. Transmission electron microscopy is the 'gold standard' for unambiguously identifying apoptotic changes in an embryo. However, this technique is technically challenging and time consuming, and is not suitable for quantifying apoptosis in a large number of embryos (Table 2). If quantification of the extent of apoptosis is an important endpoint of the study, then differential labelling (suitable for blastocysts, Table 2), or TUNEL labelling (Table 3) are much more appropriate approaches which will provide statistically meaningful results from a large number of embryos. These techniques allow localisation of dying cells when the embryo is examined in whole mount. Differential labelling has the added advantage that apoptosis in the TE and the ICM can be quantified, in parallel with an estimate of the number of cells undergoing mitosis. While some fragmented nuclei will not be TUNEL labelled and vice versa, these two techniques give a remarkably consistent estimate of the extent of apoptosis. As there is a risk that TUNEL labelling will also label necrotic nuclei, it is recommended that TUNEL labelling is used in conjunction with nuclear labelling with a polynucleotide fluorochrome such as DAPI, confirming that the nuclei are indeed apoptotic on the basis of fragmented morphology.

Quantification of gene expression is useful for assessing whether the apoptotic machinery is present at different stages of embryo development (Table 3). However, gene activity is not only regulated by the extent of transcription and translation, but also by post-translational modifications. For example, it is thought that in the presence of growth factor stimulation Bad (a pro-apoptotic member of the Bcl-2 family) is phosphorylated and sequestered in the cytoplasm. It only becomes active in the absence of growth factor stimulation, when it becomes dephosphorylated and enters the mitochondrial membrane, initiating an apoptotic cascade.

Table 3: Advantages and disadvantages of different techniques to examine biochemical and molecular changes that occur during apoptosis

Technique	Advantages	Disadvantages
Annexin V labelling	*Technique for visualisation of externalisation of phosphatidylserine* *Should be used in conjunction with propidium iodide to assess membrane permeability, and distinguish early apoptosis from secondary necrosis*	*Non-specific autofluorescence may result in false positives* *Cell damage and a leaky membrane will result in false-positives* *Has to be used on intact cells - sections will not allow differentiation between inner and outer membrane leaflets*
TUNEL	*Provides quantitative estimate for extent of apoptosis* *Technique for* in situ *assessment of DNA fragmentation* *Can be used on specimens of only a few cells* *Can localize nuclei with fragmented DNA in embryo if confocal images are rotated. Time-consuming and requires specialist software* *Reliable method of identifying apoptotic nuclei if used in conjunction with polynucleotide-specific fluorochromes e.g. DAPI to identify fragmenting nuclei*	*Potentially also labels cells with degraded DNA undergoing necrosis* *Specimens fade, cannot be stored for more than a week or so* *Laser scanning confocal microscope required for accurate assessment of specimens*
Gene expression (RT-PCR)	*Can identify presence of mRNA for specific genes involved in apoptosis*	*Need to design specific primers* *May need to do nested PCR* *Cannot localize where the gene is expressed in the embryo Quantification possible but technically challenging* *Cell death machinery probably present anyway, and modulated by post translational modifications such as phosphorylation*
Gene expression (immuno-histochemistry)	*Can localize expression of protein* *If using DAB-labelled secondary antibodies, can store slides indefinitely* *Multichannel laser-scanning confocal microsocopy allows simultaneous visualisation of a number of ligands*	*If using FITC-labelled secondary antibodies, specimens cannot be stored indefinitely*

It is important to remember that all of the techniques described here, with the exception of timelapse approaches, only provide a 'snapshot' of the embryo at the time at which it was fixed or lysed. As it is not yet known how long dying cells persist for, and as cell death has probably confounded previous estimates of cell cycle length, it is impossible to estimate the full extent of cell death in the mammalian blastocyst. Studies in other cell types suggest that the initial stages of apoptosis are extremely rapid, with cytoplasmic fragmentation often occurring within minutes and removal of apoptotic bodies within hours (Wyllie et al., 1980). If a similar programme occurs in the mammalian blastocyst, the observation of moderate numbers of dead cells may represent an extensive rate of cell death.

6. Conclusions

Apoptosis is a natural process during mammalian preimplantation development, present in embryos in vivo. This suggests that it plays a developmental role, which is as yet unclear, but could involve elimination of unwanted or damaged cells. The incidence of apoptosis can be increased in certain pathological states or under suboptimal environmental conditions, to the extent that apoptosis becomes widespread within the embryo, leading to developmental arrest. Thus it is important to create an environment that does not compromise embryo health, but that also, conversely, does not over-ride natural checkpoints that could eliminate abnormal cells, allowing them to survive.

References

Antczak M, Van Blerkom J. 1999. Temporal and spatial aspects of fragmentation in early human embryos: possible effects on developmental competence and association with the differential elimination of regulatory proteins from polarized domains. Hum Reprod 14:429-447.

Arends MJ, Wyllie AH. 1991. Apoptosis: mechanisms and roles in pathology. Int Rev Exp Pathol 32:223-254.

Bicknell GR, Cohen GM. 1995. Cleavage of DNA to large kilobase pair fragments occurs in some forms of necrosis as well as apoptosis. Biochem Biophys Res Commun 207:40-47.

Brison DR. 2000. Apoptosis in mammalian preimplantation embryos: regulation by survival factors. Hum Fertil 3:36-47.

Brison DR, Schultz RM. 1997. Apoptosis during mouse blastocyst formation: evidence for a role for survival factors including transforming growth factor alpha. Biol Reprod 56:1088-1096.

Brison DR, Schultz RM. 1998. Increased incidence of apoptosis in transforming growth factor alpha-deficient mouse blastocysts. Biol Reprod 59:136-144.

Buster JE, Bustillo M, Rodi IA, Cohen SW, Hamilton M, Simon JA, Thorneycroft IH, Marshall JR. 1985. Biologic and morphologic development of donated human ova recovered by nonsurgical uterine lavage. Am J Obstet Gynecol 153:211-217.

Byrne AT, Southgate J, Brison DR, Leese HJ. 1999. Analysis of apoptosis in the preimplantation bovine embryo using TUNEL. J Reprod Fertil 117:97-105.

Charriaut-Marlangue C, Ben-Ari Y. 1995. A cautionary note on the use of the TUNEL stain to determine apoptosis. Neuroreport 7:61-64.

Cohen GM, Sun XM, Snowden RT, Dinsdale D, Skilleter DN. 1992. Key morphological features of apoptosis may occur in the absence of internucleosomal DNA fragmentation. Biochem J 286 (Pt 2):331-334

Collins JA, Schandi CA, Young KK, Vesely J, Willingham MC. 1997. Major DNA fragmentation is a late event in apoptosis. J Histochem Cytochem 45:923-934.

Copp AJ. 1978. Interaction between inner cell mass and trophectoderm of the mouse blastocyst. I. A study of cellular proliferation. J Embryol Exp Morphol 48:109-125.

Coucouvanis E, Martin GR. 1995. Signals for death and survival: a two-step mechanism for cavitation in the vertebrate embryo. Cell 83:279-287

Datta SR, Brunet A, Greenberg ME. 1999. Cellular survival: a play in three Akts. Genes Dev 13:2905-2927.

Devreker F, Hardy K. 1997. Effects of glutamine and taurine on preimplantation development and cleavage of mouse embryos in vitro. Biol Reprod 57:921-928.

Earnshaw WC, Martins LM, Kaufmann SH. 1999. Mammalian caspases: structure, activation, substrates, and functions during apoptosis. Annu Rev Biochem 68:383-424.

El-Shershaby AM, Hinchliffe JR. 1974. Cell redundancy in the zona-intact preimplantation mouse blastocyst: a light and electron microscope study of dead cells and their fate. J Embryol Exp Morphol 31:643-654.

Enders AC, Hendrickx AG, Binkerd PE. 1982. Abnormal development of blastocysts and blastomeres in the rhesus monkey. Biol Reprod 26:353-366.

Enders AC, Lantz KC, Schlafke S. 1990. Differentiation of the inner cell mass of the baboon blastocyst. Anat Rec 226:237-248.

Enders AC, Schlafke S. 1981. Differentiation of the Blastocyst of the rhesus monkey. Am J Anat 162:1-21.

Exley GE, Tang C, McElhinny AS, Warner CM. 1999. Expression of caspase and BCL-2 apoptotic family members in mouse preimplantation embryos. Biol Reprod 61:231-239.

Gavrieli Y, Sherman Y, Ben-Sasson SA. 1992. Identification of programmed cell death in situ via specific labeling of nuclear DNA fragmentation. J Cell Biol 119:493-501.

Giles JR, Foote RH. 1995. Rabbit blastocyst: allocation of cells to the inner cell mass and trophectoderm. Mol Reprod Dev 41:204-211.

Grasl-Kraupp B, Ruttkay-Nedecky B, Koudelka H, Bukowska K, Bursch W, Schulte-Hermann R. 1995. In situ detection of fragmented DNA (TUNEL assay) fails to discriminate among apoptosis, necrosis, and autolytic cell death: a cautionary note. Hepatology 21:1465-1468.

Handyside AH, Hunter S. 1984. A rapid procedure for visualising the inner cell mass and trophectoderm nuclei of mouse blastocysts in situ using polynucleotide-specific fluorochromes. J Exp Zool 231:429-434.

Handyside AH, Hunter S. 1986. Cell division and death in the mouse blastocyst before implantation. Roux's Arch Dev Biol 195:519-526.

Hardy K. 1997. Cell death in the mammalian blastocyst. Mol Hum Reprod 3:919-925.

Hardy K. 1999. Apoptosis in the human embryo. Rev Reprod 4:125-134.

Hardy K, Handyside AH. 1996. Metabolism and cell allocation during parthenogenetic preimplantation mouse development. Mol Reprod Dev 43:313-322.

Hardy K, Handyside AH, Winston RM. 1989. The human blastocyst: cell number, death and allocation during late preimplantation development in vitro. Development 107:597-604.

Hardy K, Spanos S, Becker D, Iannelli P, Winston RM, Stark J. 2001. From cell death to embryo arrest: Mathematical models of human preimplantation embryo development. Proc Natl Acad Sci USA 98:1655-1660.

Hardy K, Warner A, Winston RM, Becker DL. 1996. Expression of intercellular junctions during preimplantation development of the human embryo. Mol Hum Reprod 2:621-632.

Hardy K, Winston RM, Handyside AH. 1993. Binucleate blastomeres in preimplantation human embryos in vitro: failure of cytokinesis during early cleavage. J Reprod Fertil 98:549-558.

Harmon BV, Winterford CM, O'Brien BA, Allan DJ. 1998. Morphological criteria for identifying apoptosis. In: Cell biology: A Laboratory Handbook. Ed JE Celis, Academic Press, San Diego: pp 327-340.

Herrler A, Krusche CA, Beier HM. 1998. Insulin and insulin-like growth factor-I promote rabbit blastocyst development and prevent apoptosis. Biol Reprod 59:1302-1310.

Hertig AT, Rock J, Adams EC, Mulligan WJ. 1954. On the preimplantation stages of the human ovum: a description of four normal and four abnormal specimens ranging from the second to fifth day of development. Contrib Embryol 35:199-220.

Huppertz B, Frank HG, Kaufmann P. 1999. The apoptosis cascade-morphological and immunohistochemical methods for its visualisation. Anat Embryol (Berl) 200:1-18.

Hurst PR, Jefferies K, Eckstein P, Wheeler AG. 1978. An ultrastructural study of preimplantation uterine embryos of the rhesus monkey. J Anat 126:209-220.

Iwasaki S, Yoshiba N, Ushijima H, Watanabe S, Nakahara T. 1990. Morphology and proportion of inner cell mass of bovine blastocysts fertilized in vitro and in vivo. J Reprod Fertil 90:279-284.

Jamieson ME, Coutts JR, Connor JM. 1994. The chromosome constitution of human preimplantation embryos fertilized in vitro. Hum Reprod 9:709-715.

Jurisicova A, Latham KE, Casper RF, Casper RF, Varmuza SL. 1998a. Expression and regulation of genes associated with cell death during murine preimplantation embryo development. Mol Reprod Dev 51:243-253.

Jurisicova A, Rogers I, Fasciani A, Casper RF, Varmuza S. 1998b. Effect of maternal age and conditions of fertilization on programmed cell death during murine preimplantation embryo development. Mol Hum Reprod 4:139-145.

Jurisicova A, Varmuza S, Casper RF. 1995. Involvement of programmed cell death in preimplantation embryo demise. Hum Reprod Update 1:558-566.

Jurisicova A, Varmuza S, Casper RF. 1996. Programmed cell death and human embryo fragmentation. Mol Hum Reprod 2:93-98.

Lea RG, McCracken JE, McIntyre SS, Smith W, Baird JD. 1996. Disturbed development of the preimplantation embryo in the insulin-dependent diabetic BB/E rat. Diabetes 45:1463-1470.

Levy R, Benchaib M, Cordonier H, Souchier C, Guerin JF. 1998. Annexin V labelling and terminal transferase-mediated DNA end labelling (TUNEL) assay in human arrested embryos. Mol Hum Reprod 4:775-783.

Long CR, Dobrinsky JR, Garrett WM, Johnson LA. 1998. Dual labeling of the cytoskeleton and DNA strand breaks in porcine embryos produced in vivo and in vitro. Mol Reprod Dev 51:59-65.

Lopata A, Kohlman DJ, Kellow GN. 1982. The fine structure of human blastocysts developed in culture. In: Embryonic development, Part B: Cellular Aspects, Alan R. Liss, New York: pp 69-85.

Martin SJ, Reutelingsperger CP, McGahon AJ, Rader JA, van Schie RC, LaFace DM, Green DR. 1995. Early redistribution of plasma membrane phosphatidylserine is a general feature of apoptosis regardless of the initiating stimulus: inhibition by overexpression of Bcl-2 and Abl. J Exp Med 182:1545-1556.

Matwee C, Betts DH, King WA. 2000. Apoptosis in the early bovine embryo. Zygote 8:57-68.

Mohr LR, Trounson AO. 1982. Comparative ultrastructure of hatched human, mouse and bovine blastocysts. J Reprod Fertil 66:499-504.

Moley KH, Chi MM, Knudson CM, Korsmeyer SJ, Mueckler MM. 1998a. Hyperglycemia induces apoptosis in pre-implantation embryos through cell death effector pathways. Nat Med 4:1421-1424.

Moley KH, Chi MM, Mueckler MM. 1998b. Maternal hyperglycemia alters glucose transport and utilisation in mouse preimplantation embryos. Am J Physiol 275: E38-47.

Munné S, Alikani M, Tomkin G, Grifo J, Cohen J. 1995. Embryo morphology, developmental rates, and maternal age are correlated with chromosome abnormalities. Fertil Steril 64:382-391.

O'Neill C. 1998. Autocrine mediators are required to act on the embryo by the 2-cell stage to promote normal development and survival of mouse preimplantation embryos in vitro. Biol Reprod 58:1303-1309.

Pampfer S, De Hertogh R, Vanderheyden I, Michiels B, Vercheval M. 1990. Cell allocation to the inner cell mass and the trophectoderm in rat embryos during in vivo preimplantation development. Roux's Arch Dev Biol 198:257-263.

Pampfer S, Vanderheyden I, McCracken JE, Vesela J, De Hertogh R. 1997. Increased cell death in rat blastocysts exposed to maternal diabetes in utero and to high glucose or tumor necrosis factor-alpha in vitro. Development 124:4827-4836.

Papaioannou VE, Ebert KM. 1988. The preimplantation pig embryo: cell number and allocation to trophectoderm and inner cell mass of the blastocyst in vivo and in vitro. Development 102:793-803.

Pereda J, Croxatto HB. 1978. Ultrastructure of a seven-cell human embryo. Biol Reprod 18:481-489.

Pierce GB, Lewellyn AL, Parchment RE. 1989. Mechanism of programmed cell death in the blastocyst. Proc Natl Acad Sci U S A 86:3654-3658.

Plante L, King WA. 1994. Light and electron microscopic analysis of bovine embryos derived by in vitro and in vivo fertilization. J Assist Reprod Genet 11:515-529

Rinkenberger JL, Horning S, Klocke B, Roth K, Korsmeyer SJ. 2000. Mcl-1 deficiency results in peri-implantation embryonic lethality. Genes Dev 14:23-27.

Sanders EJ, Wride MA. 1996. Ultrastructural identification of apoptotic nuclei using the TUNEL technique. Histochem J 28:275-281.

Sathananthan AH, Wood C, Leeton JF. 1982. Ultrastructural evaluation of 8-16 cell human embryos cultured in vitro. Micron 13:193-203.

Spanos S, Becker DL, Winston RM, Hardy K. 2000. Anti-apoptotic action of insulin-like growth factor-I during human preimplantation embryo development. Biol Reprod 63:1413-1420.

Thornberry NA, Lazebnik Y. 1998. Caspases: enemies within. Science 281:1312-1316.

Tsujimoto Y, Shimizu S. 2000. Bcl-2 family: life-or-death switch. FEBS Lett 466:6-10.

Van Blerkom J, Davis PW. 1998. DNA strand breaks and phosphatidylserine redistribution in newly ovulated and cultured mouse and human oocytes: occurrence and relationship to apoptosis. Hum Reprod 13:1317-1324.

Van Soom A, Boerjan M, Ysebaert MT, De Kruif A. 1996. Cell allocation to the inner cell mass and the trophectoderm in bovine embryos cultured in two different media. Mol Reprod Dev 45:171-182.

Warner CM, Cao W, Exley GE, McElhinny AS, Alikani M, Cohen J, Scott RT, Brenner CA. 1998. Genetic regulation of egg and embryo survival. Hum Reprod 13 Suppl 3:178-190; discussion 191-196.

Watson AJ, De Sousa P, Caveney A, Barcroft LC, Natale D, Urquhart J, Westhusin ME. 2000. Impact of bovine oocyte maturation media on oocyte transcript levels, blastocyst development, cell number, and apoptosis. Biol Reprod 62:355-64.

Weil M, Jacobson MD, Coles HS, Davies TJ, Gardner RL, Raff KD, Raff MC. 1996. Constitutive expression of the machinery for programmed cell death. J Cell Biol 133:1053-1059.

Wuu YD, Pampfer S, Becquet P, Vanderheyden I, Lee KH, De Hertogh R. 1999. Tumor necrosis factor alpha decreases the viability of mouse blastocysts in vitro and in vivo. Biol Reprod 60:479-483.

Wyllie AH. 1997. Apoptosis: an overview. Br Med Bull 53:451-465.

Wyllie AH, Kerr JF, Currie AR. 1980. Cell death: the significance of apoptosis. Int Rev Cytol 68:251-306.

Yang HW, Hwang KJ, Kwon HC, Kim HS, Choi KW, Oh KS. 1998. Detection of reactive oxygen species (ROS) and apoptosis in human fragmented embryos. Hum Reprod 13:998-1002.

Chapter 13

ELECTRON MICROSCOPY OF PRE-HATCHING MAMMALIAN EMBRYOS

THE ULTRASTRUCTURE OF CATTLE, SWINE AND HORSE EMBRYOS

Poul Maddox-Hyttel[1] and Marleen Boerjan[2]
[1] *Department of Anatomy and Physiology, The Royal Veterinary and Agricultural University, Groennegaardsvej 7, DK-1870 Frederiksberg C, Denmark,*
[2] *PasReform, Hatchery technologies, PO Box 2, NL-7038 ZG Zeddam, the Netherlands*

1. Introduction

All researchers working with stereo microscopical grading of mammalian embryos recognize the wish of turning the magnification zoom beyond its physical limit in order to assess the nature of structures that escape the resolution of the equipment. However, non-invasive microscopical techniques for embryo evaluation are hampered by on the one hand, the limit of resolution of the stereo microscope normally working within magnifications up to a few hundred times and, on the other hand, the thickness of the three dimensional structure of the embryo of 150 μm or even more. Transmission electron microscopy bypasses both of these limitations: The microscopical principle offers a much higher resolution in itself and the third dimension is almost eliminated by examining extremely thin sections of the specimen. The most obvious disadvantage by the electron microscopical approach is the inherent invasiveness including fixation, embedding and sectioning. Nonetheless, researchers who have had even limited experiences of revealing the electron microscopical structure, i.e. the ultrastructure, of embryos by either observing pictures or even sitting at the microscope themselves also recognize that knowledge of the ultrastructure allows for a much more detailed stereo microscopical assessment. A structure, the nature of which cannot be immediately distinguished by the stereo microscope, may very well reveal itself at this simple method of observation if it has previously been examined in the electron microscope. One could say that the electron microscopical knowledge provides the alphabet for reading the stereo microscopical appearance of an embryo. In this sense, electron microscopical studies of embryos have two main purposes: Firstly, they allow for a direct description and understanding of cell biological phenomena of importance for embryo development and, secondly, they serve as a bridge to the stereo microscope providing the necessary information to "read" the low power embryo morphology.

In this chapter we firstly describe the development of pre-hatching bovine embryo at the ultrastructural level. Secondly, with reference to the bovine embryo, the development of the pre-hatching porcine and horse embryo is presented. For an introduction to the electron microscope and the techniques of specimen preparation we refer to appendix I.

2. Basic considerations on the ultrastructure of pre-hatching mammalian embryos

2.1. Introduction to the ultrastructure of the somatic cell

At the first glance, the ultrastructure of the cells, i.e. the blastomeres of the embryos resembles that of differentiated somatic cells. They are build from the same organelles and inclusions. However, upon a closer examination it is clear that the embryo presents a number of particular features and modifications which cannot be regarded as general cell biological structures. Nevertheless, in order to understand the structural cell biology of the embryo, a short basic introduction to the ultrastructure of the common organelles and inclusions of the somatic cell will be presented. For an extended review of the ultrastructure of the cell we refer to basic text books.

The nucleus of the cell is bounded by the nuclear envelope, a double membrane which isolates DNA replication and RNA transcription from RNA translation of the genetic message in the cytoplasm. The pores in the nuclear envelope are involved in exchange of information between the nuclear and cytoplasmic compartments allowing for e.g. mRNA and ribosomal subunits to be exported from the nucleus. In differentiated tissue, the nucleus is actively transcribing DNA and this can be recognized at the ultrastructural level: In the interphase nucleus, the chromatin is either organized in the electron dense heterochromatin or in euchromatin which is less electron dense because it consists of uncoiled DNA being transcribed. The most prominent nuclear organelle visible also at the light microscopical level is the nucleolus, which is mainly involved in rRNA transcription and ribosome formation. The active nucleolus consists of three components: The fibrillar component, which can typically be subdivided into the fibrillar centres and the surrounding dense fibrillar component, which, in turn, is surrounded by the granular component. Such a nucleolus is referred to as being fibrillo-granular. The fibrillar centres contain the enzymatic apparatus required for rRNA transcription, e.g. RNA polymerase I, and the rRNA-genes have also been localized to this structure and the dense fibrillar component (Wachtler and Stahl, 1993). It is believed that rRNA transcription occurs at the interphase between the fibrillar centre and the dense fibrillar component, and that the latter is formed by the newly synthesized rRNA. This material is subsequently spliced and arranged with protein and other non-nucleolar transcripts to form the ribosomal subunits that make up the granular component.

The mitochondria are also visible at the light microscopical level. They are numerous and may occupy a large volume of the cytoplasm. The mitochondria play key roles in cellular energy metabolism and their specialized function is reflected in the morphology. The mitochondrion consists of an outer membrane, an inner membrane and two internal compartments: the matrix and the intermembrane space. In somatic cells the mitochondria are easily recognized because they are elongated and their inner membrane presents typical transverse cristea. The number and structural organisation of the cristea change with cell function and activity. Another organelle usually located near the nucleus is the Golgi apparatus. It is composed of numerous disc-shaped membrane bounded cisternae, which resemble a stack of plates. The Golgi apparatus is involved in intracellular transport of membrane lipids and proteins as well as in the "packing" of substances for secretion from the cell. The Golgi apparatus also plays an important role in protein glycosylation as well as in the formation of lysosomes, which are electron-dense membrane-bounded structures containing acid hydrolytic enzymes like proteases, lipases and phosphatases.

The smooth (SER) and rough endoplasmic reticulum (RER) are also prominent membranous structures present in the cytoplasm typically forming tubules (SER) or cisternae (RER). The membranes of the endoplasmic reticulum are continuous with the outer nuclear membrane and enclose a continuous internal space. These membranes play a role in the synthesis of lipids and proteins. The RER is distinct from the SER because of the ribosomes located at the membrane facing the cytoplasm giving them a "rough" appearance. Free ribosomes and polysomes can also be observed in the cytoplasm of cells actively engaged in protein synthesis.

In somatic epithelial tissues, cell-cell interactions are mediated through intercellular junctions: Gap junctions are communicating junctions, tight junctions link adjacent cells to form an an impermeable layer and desmosomes belong to the adhering type of junctions. Gap junctions are constructed from proteins that form channels between two adjacent cells allowing for exchange of molecules. In electron micrographs they can be recognized because the gap ('channel') is filled with the electron-dense osmium-oxide during the processing of tissue. Tight junctions are also recognisable in electron micrographs due to specific proteins that make a direct contact across the intercellular space. Desmosomes are characterized by electron-dense coating of the apposed cell membranes from where bundles of intermediate filaments extend into the cytoplasm.

The plasma membrane not only encloses the cytoplasm but it is also an organelle actively involved in exocytosis and endocytosis. In electron micrographs this transport function of the cell membrane can be recognized by membranous pits (endosomes and coated pits) and small coated vesicles in the cell cortex. In epithelial tissues, the outer surface of the cell is often covered by numerous microvilli, which are extensions of the plasma membrane to maximize the total surface of the cell for e.g. absorption. Microvilli contain a core of actin filaments.

2.2. Phases of oocyte, zygote and embryo development

Among the large domestic animal species, the pre-hatching embryonic ultrastructure has received far greatest attention in cattle. Therefore, the following description will be based on this species, and some comparative aspects of embryonic ultrastructure in swine and horse will be presented at the end.

Over the past decade, reliable methods for in vitro production of bovine embryos have been developed. As a consequence, a majority of investigations on embryonic cell biology are performed on in vitro produced embryos. Such data should be taken with some precaution and whenever possible, in vivo developed embryos should be included as a reference. The ultrastructural data presented in the following refer to in vivo development unless the opposite is stated.

It is evident that the ultrastructure of the bovine embryo to a great extent reflects the ultrastructure of the oocyte and to a much lesser extent that of the fertilizing spermatozoon. The ultrastructure of the mature fertilizable oocyte is gradually built up during different phases of oocyte development. The first and longest phase comprises the probably several months of growth of the oocyte accompanying follicular development from the primordial follicle up to the tertiary, i.e. antral, follicle. The second phase is the development of the oocyte in the dominant follicle, a process lasting for one to two weeks referred to as "oocyte capacitation" (Hyttel et al., 1997). The third and last phase is constituted by oocyte maturation taking place in the ovulatory follicle during the approximately 24 h period between the peak of the LH-surge and ovulation. The ultrastructure of the zygote, i.e. the one-cell fertilized egg, strongly reflects what is gradually built during these phases, and with the development of the embryo towards hatching, the ultrastructure gradually changes along with the so-called maternal-embryonic transition where the embryonic genome is activated. Consequently, in order to understand the ultrastructure of the embryo, the ultrastructure of the oocyte during its growth, capacitation and maturation is described in the following paragraphs. The description is presented in a chronological sequence, and in order to avoid repetitions only the changes noted from one stage to the next are described.

3. Ultrastructure of the oocyte, zygote and pre-hatching embryo in cattle

3.1. The oocyte

The basic ultrastructure of the oocyte is generated during its growth accompanying follicular growth from the primordial to the tertiary follicle. When the tertiary follicles in a cohort reach a diameter of about 3-5 mm, one dominant follicle is selected, and the ultrastructure of the oocyte in this particular follicle is modified during its so-called capacitation. The oestrous cycle in cattle generally comprises 2 or 3 follicular waves, and only the dominant follicle of the last wave becomes ovulatory. In the ovulatory follicle the oocyte undergoes a third and last phase of ultrastructural changes during an approximately 24 h period between the peak of the LH-surge and ovulation. During this last phase referred to as maturation, the oocyte becomes fertilizable.

3.1.1. The oocyte growth phase

During the oocyte growth phase, the inside zona diameter of the gamete increases from less than 30 μm in the quiescent primordial follicle to more than 120 μm in the tertiary follicle. We have characterized the ultrastructure, transcriptional activity and developmental competence of bovine oocytes in relation to the sequential stages of follicular development and, for oocytes from the early antral follicles, in relation to their diameter. The general ultrastructure, transcriptional activity and nucleolar ultrastructure of oocyte growth in cattle is summarized in Table 1 and Figure 1 (Crozet et al., 1986; Fair et al., 1996, 1997 a, b).

The quiescent **primordial follicle** is 34.6 ± 3.7 μm (mean ± SD) in diameter (Hulshof, 1995). The oocyte which is 27.9 ± 3.3 μm in diameter is surrounded by a single layer of flattened granulosa cells. Gap and intermediate junctions are present between adjacent granulosa cells, and the latter type of junctions, but not gap junctions, are also present between the granulosa cells and the oocyte.

Table 1: Ultrastructural characteristics of bovine oocytes during growth, capacitation and maturation in relation to follicular growth.

Feature	Primordial	Primary	Secondary	Early tertiary			Later tertiary		
					Oocytes <100 μm	*Oocytes 100-110 μm*	*Oocytes >110 μm*	*Capacitated*	*Matured*
Meiosis	Prophase I	Prophase I	Prophase I	Prophase I	Prophase I	Prophase I	Prophase I	Prophase I	Metaphase II
Oocyte-granulosa Cell junctions	Intermediate and gap	Intermediate and gap	Intermediate and gap	Intermediate and gap	Intermediate and gap	Intermediate retraction	Intermediate	Superficial	Disconnection
Zona pellucida (ZP)	Absent	Absent	Small portions	Present	Present	Present	Present	Present	Present
Perivitelline Space (PvS)	Absent	Absent	Absent	Absent	Absent	Small	Moderate	Moderate	Wide
Microvilli	Few, bent	Few, bent	More erect	Numerous ZP embedded	Numerous ZP embedded	Numerous bent in PvS	Numerous bent in PvS	Many bent in PvS	Many erect in PvS
Mitochrondria	Round	Round	Round	Round/elongated	Round/elongated And few hooded	Round/hooded	Hooded	Hooded	Hooded
Golgi complexes	Few	Few	Few	Moderate	Moderate	Moderate	Many	Moderate	Few
Free RER	Extensive	Extensive	Extensive	Extensive	Extensive	Moderate	Absent	Absent	Absent
Free SER	Extensive	Extensive	Extensive	Extensive	Extensive	Moderate	Sparse	Sparse	Moderate
Vesicles	Few	Few	Moderate	Many	Many	Many	Abundant	Abundant	Abundant
Lipid droplets	Few	Few	Moderate	Many	Many	Many	Many	Abundant	Abundant
Cortical granules	Absent	Absent	Few	Moderate clustered	Many clustered	Many clustered	Many clustered	Many clustered	Many solitary

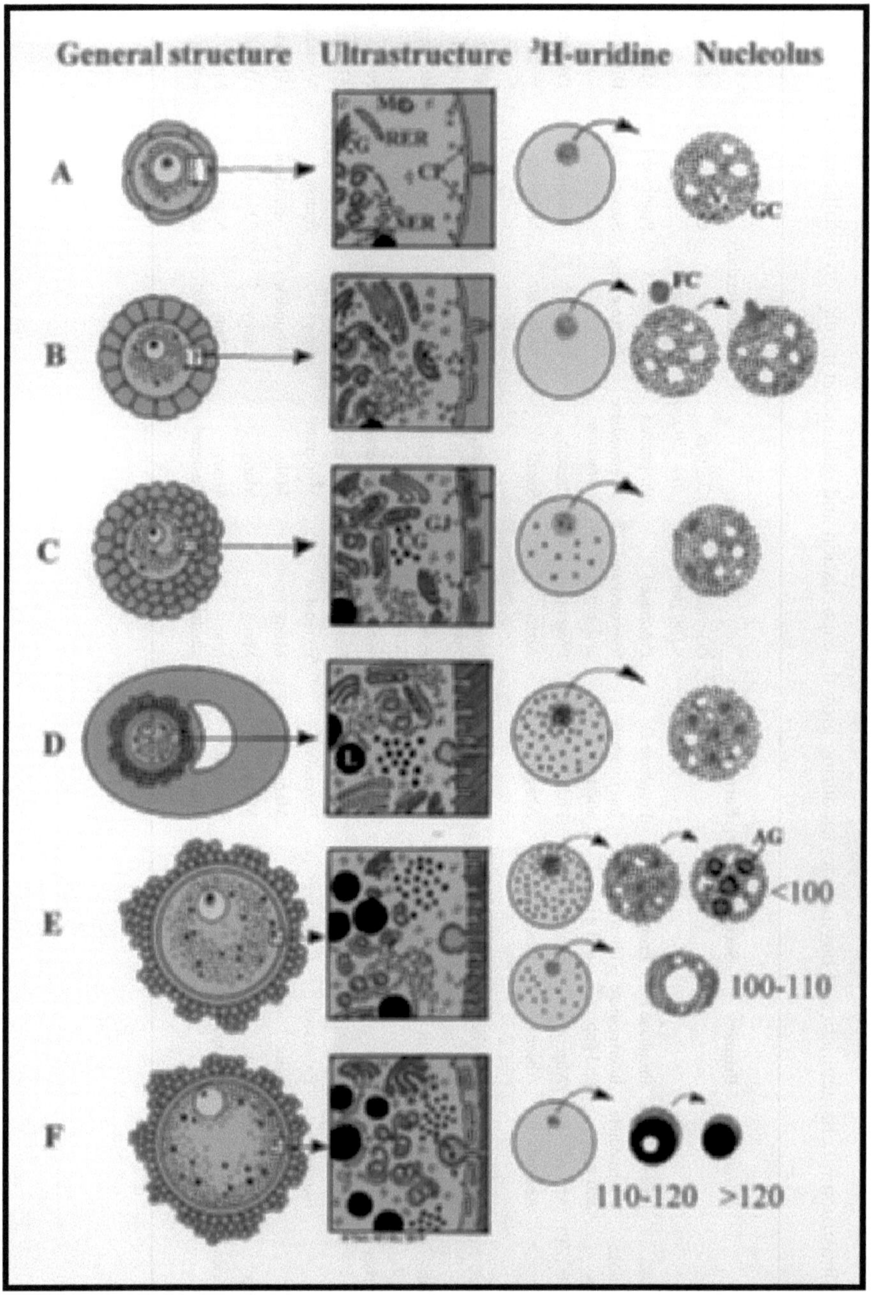

Figure 1: *Schematic drawing of aspects of bovine oocyte growth with respect to general ultrastructure, autoradiographic labelling of the nucleus on*

semithin (2µm) sections following 30 min incubation with ^{3}H-uridine, and nucleolar ultrastructure.

(A) *Primordial follicle. The oocyte is surrounded by a single layer of flattened granulosa cells. The central oocyte nucleus is surrounded by round mitochondria (M), smooth (SER) and rough (RER) endoplasmic reticulum and small Golgi complexes (G). The oocyte cortex presents numerous coated pits (CP) and vesicles. The oocyte nucleus lacks autoradiographic labelling and the nucleolus only presents the granular component (GC) interspersed with vacuoles (V).*

(B) *Primary follicle. The oocyte is surrounded by a single layer of cuboidal granulosa cells, and it presents some microvilli and some elongated mitochondria. The eccentrical oocyte nucleus lacks autoradiographic labelling. Close to the granular oocyte nucleolus fibrillar centres (FC) appear and gradually start to invade the nucleolus.*

(C) *Secondary follicle. The oocyte is surrounded by more than one layer of cuboidal granulosa cells. Small patches of zona pellucida (hatched areas) material have appeared, gap junctions (GJ) have developed between the oocyte and the granulosa cells, and the oocyte microvilli have become more erect. The number of coated pits and vesicles at the oocyte cortex have decreased. In the oocyte, the first small clusters of cortical granules (CG) have developed and some membrane bounded vesicles have appeared (not shown). The oocyte nucleus including the nucleolus displays weak autoradiographic labelling as indicated by the black dots. The fibrillar centres have become completely incorporated into the periphery of the nucleolus forming a fibrillo-granular nucleolus.*

(D) *Early tertiary follicle up to about 1mm. The antrum of the follicle has developed and the oocyte is located in the cumulus oophorus surrounded by cumulus cells and the innermost layers of corona radiata cells, which possess projections that penetrate the zona pellucida, invaginate the oolemma and make gap junctional contact to it. In the oocyte, the organelles have attained a more even distribution throughout the ooplasm, elongated mitochondria have become more numerous, lipid droplets (L) have become common, and the number and size of the cortical granule clusters have increased as has the number of vesicles (not shown). The erect microvilli have become embedded within the zona pellucida. The oocyte nucleus including the nucleolus displays distinct autoradiographic labelling. The nucleolus is fibrillo-granular and presents numerous evenly distributed fibrillar centres.*

(E) *Tertiary follicle up to about 3 mm as represented by oocytes at 80 to 110 µm in diameter. The number of lipid droplets in the oocyte has increased. In oocytes at less than 100 µm, the nucleus including the nucleolus displays distinct autoradiographic labelling and the fibrillo-*

granular nucleolus gradually develops aggregations of electron-dense granules (AG) around the fibrillar centres. In oocytes 100 to 110 μm, the nucleus displays moderate autoradiographic labelling excluding the nucleolus, and the fibrillo-granular nucleolus develops a large central vacuole and mariginalized fibrillar centres.

(F) *Late tertiary follicles as represented by oocytes at more than 110 μm in diameter. In the oocyte, the organelles have been dislocated to the periphery, the number of lipid droplets and vesicles (not shown) have increased as have the size of the Golgi complexes. The microvilli have been released from the zona pellucida and pile up in stacks in the perivitelline space. The peripheral oocyte nucleus lacks autoradiographic labelling. The nucleolus has turned into a remnant consisting of a sphere of densely packed fibrilles with a fibrillar centre attached as a halo. In oocytes at 110 to 120 μm, the nucleolar remnant may still be vacuolated.*

The oocyte plasma membrane, i.e. the oolemma, forms numerous coated pits, and coated vesicles are found in the cortical ooplasm signalling endocytosis. The oolemma also forms projections, which penetrate between adjacent granulosa cells, and few short bent microvilli lying parallel with the oocyte surface. The nucleus of the oocyte, i.e. the germinal vesicle, occupies a central or slightly off centre position and the organelles are concentrated in the perinuclear region.

The mitochondria are predominantly round with peripheral cristae. However, a limited number of elongated mitochondria with transverse cristae are also seen. Many mitochondria contain large electron-dense granules. A sparse number of lipid droplets and vesicles containing a flocculent material are seen. Both smooth and rough endoplasmic reticulum (SER and RER, respectively) are observed in the ooplasm, either as separated aggregations or as a meshwork spatially connecting mitochondria, lipid droplets and vesicles. In the fully grown oocyte (see later), the latter spatially related complexes are referred to as metabolic units and they are thought to be engaged in oocyte metabolism (Kruip et al., 1983). The ER aggregations are often found in close spatial relationship with the nuclear envelope. A few small Golgi complexes are seen and polyribosomes are present on the surface of the RER and free in the ooplasm. The nucleus contains a nucleolus which, in general, only consists of the granular component, thus lacking the fibrillar components (see later). Such a nucleolus is from an ultrastructural point of view considered as being transcriptionally inactive. Accordingly, as determined by 30 min incubation with ^{3}H-uridine followed by autoradiography, the oocyte is transcriptionally inactive. Some primordial follicles present a few cuboidal granulosa cells probably as a first sign of activation of follicular growth. The nuclei of oocytes enclosed in such follicles often present nucleoli in which tentative fibrillar centres are seen either adjacent to or in contact with the granular component. What does this particular nucleolar ultrastructure signal? During oocyte growth, transcripts are required for the synthesis of proteins for use within the oocyte or for export out of the cell. Moreover, it is believed that the storage of certain transcripts and proteins are required for building up the meiotic and developmental competence of the gamete (Szöllösi, 1993). The transcriptional activity of a cell is normally reflected in the ultrastructure of the nucleolus, which is the site of ribosomal RNA (rRNA) transcription and subsequent synthesis of the ribosomal subunits. The actively transcribing nucleolus consists of three components: The fibrillar component, which can typically be subdivided into the fibrillar centres and the surrounding dense fibrillar component, which, in turn, is surrounded by the granular component. Such a nucleolus is referred to as being fibrillo-granular. The fibrillar centres contain the enzymatic apparatus required for rRNA transcription, e.g. RNA Polymerase I and the rRNA-genes have also been localized to this structure and the dense fibrillar component (Wachtler and Stahl, 1993). It is believed that rRNA transcription occurs at the interphase between the fibrillar centre and the dense fibrillar component, and that the latter is formed by the newly synthesized rRNA. This material is subsequently spliced and arranged with protein and other non-nucleolar transcripts to form the ribosomal subunits that make up the granular component. According to this model of nucleolar biology, the invasion of the tentative fibrillar centres into the granular component of the oocyte nucleolus is thought to mark the initial step towards the assembly of a functional ribosome synthesising nucleolus in this cell.

The mean diameter of the **primary follicle** is 46.1 ± 6.1 µm (Hulshof, 1995). The oocyte which is 31.6 ± 4.3 µm in diameter is surrounded by a single layer of cuboidal granulosa cells and occasionally small portions of zona pellucida substance can be observed between the granulosa cells and the oocyte. As in the "activated" primordial follicles, the oocyte nucleoli present tentative fibrillar centres either adjacent to or in contact with the periphery of the granular component.

The **secondary follicle** is 101.7 ± 41.8 µm in diameter (Hulshof, 1995). The oocyte which is 45.6 ± 14.0 µm in diameter is surrounded by more than one layer of cuboidal granulosa cells. Portions of zona pellucida are commonly observed associated with erect microvilli and granulosa cell processes extending towards the oocyte. The continued zona formation is associated with the embedding of these processes as well as the erect oocyte microvilli into the zona pellucida. At the same time gap, junctions are established between the granulosa cell processes and the oocyte. The concomitant decrease in the frequency of coated pits and vesicles in the oocyte indicates a shift in the mode of intercellular communication between oocyte and granulosa cells from endocytotic to a directl coupling. Rarely, ooplasmic annulate lamellae are seen and small clusters of cortical granules may be found in the cortical ooplasm. The cortical granules act in establishing the block against polyspermic fertilization (see later). The oocyte nucleoli present tentative fibrillar centres more or less embedded into the granular component. At this stage of development, the first signs of oocyte transcription can be detected after 30 min incubation with ^{3}H-uridine.

The **small tertiary follicle** up to about 1 mm in diameter presents an oocyte with a complete zona pellucida traversed by numerous granulosa cells or, when the cumulus oophorus is formed, cumulus cell projections with bulbous endings invaginating the oolemma and forming gap and intermediate junctions to the oocyte. The mitochondria are distributed throughout the ooplasm and round and elongated forms are equally present. Lipid droplets and vesicles are more numerous, as are Golgi complexes in the cortical ooplasm. Clusters of cortical granules are numerous and often related to Golgi complexes. The oocyte nucleoli are typical fibrillo-granular with numerous fibrillar centres scattered within the granular component. For unknown reasons the bovine oocyte does not present a dense fibrillar component. Abundant oocyte transcription is encountered after 30 min incubation with ^{3}H-uridine.

In the **larger tertiary follicles** we have characterized the oocyte ultrastructure according to the inside zona pellucida diameter of the cell. In such oocytes <100 μm the nuclei are, in general, located slightly off centre, the mitochondria are relocated to the perinuclear and deep cortical ooplasm in many cases and the microvilli remain embedded in the zona pellucida (Figures 3 and 4). The particular hooded mitochondria unique to ruminants are observed for the first time. The oocyte nucleus presents large fibrillo-granular nucleoli which are transcriptionally active. Oocytes from 100 to 110 μm in diameter in many cases display formation of a perivitelline space, the process of which is associated with the release of the previously embedded microvilli from the zona. The oocyte nucleus is, in general, displaced further towards the periphery as are the Golgi complexes and mitochondria amongst which the hooded form becomes more numerous. The

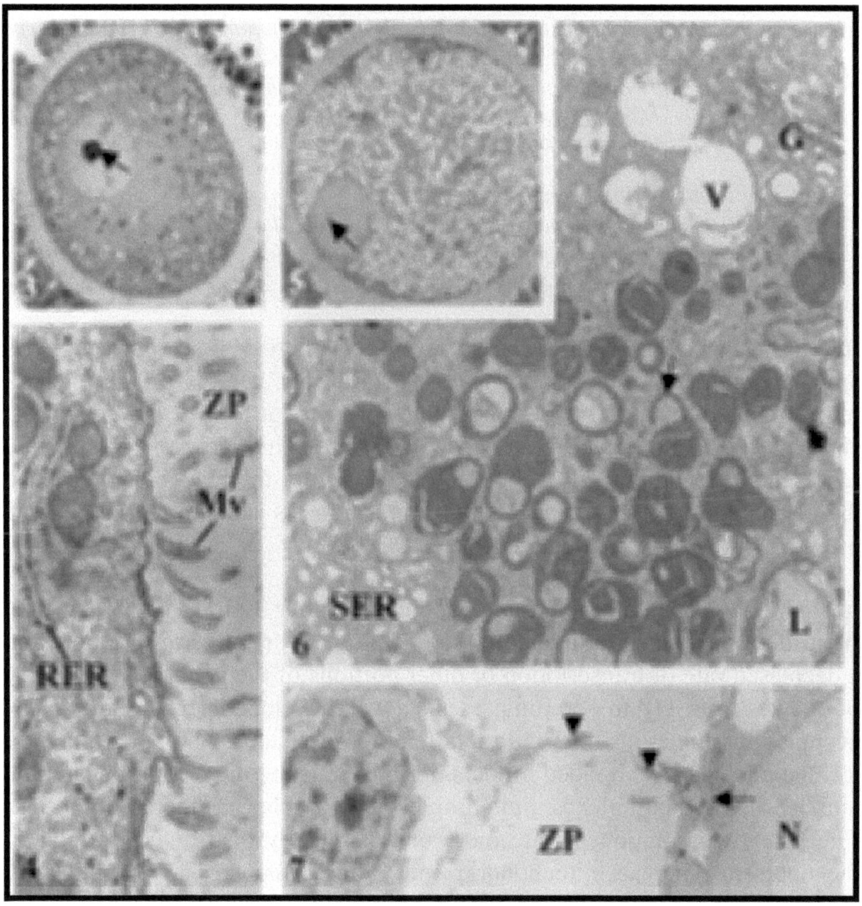

Figure 3: *Light micrograph of growing bovine oocyte. Note, in comparison with Figure 5, the more centrally located nucleus containing a large nucleolus (arrow), the relatively fewer vesicles (lucent dots in ooplasm) and the more evenly distributed mitochondria (small dark dots in ooplasm). (X180) (From Fair et al., 1997 b).*

Figure 4: *Transmission electron micrograph of growing bovine oocyte showing rough endoplamic reticulum (RER) and erect microvilli (Mv) embedded in the zona pellucida (ZP). (X14,600) (From Fair et al., 1997 b).*

Figure 5: *Light micrograph of grown bovine oocyte. Note the peripheral location of the oocyte nucleus with the small nucleolar remnant (arrow), the numerous vesicles (lucent dots in the ooplasm) evenly distributed in the ooplasm and the peripherally located mitochondria (clustered small dark dots in the ooplasm). (X200) (From Fair et al., 1997 b).*

Figure 6: *Transmission electron micrograph of mitochondria of the hooded type (arrowhead). The section results in different cut profiles of the hoods. The mitochrondria are associated with smooth endoplasmic reticulum (SER). Note the vesicles (V), Golgi complex (G) and lipid droplet (L). (X19,500) (From Fair et al., 1997 b).*

Figure 7: *Transmission electron micrograph of a cumulus cell projection (arrowheads) penetrating the zona pellucida (ZP). Note the bulbous ending of the projection forming an indentation of the oolemma (arrow) near the oocyte nucleus (N). (X3,400) (From Hyttel, 1987).*

free forms of SER and RER are less developed whereas the Golgi complexes are more numerous. In some oocytes, the fibrillar centres of the nucleoli have migrated towards the periphery and formed lentiform structures on the surface of the spherical granular component. This process presumably represents the withdrawal of the rRNA-genes from the nucleolus and, thus, the termination of rRNA and ribosome synthesis signalling that oocyte growth is more or less completed.

In such oocytes, autoradiographic examination after 30 min incubation with ^{3}H-uridine shows lack of nucleolar transcription, i.e. rRNA transcription, and a decreased rate of transcription of other portions of the genome. A majority of oocytes from 110 to 120 μm presents a well developed perivitelline space, occupied by numerous microvilli that tend to form stacks, and a peripherally located nucleus. The hooded mitochondria form is the most common and the number of Golgi complexes, lipid droplets and vesicles has increased further. The latter two components are more or less equally distributed throughout the ooplasm whereas mitochondria and Golgi complexes are peripherally located. Free forms of SER and RER are uncommon. In most oocytes, the nucleoli have expelled the granular component and only remnants consisting of a sphere of tightly packed fibrils with a fibrillar centre attached remain. Accordingly, no oocyte transcription can, in general, be detected after 30 min ^{3}H-uridine incubation. Oocytes >120 μm present even more lipid droplets and have completed nucleolar inactivation in by and large all cases. The nucleus is located peripherally in the ooplasm and the perivitelline space is evident, but the cumulus cells are still connected to the oocyte through long slender projections penetrating the zona pellucida (Figures 5 and 7). Numerous large clusters of cortical granules are present in the cortical ooplasm together with mitochondria of the hooded type (Figures 6 and 8).

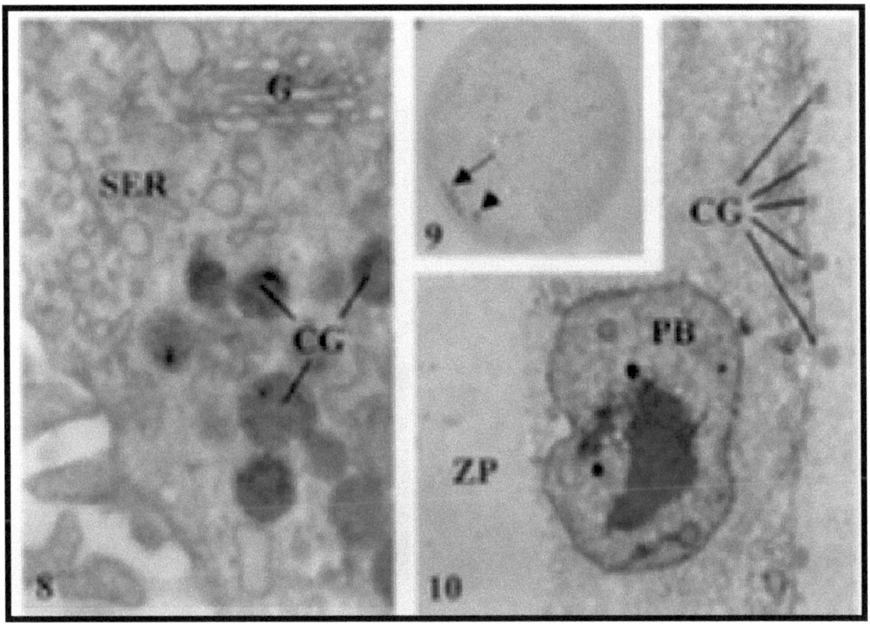

Figure 8: *Transmission electron micrograph showing a portion of a cluster of cortical granules (CG) located adjacent to a Golgi complex (G) and profiles of smooth endoplasmic reticulum (SER). (X61,000) (From Hyttel et al., 1986 a).*

Figure 9: *Light micrograph of a bovine oocyte at metaphase II. Note the metaphase plate (arrowhead) and the first polar body (arrow). (X170) (From Hyttel et al., 1986 a).*

Figure 10: *Transmission electron micrograph of the cortical granules (CG) aligned in the ooplasm along the oolemma ready for exocytosis. Note the first polar body (PB) adjacent to the zona pellucida (ZP). (X 9,400) (From Hyttel et al., 1986 a).*

Interestingly, the oocyte achieves the competence to complete meiotic maturation to metaphase II in vitro at a diameter of about 110 μm coinciding with the de-activation of its transcriptional machinery, indicating that the necessary compartment of messengers in the form of proteins and mRNAs has been set in place.

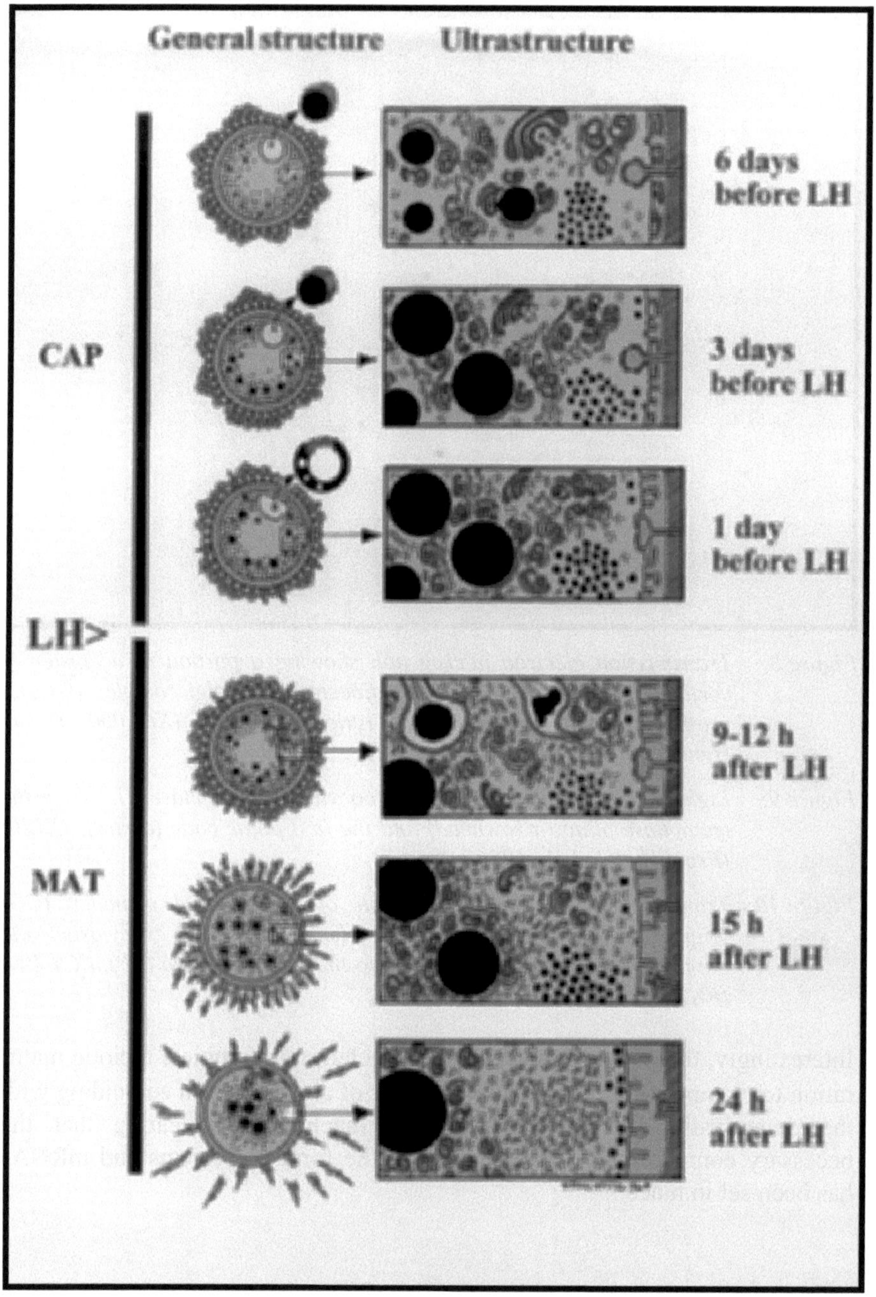

Figure 2: Schematic drawing of ultrastructural aspects of bovine oocyte capacitation (CAP) in the dominant follicle up to the LH peak and final oocyte maturation (MAT) after the peak. See Figure 1 for labels.

(A) *Oocyte from a dominant follicle 6 days before the LH peak. The general ultrastructure is identical with that obtained at the end of oocyte growth (Figure 1F).*

(B) *Oocyte from a dominant follicle 3 days before the LH peak. The number of microvilli stacks has decreased as have the size of the Golgi complexes, the amount of lipid droplets has increased, and the cortical granule clusters have dislocated to a more superficial location.*

(C) *Oocyte from a dominant follicle on the day before the LH peak. Some individual corona cells display elongation and the corona cell projections have been retracted to a more superficial location, the perivitelline space has enlarged, the microvilli have become more erect, and the size of the Golgi complexes has been further reduced. Moreover, the envelope of the oocyte nucleus has become undulating and the nucleolar remnant has transformed into a ring-like structure including the fibrillar centre with a central and several secondary vacuoles.*

(D) *Oocyte at "germinal vesicle breakdown" from an ovulatory follicle at 9-12 h after the LH peak. The perivitelline space develops further and in the oocyte the mitochondria tend to arrange around the lipid droplets and the nuclear envelope is disolved into tubules of SER and microtubules appear adjacent to the condensing chromosomes.*

(E) *Oocyte at metaphase I from an ovulatory follicle at about 15 h after the LH peak. The number and size of the lipid droplets has increased and mitochondria have assembled around the droplets and these conglomerates have attained a more even distribution throughout the ooplasm. Numerous ribosomes have appeared especially around the chromosomes and the size of the Golgi complexes has decreased further.*

(F) *Oocyte at metaphase II from an ovulatory follicle at about 24 h after the LH peak. The bulk of the cortical granules are distributed at solitary positions along the oolemma. The lipid droplets and mitochondria have attained a more central location in the ooplasm leaving a rather organelle free peripheral zone in which the most prominent features are large clusters of SER. Golgi complexes are practically absent.*

3.1.2. Oocyte capacitation

With the growth of the dominant follicle, the ultrastructure of the fully grown oocyte is modified during its capacitation. We have examined this process in a model, where the dominant follicle of the first follicular wave of the oestrus cycle in cows was induced to ovulate by treatment with a prostaglandin $F_{2\alpha}$ analogue on day 7 after ovulation (Assey et al., 1994a). This model allowed for collection of oocytes for ultrastructural analyses from the dominant follicle throughout the entire phase of dominance. The results are summarized in Table 1 and Figure 2. During the days approaching the regression of the corpus luteum, i.e. the final period of the luteal phase, the frequency of microvilli stacks on the oocyte surface decreases as does the size of the oocyte Golgi complexes whereas that of the lipid droplets increases. The cortical granule clusters are dislocated to more superficial locations and some granules migrate to solitary positions along the oolemma. During the period between luteolysis and the LH-surge, i.e. the initial period of the follicular phase, individual cumulus cells exhibit elongation and some of the cumulus cell process endings are retracted to a more superficial location on the surface of the oolemma. The perivitelline space enlarges and the size of the Golgi complexes is further reduced. Moreover, the oocyte nuclear envelope becomes undulating, especially in the regions facing the zona pellucida, and the nucleolar remnant displays vacuolisation. Both of these phenomena are presumably related to the following breakdown of the oocyte nucleus. Interestingly, data reported by Hendriksen et al. (2000) indicate that the competence of the oocyte to produce blastocysts in vitro increases with completion of capacitation. Superovulation with exogenous gonadotrophins may have an adverse effect on oocyte capacitation as indicated by a lack of at least the vacuolisation of the nucleolar remnant (Assey et al., 1994b).

3.1.3. Oocyte maturation

The maturation of the oocyte during the approximately 24 h period from the peak of the LH-surge to ovulation comprises the progression of meiosis from the diplotene stage of prophase I to metaphase II accompanied by a series of ultrastructural and molecular changes of the ooplasm. The ultrastructural changes have been described in detail in relation to the time of the LH-peak in unstimulated (Kruip et al., 1983) as well as gonadotrophin stimulated cattle (Hyttel et al., 1986a). The results are summarized in Table 1 and Figure 2. Other researchers have described certain ultrastructural aspects of the immature or maturing oocyte (Flemming and Saacke, 1972; de Loos et al., 1989, 1992; Van Blerkom et al., 1990). The breakdown of the oocyte nucleus, i.e. the germinal vesicle, occurs approximately 9 to 12 h after the LH-peak and studies of oocyte maturation in vitro have demonstrated that the nuclear envelope appears to be dissolved into SER (Hyttel et al., 1987). This process is accompanied by dissolution of the nucleolar remnant and the gradual decoupling of the cumulus cell endings from the oocyte as detailed during oocyte maturation in vitro (Hyttel, 1987). Moreover, the Golgi complexes are reduced in size and the mitochondria tend to accumulate around the lipid droplets emphasising the structure of the metabolic units of the oocyte. At about 15 h after the LH peak, most oocytes have reached metaphase I and the mitochondria move from the peripheral location to a more even distribution throughout the ooplasm. Concomitantly, the mitochondrial clustering around the lipid droplets is enforced and the Golgi compartment is further reduced. At about 20 h after the LH-peak, most oocytes have reached metaphase II and the first polar body is abstricted and at about 24 h ovulation occurs (Figure 9). During the last hours of maturation, the lipid droplets and mitochondria attain a more central location in the ooplasm, the SER tends to form large clusters in the cortical ooplasm and the cortical granules migrate to solitary positions along the oolemma (Figure 10). The peripheral migration of the cortical granules appears to be compromised to a certain degree during oocyte maturation in vitro rendering such oocytes more susceptible to polyspermic fertilization (Hyttel et al., 1986b). The large SER clusters may play a role as intracellular deposits of Ca^{++} ions as in skeletal muscle cells allowing for the intracellular increase in the Ca^{++} concentration in conjunction with oocyte activation at fertilization. Moreover, they may serve as a membrane reservoir for the formation of the envelopes of the pronuclei (see later). At both metaphase I and II, the chromosomes are contained within a spherical organelle-free area displaying increased electron-density of the cytoplasmic matrix. The density of this area is very similar to that observed in the matrix of the oocyte nucleus before its breakdown. Hence, it is tempting to speculate that substances from the nuclear matrix remain in contact with the metaphase chromosomes without being delineated by the nuclear envelope. Interestingly, spherical areas of similar density are found adjacent to the sperm head during its decondensation in the ooplasm after fertilization and, moreover, the prophase chromosomes of the first mitosis located centrally in the zygote are also embedded in such material (see later).

Through the phases of growth, capacitation and maturation, the oocyte has now reached the stage where the ultrastructural architecture for fertilization and initial embryonic development is established and the necessary messengers for the initiation of these processes are stored in the form of proteins and RNAs.

3.2. Fertilization and development of the zygote

Fertilization occurs in the - at least in cattle - poorly defined transition between the ampulla and the isthmus of the oviduct resulting in the formation of the zygote, i.e. the one-cell fertilized egg, progressing through the first post fertilization cell cycle being terminated by the first post-meiotic mitosis. During the process of fertilization, the ultrastructural components architectured during oocyte growth, capacitation and maturation are re-utilized for a number of specific purposes.

We have described the ultrastructural changes associated with fertilization in relation to the estimated time of ovulation as determined by timing of the LH-peak in gonadotrophin stimulated cows (Hyttel et al., 1988 a). The results are summarized in Figure 11. Other researchers have described more specific ultrastructural features of fertilization in unstimulated (Crozet, 1984) or stimulated cattle (Brackett et al., 1980) but without relating the findings to endocrine data. Also, the ultrastructural investigations of bovine in vitro fertilization have added to the understanding of the process of fertilization (Hyttel et al., 1988 b,c). The oocyte sheds the accompanying expanded cumulus cells during or within few hours after ovulation. The spermatozoa directly contact a major surface area of the zona pellucida (Figure 12). After the fertilizing spermatozoon has attached to the zona, the acrosome reaction occurs. This process is characterized by multiple fusion between the sperm plasma membrane and the outer acrosome membrane resulting in formation of tiny vesicles.

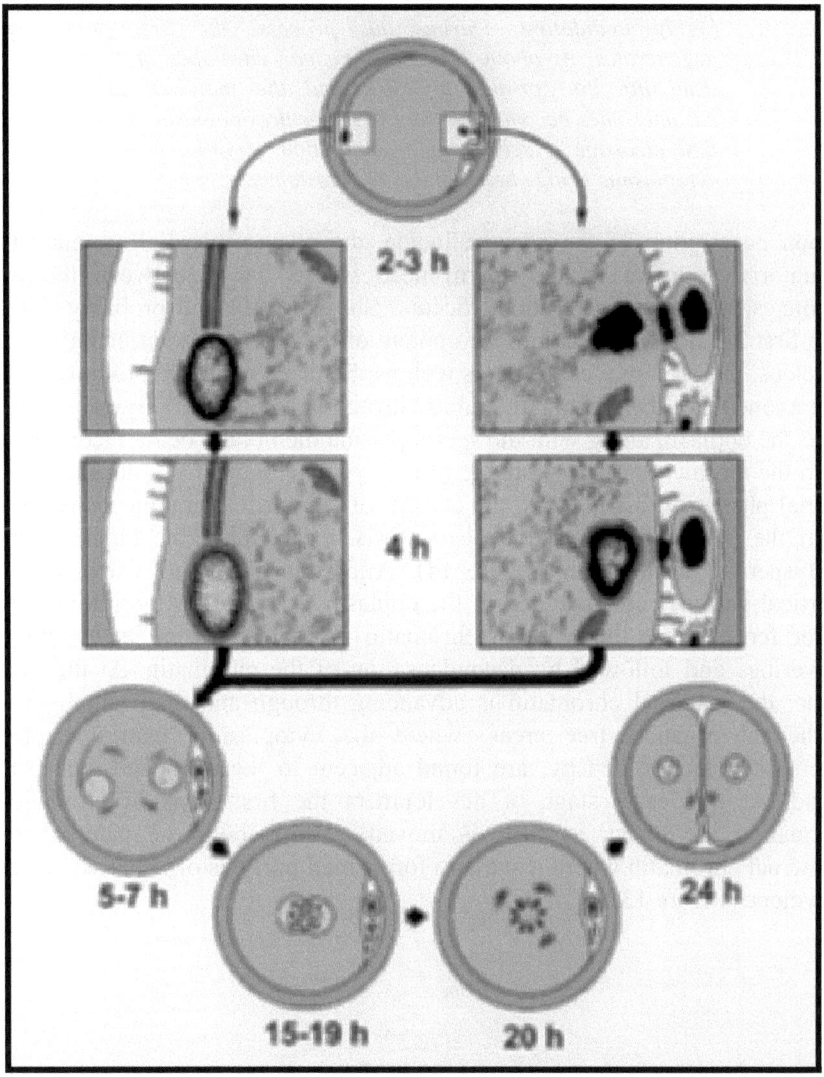

Figure 11: Schematic drawing ultrastructural aspects of pronucleus development and first cleavage in the bovine zygote. See Figure 1 for labels.

At 2-3 h after ovulation, SER extends towards the decondensing sperm head and the anaphase chomatin of the oocyte and forms fragments of nuclear envelope. At about 4 h, the chromatin is surrounded by a complete nuclear envelope and the second polar body is almost abstricted. At 5-7 h, the pronuclei swell to their characteristic spherical appearance, and in the subsequent period they migrate to a close apposition and the apposing portions of the nuclear envelopes become highly undulating. During this process, the first polar body degenerates. At about 20 h, the nuclear envelopes of the pronuclei dismantle into profiles of SER and the maternal and paternal chromosomes become localized in the central ooplasm. About 24 h, the first cleavage is seen with accumulation of Golgi complexes at the cytoplasmic bridge between the two blastomeres.

Upon penetration of the zona pellucida, the oocyte microvilli contact the equatorial segment of the sperm head where fusion between the two gametes, i.e. syngamy, initially occurs. Subsequently and probably within the first 2-3 h after ovulation, the content of the spermatozoon including the nucleus, the proximal centriole as well as the distal centriole prolonged into the axoneme of the sperm tail, and its surrounding coarse fibres, are expelled into the ooplasm along with the sperm plasma membrane being incorporated into the mosaic plasma membrane of the zygote (Figure 13). Along with the initial phases of this process, the content of the cortical granules is released into the perivitelline space by exocytosis, establishing the block against polyspermic fertilization (Figure 14). After this event, only few solitary cortical granules are scattered in the ooplasm. Also within the initial 2-3 h after fertilization, the paternal chromatin is denuded from its membrane coverings and followed by decondensation of the chromatin. At the same time, the maternal chromatin is advancing through ana- and telophase II. Spherical organelle-free areas, where the cytoplasmic matrix displays increased electron-density, are found adjacent to the decondensing sperm head. At this early stage of development the first steps of pronucleus formation are taken, as SER is moved towards both the paternal and maternal chromatin where it starts to form small portions of tentative nuclear envelope (Figure 13).

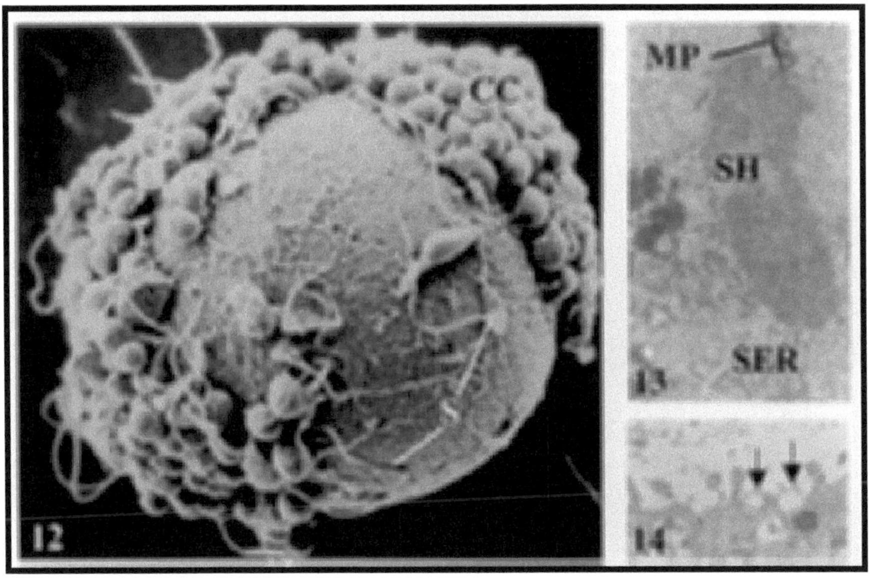

Figure 12: *Scanning electron micrograph of bovine oocyte at 1 h after insemination in vitro. Note the cumulus cells (CC) and spermatozoa (S) on the surface of the zona pellucida (X23,000).*

Figure 13: *Transmission electron micrograph of decondensing sperm head (SH) 2-3 h after ovulation. Profiles of smooth endoplasmic reticulum (SER) extend towards the sperm head and form fragments of nuclear envelope. Note that the midpiece (MP) of the spermatozoon is still associated with the sperm head. (X7,100) (From Hyttel et al., 1988 b).*

Figure 14: *Transmission electron micrograph of cortical granules liberating their content into the perivitelline space through exocytosis (arrows) (X6,700) (From Hyttel et al., 1986 a).*

Apparently within an hour or so, the two sets of chromatin are completely surrounded by nuclear envelopes establishing two tiny pronuclei. Along with this process, the abstriction of the second polar body occurs accompanied by the gradual degradation of the first polar body (Figure 15). The midpiece of the sperm tail is still closely linked to the paternal pronucleus. Apart from the importance of the sperm chromatin, the role of the proximal centriole of the spermatozoon should also be emphasized. Thus, the oocytes of the large domestic species lack centrioles, and the centrioles forming the microtubule asters of the first and succeeding mitoses arise from doublings of the proximal centriole of the fertilizing spermatozoon (Sathananthan, 1997). During these initial steps of fertilization well-developed Golgi complexes re-appear in the ooplasm.

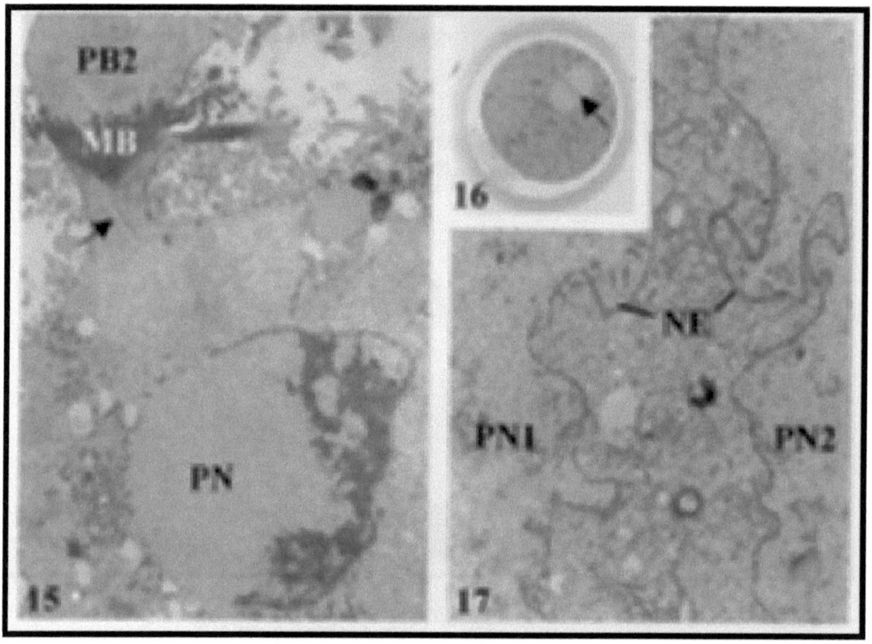

Figure 15: Transmission electron micrograph of the abstriction of the second polar body (PB2). Note the maternal pronucleus (PN) which is surrounded by a complete nuclear envelope, the midbody (MB) and the numerous microtubules extending from the pronucleus through the midbody into the polar body (arrow). (X7,500) (From Hyttel et al., 1986 a).

Figure 16: Light micrograph of a bovine zygote presenting two apposed pronuclei (arrow). (X150) (From Hyttel et al., 1988 b).

Figure 17: Transmission electron micrograph of the apposed regions of two apposed pronuclei (PN1/PN2). Note the highly undulating nuclear envelope (NE) (X6,100) (From Hyttel et al., 1988 b).

Subsequently, the pronuclei swell to their characteristic spherical appearance accompanied by completion of chromatin decondensation, and about 10 h post ovulation most zygotes exhibit spherical pronuclei (Laurincik et al., 1994). Along with this process, so-called nucleolus precursor bodies that later act as enucleation sites for nucleolus formation (see later) are formed in the pronuclei (Laurincik et al., 1996). The precursor bodies are spherical masses of tightly packed fibrils of about 1 μm in diameter that share ultrastructural similarities with the nucleolar remnants of the fully grown oocyte. With the progression of the cell cycle, they may develop a central vacuole. The precursor bodies are not active in rRNA transcription and ribosome formation. Another feature accompanying pronucleus swelling is the synthesis of numerous arrays of parallel cytoplasmic annulate lamellae connected to the SER. The functions of these lamellae is not clear, but it has been hypothesized that they are implicated in the regulation of gene expression in cells having an extended interval between transcription and translation (Kessel, 1992). This context certainly applies to the zygote and early embryo, where mRNAs transcribed during oocyte development are translated with a remarkable delay during initial embryonic development.

The two pronuclei migrate to a close apposition slightly off centre in the zygote, and about 14 h after ovulation most zygotes exhibit apposed pronuclei. In cattle, the DNA-synthetic phase (the S-phase) of the first post-fertilization cell cycle is initiated at about 12 h after ovulation and terminated at about 19 h (Laurincik et al., 1994). Along with this process, the organelles, vesicles and lipid droplets tend to accumulate centrally in the ooplasm around the pronuclei leaving a rather organelle free cortical zone. In conjunction with pronuclear apposition, pronounced undulations of the nuclear envelopes of the pronuclei are seen in the apposed regions probably preparing for breakdown of the envelopes which is seen at about 24 h after ovulation (Figures 16 and 17). This process is often referred to as synkaryosis, but it should be emphasized that the two pronuclei do not fuse, but undergo a dissolution of the nuclear envelopes similar to the one seen at the breakdown of the oocyte nucleus at maturation. Chromatin condensation and dissolution of the nucleolus precursor bodies occur in the pronuclei already before synkaryosis preparing for the first mitosis, and following synkaryosis the assembled chromatin is found in an organelle-free area displaying increased electron-density surrounded by numerous Golgi complexes. Immediately after synkaryosis, karyokinesis and cytokinesis proceed resulting in the formation of two daughter nuclei enclosed in each blastomere.

3.3. The embryo

Morphologically, the first cleavages cause repartition of the cytoplasm of the mammalian zygote into smaller and smaller compartments without changing its overall volume (McLaren, 1982). Soon after the first cleavage, a certain degree of developmental asynchrony may be seen between the individual blastomeres of the embryo. Thus, in the cattle and swine, 3-cell embryos are found. Due to this phenomenon, the ultrastructural features of a blastomere is most correctly referred to in the context of the number of the post-fertilization cell cycle that specific blastomeres have reached. Hence in the following, the 2nd cell cycle refers to the 2-cell stage, the 3rd to the tentative 4-cell stage, the 4th to the tentative 8-cell stage, and the 5th to the tentative 16-cell stage. In cattle, the embryo moves from the oviduct into the uterus about 4 days after ovulation during the 4th to 5th cell cycle.

Along with the initial cleavages the embryonic genome is gradually activated during the so-called maternal-embryonic transition. Thus, a low rate of transcription of the embryonic genome has been detected as early as during the 1st, i.e. the zygote (Hay-Schmidt et al., 2001) and 2nd post fertilization cell cycles (Plante et al., 1994; Hyttel et al., 1996; Viuff et al., 1996), and during the 4th cell cycle a major transcriptional activation occurs (Camous et al., 1986). Correspondingly, the duration of this cell cycle is about 21-30 h, which is considerably longer than both the former and the subsequent ones (Barnes and Eyestone, 1988). The activation of the embryonic genome has a profound effect on embryo physiology and it is, therefore, not surprising that this process is reflected by dramatic changes in the ultrastructural appearance of the embryo as well.

The ultrastructure of the bovine embryo developed both in vivo and in vitro has been chronologically described in detail by Plante and King (1994). Also a number of other researchers have contributed to the understanding of the general embryonic ultrastructure based on either in vivo or in vitro developed embryos (Massip et al., 1981; Mohr and Trounson, 1981; Betteridge and Flechon, 1988; Shamsuddin et al., 1992; Abe et al., 1999). These data are summarized in Table 2. We have conducted detailed chronological investigations on nuclear ultrastructure in in vitro produced bovine embryos with special emphasis on elucidating the activation of the embryonic genome as visualized by the rRNA-gene activation and the resulting nucleolus formation (Laurincik et al., 2000).

Table 2: Ultrastructural characteristics of bovine embryos during initial development.

Feature	Zygote stage	2-cell stage	4-cell stage	8-cell stage	16-cell stage	Morula	Compact morula	Early blastocyst	Expanded blastocysts
Cell cycle number	1	2	3	4	5	>5			
Microvilli	Abundant	Moderate	Moderate	Moderate	Few	Few	Abundant on TE	Abundant on TE	
Tight junctions	Absent	Absent	Absent	Absent	Appearing	Present in TE	Present in TE	Present in TE	Present in TE
Desmosomes	Absent	Absent	Absent	Absent	Absent	Absent	Present in TE	Present in TE	Present in TE
Intermediate filaments	Not seen	Not seen	Not seen	Not seen	Not seen	Not seen	Seen in TE	Seen in TE	Seen in TE
Mitochondria	Hooded	Hooded	Hooded	Hooded	Hooded	Elongated and hooded	Elongated and hooded	Elongated and few hooded	Elongated and few hooded
Nucleoli	NPB's	NPB's	NPB's	NPB's and FG	FG	FG	FG	FG	FG
Vesicles	Abundant	Abundant	Abundant	Abundant	Abundant	Moderate	Few	Few	Few
Lipid droplets	Abundant	Abundant	Abundant	Abundant	Abundant	Abundant	Abundant	Abundant	Few

TE: Trophectoderm; NPB's: Nucleolus precursor bodies; FG: Fibrillo-granular

The results are summarized in Figure 18. Certain ultrastructural data on this aspect are also available on in vivo developed embryos (Camous et al., 1986; King et al., 1988; Kopecny et al., 1989).

3.3.1. The 2nd cell cycle, i.e. the 2-cell stage

The general ultrastructure of the embryo during the 2nd cell cycle basically remains rather unchanged as compared with the first cycle. The blastomeres are organized with the nucleus in the centre surrounded by most organelles, vesicles and lipid droplets leaving a rather organelle-free cortical zone. The organelles include mitochondria with the hooded form still predominating, well-developed Golgi complexes with dilated cisternae, cytoplasmic annulate lamellae and SER, the latter of which connects mitochondria, vesicles and lipid droplets. Early during the second cell cycle, nucleolus precursor bodies resembling those described for the pronuclei are established in the nuclei (Figure 19).

As during the first cell cycle, they undergo extended vacuolisation with the progression of the cell cycle without completing the formation of fibrillo-granular nucleoli (Hyttel et al., 1996). The nucleolus precursor bodies are finally dissolved in conjunction with mitosis. Thus, the protein synthesis of the embryos must be based on the ribosome pool inherited from the oocyte, and polyribosomes are still abundant during the 2nd cell cycle. The nuclei also present small clumps of heterochromatin as well as nuclear annulate lamellae, the significance of which are unknown.

3.3.2. The 3rd cell cycle, i.e. the tentative 4-cell stage

Apart from the reduced blastomere volume, the general ultrastructure of the embryo during the 3rd cell cycle to a great extent resembles that of the previous cycle. Early during the cell cycle, nucleolus precursor bodies resembling those from the previous cell cycles are established. Again, the precursor bodies undergo vacuolisation with the progression of the cell cycle, but fibrillo-granular nucleoli are not formed. As a consequence of the lacking ribosome-synthesis, the amount of polyribosomes decreases as does the total protein synthesis of the embryo (Free et al., 1989). Also, the number of microvilli on the blastomere surface decreases slightly.

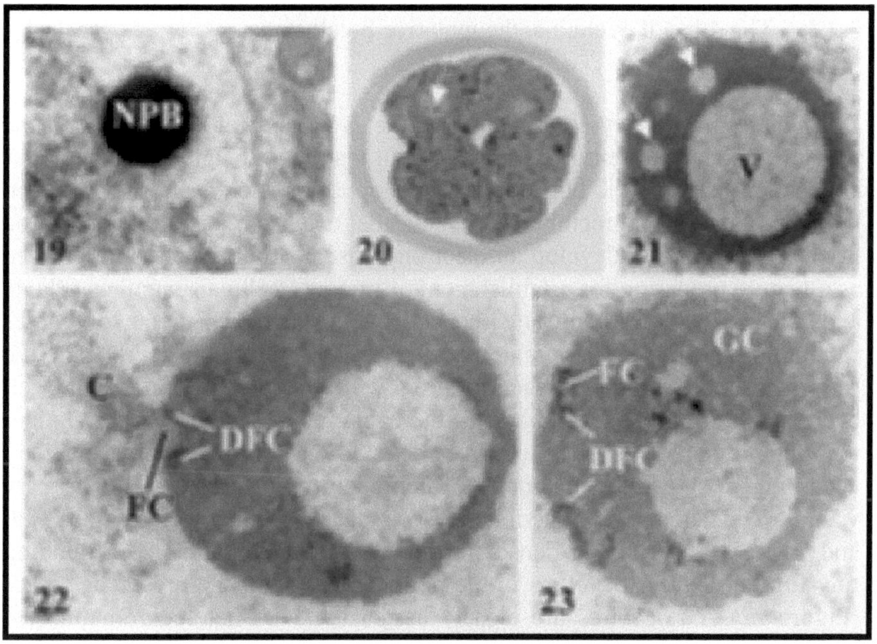

Figure 19: *Transmission electron micrograph of nucleolus precursor body (NPB) from a bovine 2-cell embryo. (X18,000).*

Figure 20: *Light micrograph of a bovine 8-cell embryo. Note the nucleolus precursor body in one of the nuclei (arrow). (X270) (From Laurincik et al., 2000).*

Figure 21: *Transmission electron micrograph of a nucleolus precursor body from an 8-cell embryo showing a central vacuole (V) and several smaller peripheral ones (arrows). (X15,600) (From Laurincik et al., 2000).*

Figure 22: *Transmission electron micrograph of the initial steps in the transformation of the nucleolus precursor body into a nucleolus. Note the chromatin (C) attached to a tiny fibrillar centre (FC) surrounded by a dense fibrillar component (DFC). X19,400 (From Laurincik et al., 2000).*

Figure 23: *Transmission electron micrograph of a fibrillo-granular nucleolus from a bovine 8-cell embryo. Note the granular component (GC) more or less embedding strands of dense fibrillar component (DFC) surrounding tiny fibrillar centres (FC). X9,200 (From Laurincik et al., 2000).*

3.3.3. The 4th cell cycle, i.e. the tentative 8-cell stage

During the 4th cell cycle, which in cattle is considerable longer than the previous and subsequent cycles, the first ultrastructural signs of intercellular connection between the blastomeres develop as short contact areas between the apposed plasma membranes. With the extension of these areas the involved membranes develop an electron-dense coating which is interpreted as a feature preparing for establishment of true intercellular junctions (Plante and King, 1994). Early during the cell cycle, nucleolus precursor bodies resembling those from the previous cell cycles are re-established. Again, the precursor bodies undergo vacuolisation with the progression of the cell cycle, and during the second half of the cycle, which is considerably longer than the previous ones, the precursors develop into fibrillo-granular nucleoli according to the following mode of action. First an eccentrically primary vacuole and later several peripheral secondary vacuoles form (Figures 20 and 21). Subsequently, a dense fibrillar nucleolar component surrounding tiny fibrillar centres develops in the peripheral zone of the precursor bodies (Figure 22). The centres are closely associated with heterochromatin. We assume that this chromatin arrangement allows for the rRNA-genes to be localized in the centres as a prerequisite for rRNA-gene transcription. Additionally, portions of dense fibrillar component form at the inner lining of the large primary vacuoles. The dense fibrillar component and the fibrillar centres expand and a presumptive granular component consisting of pre-ribosomal particles emerges and occupies the remaining portion of the precursor body (Figure 23). Thus, through this process a spherical fibrillo-granular nucleolus emerges. Concomitant with the first observation of fibrillar centres in the transmission electron microscope towards the end of the 4th cell cycle, it is, for the first time since the zygote stage, possible to localize RNA polymerase I and other important nucleolar proteins of significance for rRNA-gene transcription to the fibrillar centres by means of immunocytochemistry and confocal laser scanning microscopy (Laurincik et al., 2000). The development of the nucleolus allowing for ribosome synthesis is a prerequisite for continued embryonic development. Thus, the amount of polyribosomes decreases to a minimum during the 4th cell cycle as does the level of embryonic protein synthesis (Frei et al., 1989). At mitosis the nucleoli are dissolved as are the precursor bodies during the previous cell cycles. Along with the activation of the rRNA-genes, sets of other genes resulting in transcription of different mRNAs are activated during the 4th cell cycle.

3.3.4. The 5th cell cycle, i.e. the tentative 16-cell stage

During the 5th cell cycle, tight junctions represented by electron-dense areas at the latero-apical regions of neighbouring external blastomeres become discernible for the first time preparing for the sealing of the subsequent trophectoderm. The density of microvilli on the blastomere surfaces has decreased and the Golgi complexes present less dilated cisternae. In the nuclei, spherical fibrillo-granular nucleoli develop already from the onset of the cell cycle (Figure 23). Consequently, the amount of polyribosomes increases as does the total embryonic proteins synthesis (Frei et al., 1989). The nuclei display more clumps of heterochromatin scattered throughout.

3.3.5. The morula

Initially, the individual blastomeres of the morula are clearly discernible as individual entities protruding from the embryo surface, but with the process of compaction the profile of the individual blastomeres is lost and the surface of the embryo levels out. With this development, two cell populations are established: External cells that, in general, appear to have a lower electron-density of the cytoplasmic matrix, and internal cells that display a higher electron-density of the matrix. The external cells are connected by developing tight junctions while the internal cells are only connected by focal membrane contacts with electron-dense coating as described for the 4th cell cycle. The ultrastructural appearance of the mitochondria changes markedly at this stage of development. Thus, mitochondria of the hooded form, which were established back during the development of the oocyte in the early tertiary follicle, become fewer and elongated types with transverse cristae more numerous. Moreover, the mitochondrial matrix appears less electron-dense; a change that apparently occurs gradually from the 2nd cell cycle and onwards. In heart tissue, two populations of mitochondria have been described, an electron-lucent and an electron dense, and it has been shown that the electron-lucent population contains less RNA (Bereiter-Hahn, 1990). Taken together with the notion that in the mouse transcription of the mitochondrial genome does not occur until the blastocyst stage (Pikó, 1970, Pikó and Chase, 1973), the decreasing electron-density of the mitochondrial matrix in the developing bovine embryo may signal a reduced RNA content. RER starts to develop again for the first time since the termination of the oocyte growth phase, signalling the synthesis of proteins for export out of the cells. Phagosomes and lysosomes become a regular finding in the cells. The action of these compartments may be related to the elimination of e.g. cells undergoing apoptosis; a phenomenon that is considered as being a normal feature of initial bovine embryonic development from the morula stage and onwards (Hardy, 1997).

3.3.6. The blastocyst

In the blastocyst, the fully developed trophectoderm cells surrounds the blastocoel and the inner cell mass (Figures 24 and 25).

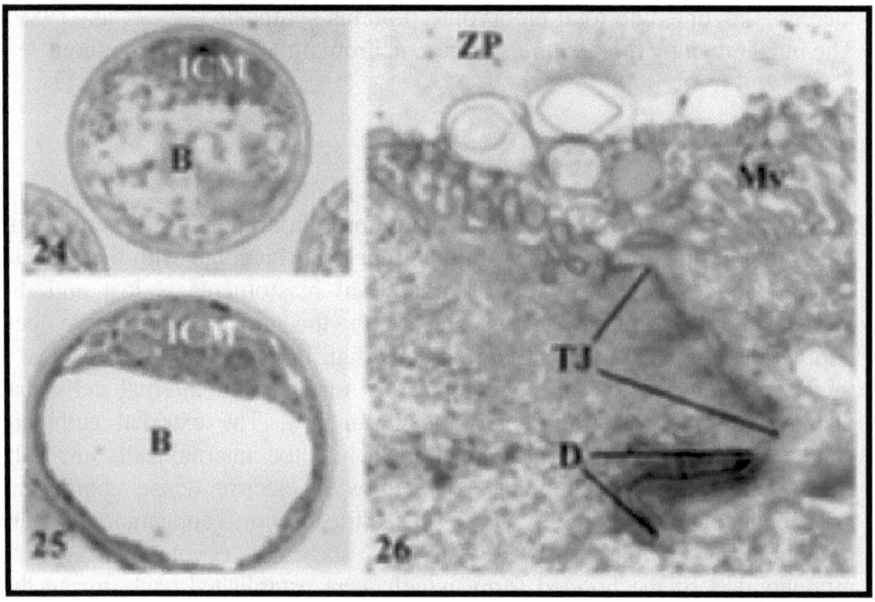

Figure 24: *Stereo micrograph of an in vitro produced bovine blastocyst. Note the blastocoel (B) and the inner cell mass (ICM). (X170) (From Vajta et al., 1997).*

Figure 25: *Light micrograph of an in vitro produced bovine blastocyst. Note the blastocoel (B) and the inner cell mass (ICM). (X250) (From Vajta et al., 1997).*

Figure 26: *Transmission electron micrograph of the junction between two adjacent trophoblast cells presenting a prominent tight junction (TJ) and a desmosome (D). Note the numerous microvilli (Mv) projecting towards the zona pellucida (ZP). (X27,000) (From Vajta et al., 1997).*

Neighbouring trophectoderm cells are connected by prominent junctional complexes presenting well developed tight junctions and desmosomes (Figure 26). In the cytoplasm, the internal face of the desmosomes are connected to a prominent meshwork of intermediate filaments. The cells of the inner cell mass remain connected with the focal electron-dense contacts as described earlier. The amount of microvilli on the surface of the trophectoderm cells increases again. Elongated mitochondria with transverse cristae and electron-lucent matrix are predominating and with the expansion of the blastocyst, the amount of lipid droplets decreases signalling that this compartment presumably constitutes an energy reserve consumed during initial embryonic development. The RER becomes more abundant whereas the SER decreases in amount.

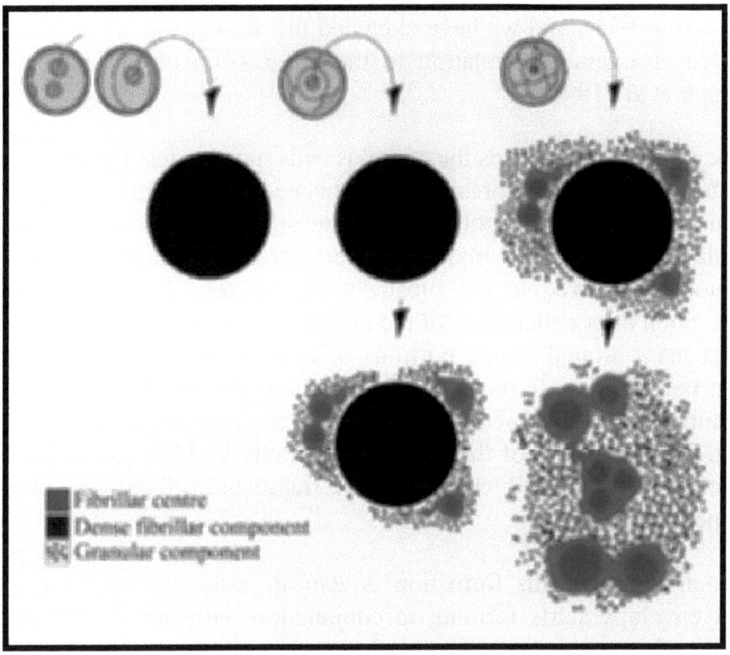

Figure 27: Schematic drawing of the ultrastructural aspects of nucleolus development in porcine embryos. During the 1st and 2nd cell cycle the nuclei display spherical fibrillar spheres. During the 3rd cell cycle, dense fibrillar component, fibrillar centres and granular component form on the surface of the sphere. During the 4th and 5th cell cycle a fibrillo-granular nucleolus emerges from the start.

4. Ultrastructure of the zygote and pre-hatching embryo in swine

4.1. Fertilization and development of the zygote

In swine, the oviduct transport is quicker and the embryos reaches the uterus already about 2 days after ovulation. The ultrastructure of fertilization is described in detail by Szöllösi and Hunter (1973) using a model where a low degree of polyspermia is induced in vivo. Additional information is provided by Norberg (1973), and we have extended this knowledge by an analysis of pronucleus formation in relation to the progression of the 1st cell cycle (Laurincik et al., 1995).

The oocyte apparently sheds the cumulus cells more or less during ovulation, and tubal oocytes only present few attached cells with their projection embedded into the zona pellucida. Some spermatozoa close to the zona pellucida may not have undergone the acrosome reaction whereas all spermatozoa observed in the substance of the zona have completed this process. Soon after penetration of the oocyte, the maternal chromatin forms a compact mass around which portions of nuclear envelope are established. Nuclear pores appear in those portions of the envelope that is closest to the chromatin. The development of intranuclear annulate lamellae is a common finding. The chromatin of the second polar body is, likewise, arranged in a compact mass around which no or only fragmentary formation of nuclear envelope is seen.

The paternal pronucleus formation is also an early phenomenon, and the nuclear envelope starts forming in conjunction with the earliest stages of chromatin decondensation. Again, the development of nuclear pores and intranuclear annulate lamellae is a consistent finding. Another characteristic is the formation of lateral projections of the forming pronucleus. It is believed that these projections contain thin sheets of chromatin. The mitochondria of the mid-piece of the fertilizing spermatozoon display swelling and degeneration. Swelling of the pronuclei to a spherical appearance is completed within 10 h after ovulation and migration of the pronuclei to a close apposition within 20 h. The S-phase of the first cell cycle is initiated at approximately 14 h after ovulation and completed at about 20 h (Laurincik et al., 1995). Abundant cisternae of SER, prominent

Golgi complexes, abundant vesicles containing flocculent material, and arrays of parallel cytoplasmic annulate lamellae are located centrally in the ooplasm around the pronuclei, and this complex of tightly packed elements is, in turn, surrounded by a zone rich in numerous large lipid droplets still leaving the most peripheral zone of the ooplasm rather organelle-free. The lipid compartment is markedly higher than in cattle. Mitochondria are found scattered all around in the ooplasm and are mostly spherical or ovoid presenting only few peripheral cristae. The vesicles and to some degree the lipid droplets are surrounded by profiles of SER.

The pronuclei contain large nucleolus precursor bodies of up to about 4μm in diameter, i.e. about 4 times as large as in cattle, appearing as spherical masses of tightly packed fibrils (Figure 28). With the apposition of the pronuclei, the precursors become localized together with condensed chromatin in the apposed regions of the pronuclei where marked undulations of the nuclear enveloped develop. Vacuolisation of the nucleolus precursor bodies is not observed.

4.2. The embryo

In swine, the major activation of the embryonic genome takes place already during the 3rd cell cycle (Tomanek et al., 1989; Hyttel et al., 2000). Accordingly, this cell cycle is of a duration of about 50 h, which is considerably longer than the previous and subsequent ones (Schoenbeck et al., 1992). There is no knowledge of whether the major transcriptional activation is preceded by a minor activation as demonstrated in cattle.

The general ultrastructure of the pre-hatching porcine embryo has been described by Norberg (1973), Szöllösi and Hunter (1973), Barends et al. (1989), Stroband et al. (1984), Stroband and van der Lende (1990) and Antalíková et al. (1992). We have extended this knowledge by a chronological ultrastructural investigation of embryonic development in vivo and in vitro (Hyttel and Niemann, 1990). Furthermore, we have conducted a detailed chronological investigation on nuclear ultrastructure in in vivo developed embryos with special emphasis on elucidating the activation of the embryonic genome as visualized by the rRNA-gene activation and the resulting nucleolus formation (Hyttel et al., 2000). The results are summarized in Figure 27. Certain ultrastructural data on the nucleolus formation were also previously available (Norberg, 1970; Tomanek et al., 1989; Laurincik et al., 1995).

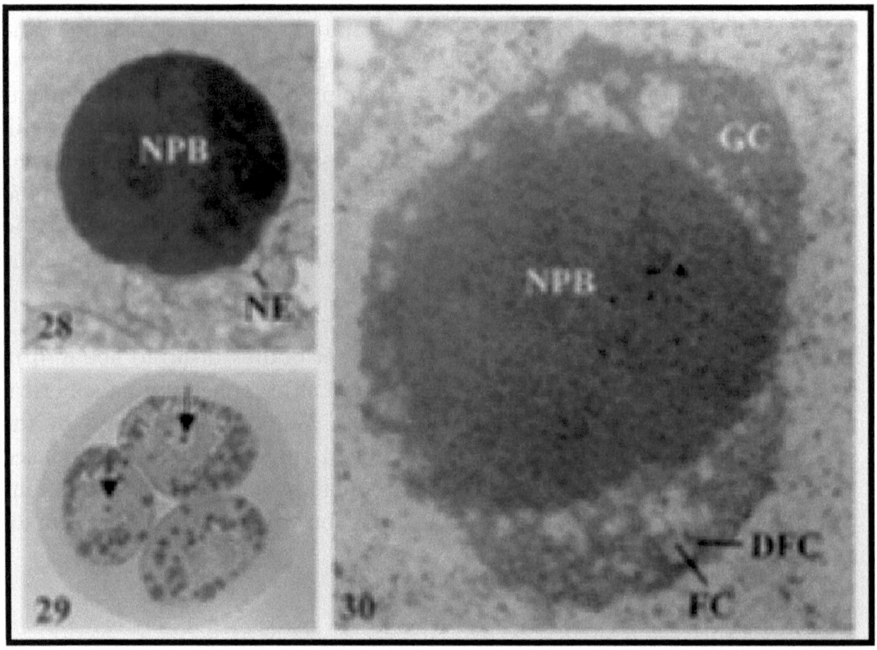

Figure 28: *Transmission electron micrograph showing a nucleolus precursor body (NPB) located adjacent to the nuclear envelope (NE) in the pronucleus of a porcine zygote. (X11,100) (From Hyttel et al., 2000).*

Figure 29: *Light micrograph showing a porcine 4-cell embryo. Note the nucleolus precursor bodies in two of the nuclei (arrows). (X290) (From Hyttel and Niemann, 1990).*

Figure 30: *Transmission electron micrograph showing the transformation of the nucleolus precursor body into a nucleolus. Note the tiny fibrillar centre (FC) surrounded by dense fibrillar component (DFC) and granular component (GC) attached to the nucleolus precursor body (NPB) as lentiform structures. (X17,600) (From Hyttel et al., 2000).*

4.2.1 The 2nd cell cycle, i.e. the 2-cell stage

The blastomeres maintain the same overall organisation as observed for the zygote: The nucleus is located in the centre surrounded by a zone rich in tightly packed SER, Golgi complexes, vesicles and cytoplasmic annulate lamellae which, in turn, is surrounded by a zone occupied by abundant large lipid droplets. Mitochondria, still of the spherical or ovoid type, are distributed throughout the cytoplasm and are often located adjacent to lipid droplets. In the nuclei, large nucleolus precursor bodies are seen, but again no vacuolisation occurs with the progression of the cell cycle.

4.2.2 The 3rd cell cycle, i.e. the tentative 4-cell stage

At the onset of the 3rd cell cycle, the blastomeres are round and present the same internal organisation as during the previous cycles. Later during the cycle, however, they assume a more polyhedral outline leading to increased appositions of their plasma membranes resulting in a decrease in the intercellular spaces. This phenomenon is regarded as the first phase of compaction (Boerjan and Te Kronnie, 1993).

Early during the 3rd cell cycle, nucleolus precursor bodies resembling those from the previous cell cycles are established again. However, from around 20 h post cleavage to the 3rd cell cycle, the nuclei of the blastomeres display different stages of nucleolus formation, ranging from the precursor bodies to fibrillo-granular nucleoli presenting fibrillar centres, a dense fibrillar component and a granular component. Different stages of nucleolus development are observed even within the same nucleus. Apparently, nucleolus formation is initiated by the formation of the dense fibrillar component and the granular component and later the fibrillar centres on the surface of, but not in, the nucleolus precursor body. Throughout this process, semilunar fibrillo-granular nucleoli more or less encapsulate precursor body (Figures 29 and 30). Soon after the observation of fibrillar centres in the transmission electron microscope, it is, for the first time since fertilization, possible to localize at least RNA polymerase I to the fibrillar centres by means of immunocytochemistry and confocal laser scanning microscopy (Hyttel et al., 2000).

4.2.3. The 4th cell cycle, i.e. the tentative 8-cell stage

During the 4th cell cycle, the next phase of compaction occurs. The blastomeres tend to become polarized: The nucleus is located towards one pole adjacent to a zone rich in organelles and vesicles with a rather organelle-free zone rich in large lipid droplets towards the other pole. Moreover, the apposition of the plasmamembranes of the blastomeres develops to such a degree that the intercellular spaces are no longer visible by light microscopy. Fibrillo-granular nucleoli, as described for the previous cycle, are already formed at the onset of this cell cycle. Again, the fibrillo-granular nucleoli and inactive nucleolus precursor bodies can be seen within the same nuclei.

4.2.4. The morula

Due to the early compaction of the porcine embryo, it is fair to regard the embryo as a morula already at the tentative 16-cell stage. At this stage of development a peripheral layer of trophectoderm cells and a group of internal cells is more or less obvious. The outer cells may still posses the polarized appearance. The mitochondria are more filamentous and their inner membranes form more cristae and tubules.

In the morula, the outer cells have a bean-like shape and they are connected by tight junctions, thus, forming a seal to create an internal environment for the inner cells.

4.2.5. The blastocyst

4.2.5.1. The early blastocyst

In the early blastocyst the blastocoel is small and the mural trophoblast only consists of one or two cells. The inner cell mass still have a compacted organisation. At the blastocoelic side of the inner cell mass and the trophectoderm cells, cytoplasmic extensions rich in microfilaments, free ribosomes and desmosomes are formed (Figure 31). The flattened outer cells are covered with microvilli towards the zona pellucida and the vesicles, mitochondria and SER assume a polarized localisation in the cytoplasm.

These cells are connected by well developed tight junctions and desmosomes (Figure 32).

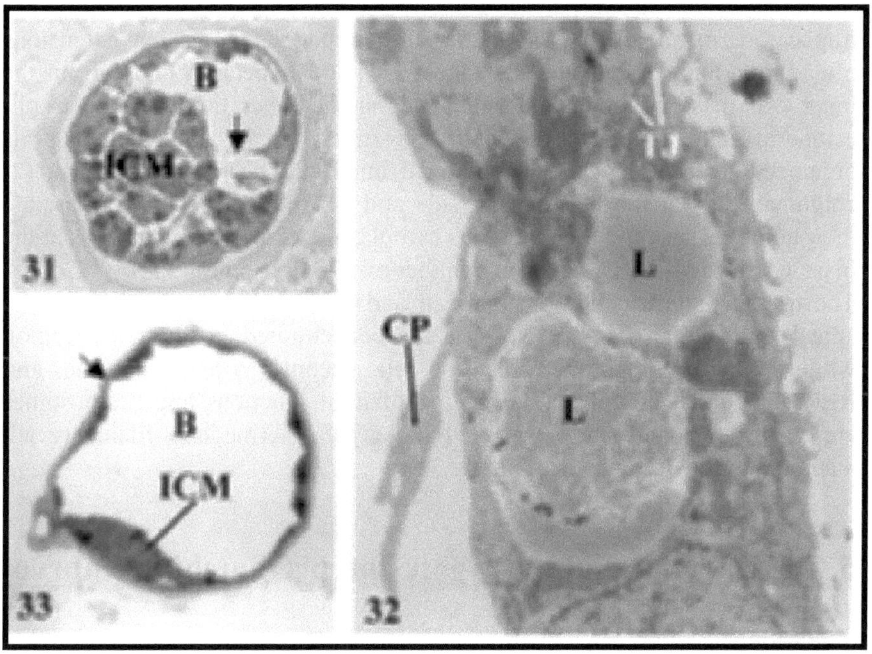

Figure 31: Light micrograph of early porcine blastocyst. Note the blastocoel (B), the inner cell mass (ICM) and the long projections extending form the trophectoderm cells (arrow). (X340) (From Hyttel and Niemann, 1990).

Figure 32: Transmission electron micrograph of the junction between two adjacent trophectoderm cells presenting a tight junction (TJ). Note the large lipid droplets (L) and the cytoplasmic projection (CP) extending into the blastocoel cavity. (X 10,000)

Figure 33: Light micrograph of expanded porcine blastocyst. Note the thin zona pellucida (arrow, as compared with Figure 31), the blastocoel (B) and the inner cell mass (ICM).(X170) (From Hyttel and Niemann, 1990).

The organelle rich cytoplasm is gradually replaced by cytoplasm devoid of SER and Golgi complexes but rich in swollen RER and free ribosomes. These cytoplasmic changes occur finally in all cells although some of the internal cells, with symmetry of cell contacts, preserve their organelle rich plasm longest.

4.2.5.2. The expanded blastocyst

In the expanded blastocyst, the cells of the inner cell mass as well as the trophectoderm cells display an increasing number of cytoplasmic extensions often extending far into the blastocoelic cavity. Especially those from the inner cell mass are long and extend along the inner lining of the trophectoderm cells. In blastocysts collected at day 7, the inner cell mass is organized in three layers: (1) polarized microvilli bearing outer cells, (2) columnar cells at the blastocoelic side and (3) internal cells with variable morphology caused by the numerous cytoplasmic extensions. The cells at the blastocoelic side show increasing number of free ribosomes.
The trophectoderm cells are polarized and microvilli are abundant on the apical side. Organelles and vesicles are concentrated in the apical region while the basal compartment of these cells becomes rich in ribosomes and RER. Junctional complexes consisting of tight junctions and desmosomes are well developed and massive bundles of intermediate filaments are anchored to the desmosomes (Figure 33).

5. Ultrastructure of the zygote and pre-hatching embryo in horse

Fertilization and initial embryonic development in the horse present several unique features. First of all, the oviduct transport is delayed and the embryo does not reach the uterus until by and large 6 days after ovulation (Battut et al., 1997). At this time development has progressed to the morula or early blastocyst stage. Moreover, the term pre-implantation covers a much more extended period in the horse than in cattle and swine. Thus, whereas bovine and porcine embryos undergo initial attachment preparing for placentation already 16-17 and 12-13 days after ovulation, respectively, the equine embryo, its foetal membranes and its circulatory system develop to an extraordinary extent before the first signs of a functional yolk sac placenta or the succeeding allantochorionic placenta are recognized at Days 22 and 40, respectively (Enders et al., 1993). In spite of this phenomenon, the present text will only consider development up to hatching. In the equine embryo, major activation of the embryonic genome included in the maternal-embryonic transition occurs during the 4th cell cycle (Ball et al., 1993; Brinsko et al., 1995; Grøndahl and Hyttel, 1996). Whether this major transcriptional activation is preceded by a minor activation as in cattle has not been reported.

The general ultrastructure of the equine zygote and embryo has been reviewed by Betteridge (1995). We have conducted chronological investigations on nuclear ultrastructure in in vivo developed embryos with special emphasis on elucidating the activation of the embryonic genome as visualized by the rRNA-gene activation and the resulting nucleolus formation (Grøndahl and Hyttel, 1996).

5.1. Fertilization and development of the zygote

The equine zygote is characterized by a pronounced polarisation inherited from the oocyte, which as in the former species is ovulated in metaphase II (Grøndahl et al., 1995). Thus, more or less one half of the ooplasm is occupied by organelles and vesicles while the other half is occupied by large lipid droplets. The pronuclei present nucleolus precursor bodies of similar ultrastructure and diameter as described for cattle. No vacuolisation of the precursor bodies is seen at any stage of development.

5.2. The embryo

During the 2nd cell cycle, i.e. the 2-cell stage, the blastomeres maintain the polarized appearance. The nuclei contain nucleolus precursor bodies. During the 3rd cell cycle, i.e. the tentative 4-cell stage, the polarisation of the blastomeres is still appreciated and, again, the nuclei contain nucleolus precursor bodies. During the 4th cell cycle, i.e. the tentative 8-cell stage, the polarisation of the blastomeres is still evident. Initially during this cell cycle nucleolus precursor bodies are observed again, but later these entities are transformed into fibrillo-granular nucleoli. Apparently, the precursor body is decondensed and the tentative dense fibrillar component of the nucleolus appear at multiple foci in it surrounded by a shell of granular component. The most striking feature of the subsequent embryonic development in the horse is the formation of the embryonic capsule. Already before the embryo hatches from the zona pellucida by and large 7 days after ovulation, the synthesis of the capsule consisting of glycoproteins is in progress, and upon hatching the capsule completely encloses the expanding blastocyst during the period up to about 21 days after ovulation.

Acknowledgements

The authors acknowledge the skilled help of Ms. Jytte Nielsen for preparing the microscopical illustrations. Figures 3, 4, 5, 6 and 7 are reproduced from Anatomy and Embryology with kind permission from Springer-Verlag, Figures 8, 9, 10, 13, 14, 15, 16 and 17 from Journal of Reproduction and Fertility with kind permission from Journals of Reproduction and Fertility, Ltd., Figures 20, 21, 22, 23, 28 and 30 from Biology of Reproduction with kind permission from Society of for the Study of Reproduction, Inc., and Figures 24, 25, 26, 29, 31 and 33 from Molecular Reproduction and Development with kind permission from Wiley-Liss, Inc.

References

Abe H, Otoi T, Tachikawa S, Yamashita S, Satoh T, Hoshi H. 1999. Fine structure of bovine morulae and blastocysts in vivo and in vitro. Anat Embryol 199:519-527.

Antalíková L, Tománek M, Jendrulek T. 1992. Ultrastructural and morphometric analysis of cytoplasmic structures in preimplantation pig embryos. Folia Biologica (Praha) 38:113-120.

Assey RJ, Hyttel P, Greve T, Purwantara B. 1994a. Oocyte morphology in dominant and subordinate follicles. Mol Reprod Dev 37:35-344.

Assey RJ, Hyttel P, Roche JF, Boland M. 1994b. Oocyte structure and follicular steroid concentrations in superovulated versus unstimulated heifers. Mol Reprod Dev 39:8-16.

Ball BA, Ignotz GG, Brinsko SP, Thomas PGA, Miller PG, Ellington JE, Currie WB. 1993. The in vitro block to development and initiation of transcription in early equine embryos. Equine Vet J (Suppl) 15:87-90.

Barends PMG, Stroband HWJ, Taverne N, te Kronnie G, Leën MPJM, Blommers PCJ. 1989. Integrity of the preimplantation pig blastocyst during expansion and loss of polar trophectoderm (Rauber cells) and the morphology of the embryoblast as an indicator for developmental stage. J Reprod Fertil 87:715-726.

Barnes FL, Eyestone, WH. 1988. Early cleavage and the maternal zygotic transition in bovine embryos. Theriogenology 33:141-151.

Battut I, Colchen S, Fieni F, Tainturier D, Bruyas JF. 1997. Success rates when attempting to nonsurgically collect equine embryos at 144, 156 or 168 hours after ovulation. Equine Vet J (Suppl) 25:60-62.

Bereiter-Hahn J. 1990. Behavior of mitochondria in the living cell. Int Rev Cytol 122:1-63.

Betteridge KJ, Fléchon J-E. 1988. The anatomy and physiology of pre-attachment bovine embryos. Theriogenology 29:155-187.

Betteridge KJ. 1995. Equine pregnancy: the road from Caxambu? Biol Reprod Mono Ser 1:115-123.

Boerjan M, te Kronnie G. 1993. The segregation of inner and outer cells inporcine embryos follows a different pattern compared to the segregation in mouse embryos. Roux's Arch Dev Biol 203:113-116.

Brackett BG, Yon KO, Evans JF, Donawick WJ. 1980. Fertilization and early development of cow ova. Biol Reprod 23:89-205.

Brinsko SP, Ball BA, Ignotz GG, Thomas PG, Currie WB, Ellington JE. 1995. Initiation of transcription and nucleologenesis in equine embryos. Mol Reprod Dev 42:298-302.

Camous S, Kopecny V, Flechon JE. 1986. Autoradiographic detection of the earliest stage of [3H]-uridine incorporation into the cow embryo. Biol Cell 58:195-200.

Crozet N, Kanka J, Motlik J, Fulka, J. 1986. Nucleolar fine structure and RNA synthesis in bovine oocytes from antral follicles. Gamete Research 14:65-73.

Crozet N. 1984. Ultrastructural aspects of in vivo fertilization in the cow. Gamete Res 10:241-251.

de Loos F , Van Maurik P, Van Beneden T, Kruip TAM. 1992. Structural aspects of bovine oocyte maturation. Mol Reprod Dev 31:208-214.

de Loos F, Van Vliet C, Van Maurik P, Kruip Th.A.M. 1989. Morphology of immature bovine oocytes. Gamete Res 24:197-204.

Enders AC, Schlafke S, Lantz KC, Liu IK. 1993. Endoderm cells of the equine yolk sac from day 7 until formation of the definitive yolk sac placenta. Equine Vet J (Suppl) 15:3-9.

Fair T, Hulshoff SCJ, Hyttel P, Greve T, Boland M. 1997b. Oocyte ultrastructure in bovine primordial to tertiary follicles. Anat Embryol, 195:327-336.

Fair T, Hyttel P, Greve T, Boland M. 1996. Nucleolus structure and transcriptional activity in relation to oocyte diameter in cattle. Mol Reprod Dev 43:503-512.

Fair T, Hyttel P, Hulshof SCJ, Boland M, Greve T. 1997a. Nucleus ultrastructure and transcriptional activity of bovine oocytes in preantral and early antral follicles. Mol Reprod Dev 46:208-215.

Flemming WN, Saacke RG. 1972. Fine structure of the bovine oocyte from the mature graafian follicle. J Reprod Fertil 29:203-213.

Frei RE, Schultz GA, Church RB. 1989. Qualitative and quantitative changes in protein synthesis occur at the 8-16-cell stage of embryogenesis in the cow. J Reprod Fert 86:637-641.

Grøndahl C, Hyttel P, Grøndahl ML, Eriksen T, Godtfredsen P, Greve T. 1995. Structural aspects of equine oocyte maturation in vivo. Mol Reprod Dev 42:94-105.

Grøndahl C, Hyttel P. 1996. Nucleologenesis and RNA synthesis in preimplantation equine embryos. Biol Reprod 55:769-774.

Hardy K. 1997. Cell death in the mammalian blastocyst. Mol Hum Reprod 3:919-925.

Hay-Schmidt A, Viuff D, Greve T, Hyttel P. 2001. Transcriptional activity in in vivo developed early cleavage stage bovine embryos. Theriogenology, 56:167-176.

Hendriksen PJM, Vos PLAM, Steenweg WNM, Bevers MM, Dieleman SJ. 2000. Bovine follicular development and its effect on the in vitro competence of oocytes. Theriogenology 53:11-20.

Hulshof SCJ. 1995. Bovine preantral follicles and their development in vitro. PhD Thesis, Utrecht University, The Netherlands.

Hyttel P, Callesen H, Greve T. 1986a. Ultrastrucural features of preovulatory oocyte maturation in superovulated cattle. J Reprod Fert 76:645-656.

Hyttel P, Fair T, Callesen H, Greve T. 1997. Oocyte growth, capacitation and final maturation in cattle. Theriogenology 47:23-32.

Hyttel P, Greve T, Callesen H. 1988a. Ultrastructure of in vivo fertilization in superovulated cattle. J Reprod Fert 82:1-13.

Hyttel P, Laurincik J, Rosenkranz C, Rath D, Niemann H, Ochs RL, Schellander K. 2000. Nucleolar proteins and ultrastructure in pre-implantation porcine embryos developed in vivo. Biol Reprod 63:1848-1856.

Hyttel P, Niemann H. 1990. Ultrastrucure of porcine embryos following development in vitro versus in vivo. Mol Reprod Dev 27:136-144.

Hyttel P, Viuff D, Avery B, Greve T. 1996. Transcription and cell cycle dependent development of intranuclear bodies and granules in 2-cell bovine embryos. J Reprod Fert 108:263-270.

Hyttel P, Xu KP, Greve T. 1988b. Scanning electron microscopy of in vitro fertilization in cattle. Anat Embryol 178:41-46.

Hyttel P, Xu KP, Greve T. 1988c. Ultrastructural abnormalities of in vitro fertilization of in vitro matured bovine oocytes. Anat Embryol 178:47-52.

Hyttel P, Xu KP, Smith S, Callesen H, Greve T. 1987. Ultrastructure of the final nuclear maturation of bovine oocytes in vitro. Anat Embryol 176:35-40.

Hyttel P, Xu KP, Smith S, Greve T. 1986b. Ultrastructure of in vitro oocyte maturation in cattle. J Reprod Fert 78:615-625.

Hyttel P. 1987. Bovine cumulus-oocyte disconnection in vitro. Anat Embryol 176:41-44.

Kessel RG. 1992. Annulate lamellae: a last frontier in cellular organelles. Int Rev Cytol 133:43-120.

King WA, Niar A, Chartrain I, Betteridge KJ, Guay P. 1988. Nucleolus organizer regions and nucleoli in preattachment bovine embryos. J Reprod Fert 82:87-95.

Kopecny V, Flechon JE, Camous S, Fulka J Jr. 1989. Nucleologeneis and the onset of transcription in the eight-cell bovine embryo: Fine-structural autoradiographic study. Mol Reprod Dev 1:79-90.

Kruip TAM, Cran DG, van Beneden TH, Dieleman SJ. 1983. Structural changes in bovine oocytes during final maturation in vivo. Gamete Res 8:49-47.

Laurincik J, Hyttel P, Kopecny V. 1995. DNA synthesis and pronucleus development in pig zygotes obtained in vivo: An autoradiographic and ultratstructural study. Mol Reprod Dev 40:325-332.

Laurincik J, Kopecny V, Hyttel P. 1994. Pronucleus development and DNA synthesis in bovine zygotes in vivo. Theriogenology 42:1285-1293.

Laurincik J, Kopecny V, Hyttel P. 1996. Dertailed analysis of pronucleus development in bovine zygotes in vivo: Ultrastructure and cell cycle chronology. Mol Reprod Dev 43:62-69.

Laurincik J, Thomsen PD, Hay-Schmidt A, Avery B, Greve T, Ochs RL, Hyttel P. 2000. Nucleolar proteins and nuclear ultrastructure in pre-implantation bovine embryos produced in vitro. Biol Reprod 62:1024-1032.

Massip A, Mulnard J, Huygens R, Hanzen C, Van der Zwalmen P, Ectors F. 1981. Ultrastructure of the cow blastocyst. J Submicrosc Cytol 13:31-40.

McLaren, A. 1982. The embryo. In Reproduction in Mammals, Vol. 2, Austin CR and Short RV (eds.). Second edition. Cambridge; Cambrigde University Press, 1-25.

Mohr LR, Trounson AO. 1981. Structural changes associated with freezing of bovine embryos. Biol Reprod 25:1009-1025.

Norberg HS. 1970. Nucleosphaeridies in early pig embryos. Z Zellforsch 110:61-71.

Norberg HS. 1973. Ultrastructural aspects of the preattached pig embryo: cleavage and early blastocyst stages. Z Anat Entwickl-Gesch 143:95-114.

Pikó L, Chase DG. 1973. Role of the mitrochondrial genome during early development in mice. J Cell Biol 58:357-378.
Pikó L. 1970. Synthesis of macromolecules in early mouse embryos cultured in vitro: RNA, DNA, and a polysaccharide component. Biol Reprod 21:257-279.
Plante L, King WA. 1994. Light and electron microscopic analysis of bovine embryos derived by in vitro and in vivo fertilization. J Ass Reprod Gen 11:515-529.
Plante L, Plante C, Shepherd DL, King WA. 1994. Cleavage and 3H-uridine incorporation in bovine embryos of high in vitro developmental potential. Mol Reprod Dev 39:375-383.
Sathanathan AH. 1997. Sperm centrioles in human development: Its role in mitoses. In Microscopy of reproduction and development: A dynamic approach, Motta PM (ed.), Rome, Antonio Delfino Editore, 51-66.
Schoenbeck RA, Peters MS, Rickords LF, Stumpf TT, Prather RS. 1992. Characterisation of deoxyribonucleic acid synthesis and the transition from maternal to embryonic control in the 4-cell porcine embryo. Biol Reprod 47:1118-1125.
Shamsuddin M, Larsson B, Gustafsson H, Gustrari H, Bartolome J, Rodriguez-Martinez H. 1992. Comparative morphological evaluation of in vivo and in vitro produced bovine embryos. Proc. 12th International Congress on Animal Reproduction, The Hague, The Netherlands, 1333-1335.
Stroband HWJ, Taverne N, Van den Bogaard M. 1984. The pig blastocyst: its ultrastructure and the uptake of protein macromolecules. Cell Tissue Res 235:347-356.
Stroband HWJ, van der Lende T. 1990. Embryonic and uterine development during early pregnancy in pigs. J Reprod Fertil Suppl 40:261-277.
Szöllösi D, Hunter RHF. 1973. Ultrastructural aspects of fertilization in the domestic pig: sperm penetration and pronucleus formation. J Anat 116:181-206.
Szöllösi D. 1993. Oocyte maturation. In: Thibault C, Levasseur MC, Hunter RHF (eds) Reproduction in mammals and man. Paris: Ellipses, 307-325.
Tománek M, Kopecný V, Kanka J. 1989. Genome reactivation in developing early pig embryos: an ultrastructural and autoradiographic analysis. Anat Embryol 180:309-316.
Vajta G, Hyttel P, Callesen H. 1997. Morphological changes of in vitro produced blastocysts after vitrification, in-straw direct rehydration and culture. Mol Reprod Dev 48:9-17.
Van Blerkom J, Bell H, Weipz D. 1990. Cellular and developmental biological aspects of bovine meiotic maturation, fertilization and preimplantation embryogenesis in vitro. J Electron Microsc Tech 16:298-323.
Viuff D, Avery B, Greve T, King WA, Hyttel P. 1996. Transcriptional activity in in vitro produced bovine 2- and 4- cell embryos. Mol Reprod Dev 43:171-179.
Wachtler F, Stahl A. 1993. The nucleolus: A structural and functional interpretation. Micron 24:473-505.

Chapter 14

DIFFERENCES IN GENE EXPRESSION PATTERNS RELATED TO ORIGIN OF PREIMPLANTATION EMBRYOS

Christine Wrenzycki and Heiner Niemann
Department of Biotechnology, Institute of Animal Science, Federal Agricultural Research Center (FAL), Höltystr. 10, D-31535 Neustadt, Germany

1. Introduction

The early stage of mammalian development is characterized by a fertilized oocyte, which divides continuously followed by a sequence of distinct morphological changes such as compaction, cavitation and blastocoel formation. These rapid changes in morphology require the well-orchestrated expression of developmentally important genes expressed from the maternal and/or embryonic genome (Kidder, 1992). The early stages of mammalian preimplantation development are regulated by maternally inherited mRNAs and proteins that are present in oocyte cytoplasm. After one to three cleavage divisions, dependent on the species, control of development is taken over by expression from the embryonic genome, and the maternally derived transcripts and proteins are gradually degraded (Telford et al., 1990). This transition from maternal to embryonic control of development is referred to as the maternal-embryonic transition (MET) or activation of the embryonic genome (embryonic gene activation, EGA).

Information on the regulation and time course of gene expression during oocyte maturation and preimplantation development is crucial for a better understanding of embryonic development and refinement of reproductive technologies. To achieve this, a set of "marker genes" important for the assessment of embryonic health and viability has to be identified. Therefore, detailed analysis of gene expression patterns during preimplantation development is a valuable tool for (1) understanding basic cellular and molecular mechanisms in the control of embryonic gene expression, (2) development of improved embryo culture systems as well as (3) to enhance strategies for the production of transgenic animals and cloning.

In vitro production (IVP) of bovine embryos is a useful tool for basic embryological studies as well as for commercial applications (Bavister, 1995). However, despite the recent improvements in bovine IVP, in vitro generated embryos display a number of marked differences compared to their in vivo counterparts (for review see Thompson, 1997; Holm and Callesen, 1998; Niemann and Wrenzycki, 2000) including different gene expression patterns. It has been suggested that persistent alterations from the normal expression patterns may have detrimental effects on embryonic development and contribute to the "Large offspring syndrome", frequently observed in cattle and sheep derived from IVP or cloning (Walker et al., 1996; Young et al., 1998; Niemann and Wrenzycki, 2000).

Previous studies in mice have shown that in vitro culture and manipulation of oocytes and preimplantation embryos can lead to reduced viability and growth, as well as developmental abnormalities (Bowmann and McLaren, 1970; Dean et al., 1998; Reik et al., 1993). Currently, human embryos are cultured for short periods prior to transfer. However, with the advances in assisted reproductive technologies (ART), human embryos will be cultured for longer periods to obtain blastocyst stages and to avoid multiple pregnancies. The assumption with such strategy is that embryos that are able to develop to the blastocyst stage are of superior quality.

It has been shown that the incidence of babies that are too small for their gestational age was significantly greater in singleton IVF pregnancies than in the normal obstetric population (Wang et al., 1994; Tanbo et al., 1995). This IVF related anomaly emphasizes the need for an in-depth understanding of preimplantation development at the molecular level. The bovine embryo could serve as a useful model. There is growing evidence that the bovine is, compared to the mouse, a better model for human preimplantation development; especially with regard to timing of genome activation, intermediate metabolism, interaction with the culture medium (Anderiesz et al., 2000; Menezo et al., 2000; Neuber and Powers, 2000; Niemann and Wrenzycki, 2000).

Over the past years, we have done extensive studies analyzing the modulation of embryonic mRNA expression patterns using the bovine in vitro produced embryo as a model. To unravel effects from the culture system on embryonic transcription, embryos derived from superovulated oocytes and fertilized in vivo were analysed in these studies (in vivo derived embryos). The following is a summary of these experiments. It also includes the current state of knowledge in rodent and human embryos.

2. Methodological aspects of mRNA expression analysis

The classical methods to detect gene transcripts like Northern Blotting require pools of hundreds of embryos. With the advent of the sensitive RT-PCR technology, it has now become possible to detect mRNAs of low abundance in few or even single embryos at the *qualitative* level. By simultaneous amplification of an internal standard it is feasible to determine the relative abundance of the desired transcript. The key steps of this *semi-quantitative* RT-PCR assay are shown in the following scheme (Figure 1).

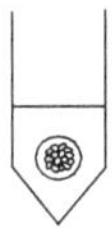

- Addition of a known amount of globin RNA as internal
  standard to a single embryo or pools of embryos
- Preparation of RNA

- Reverse transcription (RT) of RNA in cDNA
- Polymerase chain reaction (PCR)

- Agarose gel electrophoresis and ethidium bromide staining
- Digitalisation and quantification of the gel photograph employing a computer-assisted image analysis system
- Calculation of the relative abundance of each product (ratio between the intensity of the gene specific fragment and that of the globin band)

Figure 1: Schematic diagram of the semi-quantitative RT-PCR assay employed in experiments on bovine embryonic mRNA expression

With this technique, the sensitivity to detect differences in mRNA expression even in small numbers or individual oocytes or embryos is greatly improved. However, care must be taken in optimising RNA isolation procedure and RT-PCR conditions. Extracting very small amounts of RNA present within the preimplantation embryo(s) may result in the loss of RNA. Even small losses during the extraction procedure may lead to false results. Therefore rabbit globin RNA is added as an internal standard to cope with this difficulty. Priming the RT reaction with random hexamers to get the widest array of cDNAs instead of oligo dT rules out the possibility that the increase in the amount of a PCR product is attributed to an increased length of the poly(A) tail that results in a higher efficiency of reverse transcription employing oligo dT (Temeles and Schultz, 1997). Furthermore, the use of random hexamer primers rather than oligo dT minimizes the effects of sequence complexity and mRNA secondary structure. However, the highest sensitivity will be achieved when the 3'-specific primer was used during reverse transcription. The disadvantage of this approach is that only one transcript can be analysed within one sample. PCR requires the application of specific primer pairs and thus knowledge of homologous or heterologous gene sequences. If only heterologous sequences are available, a primer pair is first designed from the given heterologous sequence. Then the product is sequenced, and the resulting bovine-specific sequence is used to create the primer pair employed to detect the bovine transcript of interest. For each pair of gene specific primers, semilog plots of the fragment intensity as a function of cycle number are used to determine the range of cycle number over which linear amplification occurred and the number of PCR cycles is kept within this range. To ensure specific amplification, a hot start PCR is employed by adding the DNA polymerase at 72°C.

To circumvent the problem that differences in the relative abundance of the transcripts are due to different cell numbers of the blastocysts analysed, the number of replicates is calculated to get a high enough repeatability of the assay. The average repeatability (precision) of the assay varies from 0.60 to 0.70. A total of 6-8 replicates has to be performed for each specific gene transcript from an individual embryo. This allows calculation of statistically significant differences between treatment groups for each transcript.

The semi-quantitative RT-PCR assay can be used to compare the relative amounts of one mRNA among different samples but not the absolute level of one mRNA to that of another since the total efficiency of amplification for each set of primers during each cycle is not known (Ho et al., 1995; Temeles et al., 1997).

This methodology and various modifications thereof have been applied to investigate the effects of different culture systems and medium supplementation's as well as different nuclear transfer protocols on the mRNA expression patterns of in vitro produced embryos. These were compared with their in vivo counterparts.

Expression profiles of different genes within one group of embryos have been reviewed recently (De Sousa et al., 1998a; Memili and First, 2000; Niemann and Wrenzycki, 2000). Here, we focus on the differences in gene expression patterns in preimplantation embryos as induced by variations in their environment(s) and/or manipulation procedures. Studies comparing the temporal and quantitative gene expression patterns in in vitro or cloned versus in vivo generated bovine embryos should reveal the "normality" of in vitro produced or manipulated embryos.

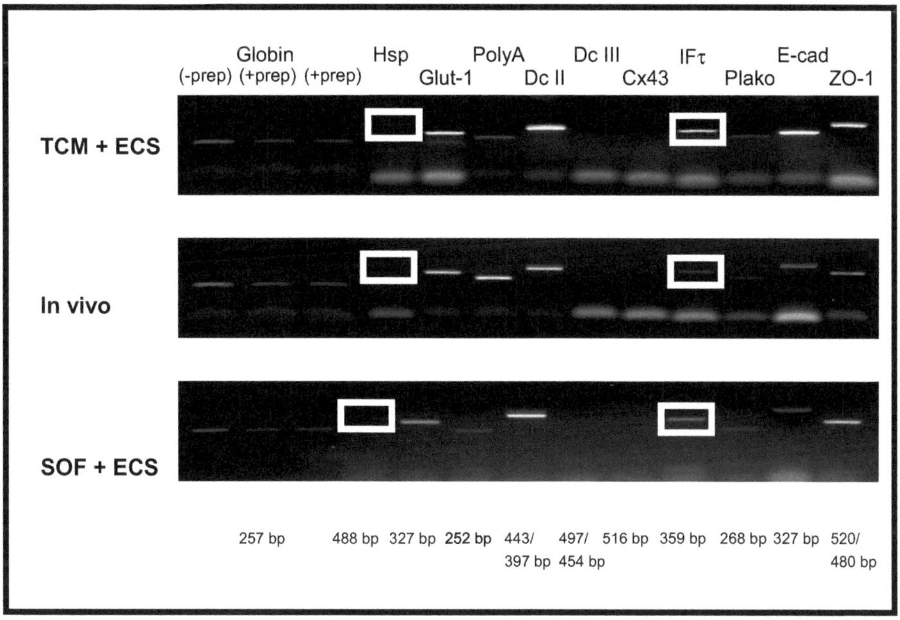

Figure 2: Representative gel photographs showing differences in the relative amount of Hsp and IF τ transcripts (white boxes) in bovine blastocysts generated in TCM supplemented with serum (TCM + ECS), in vivo, or in SOF plus serum (SOF + ECS).

The products of the analysed genes play important roles during preimplantation and postimplantation development and, hence, have the potential to be used as genetic markers (referred to as 'marker genes') for embryo viability. They are thought to be predominantly involved in the following physiological processes: compaction/cavitation, metabolism, transcription/translation, DNA methylation, (oxidative) stress, growth factor/cytokine signalling, cell cycle regulation, and growth arrest.

Figure 2 shows representative gel photographs illustrating the relative abundance of transcripts from several genes in bovine blastocysts generated in vivo (in vivo) or in vitro employing two different culture systems [TCM supplemented with serum (TCM + ECS) or SOF plus serum (SOF + ECS)].

3. Gene expression patterns in bovine preimplantation embryos

3.1. In vitro production (IVP) and nuclear transfer (NT)

Information on gene expression of domestic animal embryos predominantly stems from in vitro produced cattle embryos due to the improvement of in vitro production (IVP) techniques during the past fifteen years.

Bovine embryos generated in vitro are able to develop up to the hatching blastocyst in various formulations of media and under different culture conditions. Blastocyst developmental rates up to 40% can be obtained with optimized IVP systems (Bavister, 1995). The basic culture media widely used for bovine IVP differ in essential components; i.e. tissue culture medium (TCM) belongs to the "complex" media and was designed for cell culture in general, whereas SOF (synthetic oviduct fluid; Tervit et al., 1972) is a "simple" medium, specifically designed for the demands of the preimplantation embryo. Most media used for embryo production are supplemented with serum or serum albumin as protein sources, frequently together with somatic cells as coculture (Bavister, 1995). However, serum and BSA are complex undefined mixtures containing hormones, growth factors, vitamins, and numerous other unknown factors (Gardner and Lane, 1993). Both BSA and serum are known to be internalized or depleted of growth-promoting factors by embryos cultured in vitro. To standardize culture media, these commonly used, but "undefined" components are replaced by synthetic macromolecules, such as polyvinyl alcohol (PVA) resulting in chemically defined media (Pinyopummintr and Bavister, 1991; Eckert and Niemann, 1995; Keskintepe and Brackett, 1996).

Cloning of bovine embryos via nuclear transfer (NT) employing foetal or adult somatic donor cells has also made great progress during the past few years. The production of viable embryos and offspring derived from NT depends upon the successful combination of a great variety of factors including oocyte quality and age, the reconstruction process, the culture conditions as well as the cell cycle stages of both the donor and recipient cells (Campbell, 1999). For successful development of the reconstructed embryo, the transferred nucleus must be reprogrammed to establish the temporal, spatial and quantitative well-orchestrated expression pattern of genes correlated with normal development. This reprogramming of a nucleus constitutes a structural remodelling of the chromatin and associated histone proteins and is associated with fundamental changes in genomic activity (Gurdon, 1974).

3.2. Effects of different culture systems on gene expression patterns

The first observation that bovine embryos produced in vitro differ from their in vivo counterparts was related to the expression of the gap junction protein Connexin43 (Cx43) (Prather and First, 1993) which is necessary for the maintenance of compaction (Lee et al., 1987). Cx43 transcripts were detected in bovine morulae and blastocysts grown in vivo. In contrast, blastocysts and hatched blastocysts produced in vitro did not show detectable amounts of this gene. It has been hypothesized that under in vitro conditions the Cx43 gene either had never been activated, which would require that the maternal transcript was stable through early development, or embryonic gene transcription that had been active was terminated prematurely (Wrenzycki et al., 1996; 1998).

In another study from our lab, expression pattern of the LIF-LIF-receptor system was investigated in bovine IVP embryos and was compared with that for in vivo derived embryos. Leukaemia inhibitory factor (LIF) plays an essential role during cell differentiation and implantation in the mouse (Stewart, 1994) exerting its effects via the two receptor subunits glycoprotein 130 (gp130) and LIF-receptor β (LIFR-β) which form dimers.

Glycoprotein 130 and LIFR-β mRNA were detected throughout embryo development in vitro from the immature oocyte to the hatched blastocyst. Bovine LIF ligand transcripts were absent in immature oocytes, inconsistently expressed from the matured oocyte up to the 16-cell stage as well as in blastocysts. LIF ligand mRNA was not found in morulae, but was present again in hatched blastocysts. In contrast, in in vivo derived embryos, no LIF ligand transcripts were found in morulae and blastocysts. LIFR-β was not detected at the morula to blastocyst transition while gp130 transcripts were observed from the morula to the hatched blastocyst stage. These perturbations of the mRNA expression pattern of the LIF-LIF-receptor system could be involved in abnormal differentiation of the cell compartments forming the blastocyst (Eckert and Niemann, 1998). If the bovine LIF-LIF-receptor system plays a similar essential role in implantation as in the mouse needs further investigation. It is conceivable that either the LIF-LIF-receptor system is switched on again upon implantation or that other cytokines take over its role in bovine implantation which is supported by the lack of LIF-effects to enhance pregnancy rates upon embryo transfer (Fry, 1992).

Analysis of the Cu/Zn-SOD (superoxide dismutase) and Mn-SOD expression revealed that embryonic expression of Mn-SOD is affected by in vitro culture conditions. In the presence of serum, Mn-SOD transcripts were detected in earlier stages of development than in serum-free culture (Lequarré et al., 2001). In approximately 50% of the morulae the transcript was detectable when cultured with 5% serum while it was never found in embryos cultured under serum-free conditions. As oxidative metabolism increases sharply at the morula stage (Thompson et al., 1996) this finding could explain why transcription onset of Mn-SOD started at the earliest at that stage. Moreover, Mn-SOD expression is associated with cellular differentiation (St Clair et al., 1994) and compaction at the morula stage correlates with the first differentiation event (Shehu et al., 1996).

We have analysed the relative abundance of a set of "marker genes" in embryos cultured in medium supplemented with either serum or PVA (Wrenzycki et al., 1999). This set of developmentally important gene transcripts characterizes several physiological functions and the marker genes are involved in compaction/cavitation [gap junction protein connexin43 (Cx43), desmosomal protein plakophilin (Plako), desmosomal glycoproteins desmocollins II, III (Dc II, Dc III)], in energy metabolism [glucose transporter-1 (Glut-1), in RNA processing [poly(A) polymerase (PolyA)], in stress [heat shock protein 70.1 (Hsp)], and maternal recognition of pregnancy [interferon tau (IFN-τ)]. Transcription of the majority of genes

(Cx43, Dc II, Dc III, Glut-1, PolyA, IF τ; with the exception of Hsp) was increased in PVA-supplemented/serum-free medium derived embryos compared to their serum-generated counterparts beginning mainly at the 8-16 cell stage of development. The higher levels of mRNA under protein-free conditions could indicate the embryo's efforts to compensate for suboptimal culture conditions. Hsp transcripts were strongly increased in embryos generated in the presence of serum in the medium which might also be an indication of stress caused by non-physiological culture conditions. Expression of Hsp has been related to cellular stress (Morimoto and Santoro, 1998).

A comparison of the effects of two basic culture systems (TCM with 5% CO_2 in air vs SOF with 7% O_2, 88% N_2, 5% CO_2), each supplemented with various proteins (serum vs BSA vs PVA), on the expression of the above described set of "marker genes" revealed that the basic culture system including the medium composition and oxygen tension had profound effects on the level of specific transcripts, whereas the protein source was of minor importance (Wrenzycki et al., 2001a; see Figure 3).

With regard to the relative amounts of mRNA of the various genes of the set of “marker genes” a greater number of differences were found between embryos produced in the TCM-system and in vivo derived embryos than in SOF-generated embryos and their in vivo counterparts (see Figure 3). The SOF-system may provide an environment in which preimplantation development is more similar to that occurring in vivo than in the TCM-system. One factor contributing to these results may be the reduced O_2 tension of the SOF-system via the reduction of deleterious effects of reactive oxygen species on early development.

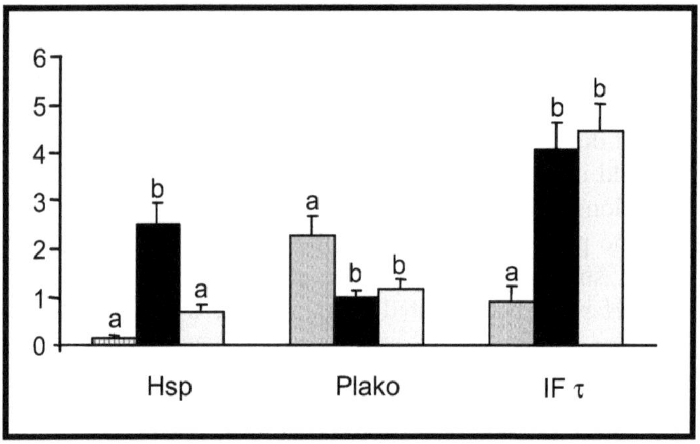

Figure 3: *Effects of basic culture system [either TCM (black bars) or SOF (open bars)] on relative mRNA abundance (means ± S.E.M.) in bovine blastocysts compared to their in vivo (lined bars) generated counterparts (for further information see text). Bars with different superscripts within each gene transcript differ significantly (ANOVA a:b $p \leq 0.05$). Experiments employing pools of embryos were repeated with at least five different embryo batches each derived from different IVP runs for each mRNA.*

No differences with respect to transcriptional activity of the above mentioned genes in embryos generated under chemically defined conditions in two different laboratories were detected. This finding provides evidence that a chemically defined simple medium can significantly reduce laboratory variability at the mRNA level of in vitro produced bovine embryos.

Recently, we analysed the expression pattern of insulin-like growth factors (IGF-I, -II) and their cognate receptors (IGF-IR, -IIR) which play important roles in the regulation of embryonic and foetal development in bovine embryos employing two different culture systems. Embryos cultured in TCM plus serum showed a significant increase of IGF-IIR mRNA at the blastocyst stage compared to those cultured in SOF plus PVA (Yaseen et al., 2001). IGF-I expression was not detected at any developmental stage. This is in contrast to previous studies in the bovine (Lonergan et al., 2000; Watson et al., 1992). The same controversy has been reported for mouse and human embryos (Rappolee et al., 1992; Doherty et al., 1994; Lighten et al., 1997). The conflicting results with regard to IGF-I expression in bovine embryos might be due to differences in culture systems or interaction of RT-PCR with specific components used in the experiments, specially the design of primer pairs. Our results indicate that it is preferentially the paracrine rather than the autocrine mechanism which is mode for action of IGF-I. Embryonic expression of IGF-I most likely does not begin until post-implantation stages.

When cultured in TCM in the presence of serum or SOF supplemented with citrate, no difference in the relative abundance of mRNA for the eukaryotic translation initiation factor eIF-1A was observed in 8-cell bovine embryos (De Sousa et al., 1998b). This is the time point when the major activation of the embryonic genome occurs suggesting that the presence of eIF-1A transcripts represents a fundamental feature of early development under conditions that support growth.

So far, only one paper reported differences in gene expression between in vivo and in vitro generated porcine embryos. No difference was observed in the quantity of cyclin B1 transcripts in 4-cell embryos (Anderson et al., 1999).

3.3. The effects of in vitro maturation on oocyte transcript levels

To characterize specific requirements for oocyte maturation, the impact of in vitro conditions during the maturation period on oocyte transcript levels has been determined in a few studies. For this, the effects of four different maturation media (TCM plus serum, SOF plus serum, SOF with or without amino acids; Watson et al., 2000) were analysed. The greatest deficit with regard to mRNA expression was observed using the amino acid-free SOF-medium indicated by decreased levels of Na/K-ATPase and cyclin A mRNAs. These alterations may reflect the acceleration of mechanisms that direct the degradation of maternal mRNAs prior to an eventual activation of the embryonic genome. A comparative analysis of the effects of in vitro versus in vivo maturation on gene expression in single bovine blastocysts revealed no significant differences in the relative abundance of Glut-1, Dc II, Plako, and E-cadherin mRNA (Knijn et al., 2001). This suggests that maturation is not the major step in the IVP process affecting expression of these genes in the preimplantation embryo.

3.4. Effects of culture conditions on foetal development

Altered gene expression patterns probably caused by different culture conditions has also been determined in foetal development. Expression of IGF-II-mRNA has been analysed in foetal bovine tissues at day 70 of gestation (63 days after transfer; Blondin et al., 2000). Livers from male foetuses originating from IVP embryos possessed approximately 2-fold higher levels of mRNA of IGF-II than those from in vivo derived males. Levels of mRNA for IGF-II tended to be lower in skeletal muscle of foetuses originating from the IVP groups compared to their in vivo counterparts. In sheep, no differences were observed in the steady-state transcript levels of IGF-II in liver, kidney, forelimb extensor carpi radialis muscle and heart between ovine control and large offspring foetuses. By contrast, the expression of IGF-IIR in large offspring foetuses was reduced by 30-60% relative to the control group (Young et al., 2001). This reduced expression was due to an epigenetic effect, as a complete loss of methylation in 9 of 12 large offspring individuals was detected.

At the protein level, no differences between in vivo and in vitro derived embryos could be detected with regard to the cell surface antigens (TEC antigens) TEC-1, -2, -3, -4 (Dowsing et al., 1998) and the transcription factor Oct-4 (Kirchhof et al., 2000).

Table 1 summarizes qualitative and quantitative differences in gene expression patterns of bovine preimplantation embryos due to different culture systems. If available, information of their in vivo counterparts is included.

Table 1: Gene expression in in vivo or in vitro derived bovine embryos

Gene	Developmental stage	Expression patterns * in vitro	in vivo	Reference
A. Qualitative differences:				
Cx43	Blastocyst	-	+	Wrenzycki et al., (1996; 1998)
LIF	Blastocyst, Hatched blastocyst	+	-	Eckert and Niemann, (1998)
LR-β	Morula, blastocyst, hatched blastocyst	+	-	Eckert and Niemann, (1998)
Mn SOD	Morula	+ / - (+ / - serum)	+	Lequarré et al., (2001)
B. Quantitative differences:				
Cx43, Glut-1, PolyA, Dc II, Dc III, IF τ	Various stages of early development, depending on the transcript	↓ / ↑ (serum / PVA)	n.a.	Wrenzycki et al., (1999)
Hsp	Matured oocyte to hatched blastocyst	↑ / ↓ (serum / PVA)	n.a.	Wrenzycki et al., (1999)
Hsp, Dc III	Morula	∅ / ↑ (SOF) / (TCM)	∅	Wrenzycki et al., (2001a)
Dc II, Plako, Glut-1, E-cad	Morula	↓	∅	Wrenzycki et al., (2001a)
Hsp	Blastocyst	∅ / ↑ (SOF) / (TCM)	∅	Wrenzycki et al., (2001a)
Plako	Blastocyst	↓	∅	Wrenzycki et al., (2001a)
IF τ	Blastocyst	↑	∅	Wrenzycki et al., (2001a)
IGF-IIR	Blastocyst	↑ / ∅ (TCM+ECS / SOF+PVA)	n.a.	Yaseen et al., (2001)
Na/K-ATPase, cyclin B	Matured oocyte	∅ / ↓ (SOF + / - aa)	n.a.	Watson et al., (2000)

* + : transcript present
  - : transcript absent
∅: "normal" pattern
↑ : up regulated
↓ : down regulated
n.a.: not analysed

3.5. Effects of feeding regime of the donor animal on mRNA expression patterns in recovered (in vivo) embryos

Embryonic transcription can also be modified by the feeding regimen of the donor animal (Wrenzycki et al., 2000). In embryos derived from heifers fed a pulp-based diet, the relative abundance of Na/K-ATPase and Cu/Zn-SOD transcripts were not affected either by day of collection or quantity of diet. Similarly, in embryos derived from heifers fed a barley-based diet, the relative abundance of Na/K-ATPase transcripts were also not changed by these main effects. However, the relative abundance of Cu/Zn-SOD transcripts were affected by day of collection and by the quantity of the diet. Differences in embryos collected at different developmental days may be attributed to varying cell numbers, whereas alterations in the transcriptional activity in embryos collected at the same day were likely to be due to alterations in metabolic end products that accumulate in reproductive tract fluids. The results of this study show for the first time that molecular variations may exist in embryos derived in vivo which developed in donor heifers on nutritional regimens differing in type and quantity. Furthermore, these findings will help to determine conditions conducive to the production of good quality embryos.

3.6. Effects of nuclear transfer protocol on mRNA expression patterns

It has been shown that heterogeneous RNA (hnRNA) production was higher in NT-derived embryos than in in vitro produced embryos (Lavoir et al., 1997). Qualitative alterations of specific gene transcripts have been detected for several growth factors (bFGF, IGF-II, IGF-Ir, IGF-I, TGF-α; Schultz et al., 1996). Aberrant transcription patterns were also observed for IL6, FGF4, and FGFr2; while similar patterns were seen for Oct4, FGF2, and gp130 (Daniels et al., 2000). Furthermore, three metabolic enzymes (LDH, CS, PFK) maintained normal expression patterns and therefore must have been properly reprogrammed following nuclear transfer (Winger et al., 2000).

Recently, we have determined the relative abundance of eight [Dc II, E-cad (cell adhesion protein E-cadherin), Glut-1, Hsp, IF τ, Mash2 (mammalian achaete-scute homologue), IGF-IIR, and DNMT (DNA methyltransferase)] specific transcripts in reconstructed blastocysts derived from distinct modifications of the nuclear transfer protocol as well as in in vivo or in vitro generated embryos (Wrenzycki et al., 2001b). Employing either FBA (fusion before activation) or AFS (fusion and activation simultaneously), Hsp mRNAs were not found in NT embryos from either protocol, whereas Hsp transcripts were detectable in IVP embryos analysed in parallel. The relative abundance (RA) of IF transcripts was significantly increased in the AFS and IVP groups compared to the FBA treatment. The use of either G_0 or G_1 donor cells significantly reduced the relative amount of DNMT transcripts and significantly increased the RA of Mash2 compared to the IVP embryos. In addition, IF transcript levels were significantly elevated in NT-blastocysts employing G_1 donor cells for NT compared to IVP embryos and those generated using G_0 cells (Figure 4). Identical results were obtained when donor cells from passage no. 5/6 or 8 were employed in nuclear transfer. Depending on steps of the cloning procedure NT-derived embryos display marked differences in their expression pattern from their IVP and in vivo derived counterparts. An aberrant expression pattern in NT-embryos was found with respect to stress adaptation, trophoblastic function and DNA methylation during preimplantation development (Wrenzycki et al., 2001b).

Figure 4: Alterations in mRNA abundance (means ± S.E.M.) attributed to two

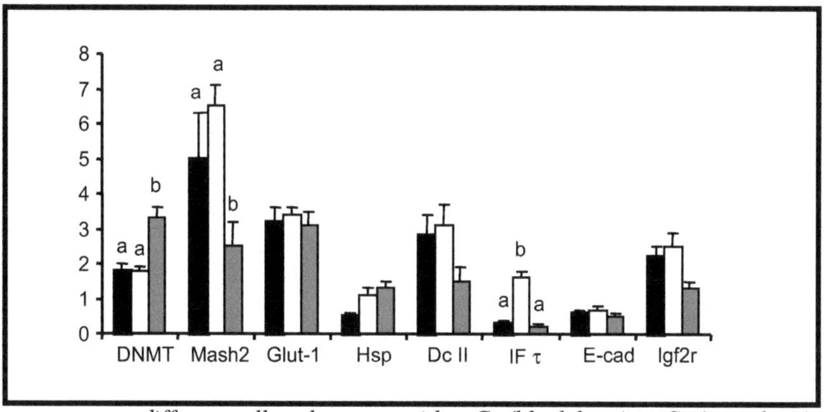

different cell cycle stages, either G_0 (black bars) or G_1 (open bars), of donor cells used for NT compared to IVP embryos (grey bars; for further information see text). Bars with different superscripts within each gene transcript differ significantly (ANOVA a:b $p \leq 0.05$). Experiments employing single embryos were repeated with eight different embryos.

At the protein level, TEC-3 antigen disappeared after fusion and was expressed again at the correct time in development when the NT embryos developed to the morula and blastocyst stage indicating complete reprogramming with respect to this gene (Van Stekelenburg-Hamers et al., 1994).

The following table summarizes qualitative and quantitative differences in gene expression patterns of bovine preimplantation embryos due to cloning. If available, information of their in vivo and in vitro counterparts is included.

Table 2: Gene expression in cloned, in vitro or in vivo generated bovine embryos

Gene	Developmental stage	Expression patterns cloned	In vitro	in vivo	Reference
A. Qualitative differences:					
bFGF	Blastocyst	+ +	-	-	Schultz et al., (1996)
IGF-II, IGF-IR	Blastocyst	+ +	(+)	(+)	Schultz et al., (1996)
IGF-I	Blastocyst	-	(+)	(+)	Schultz et al., (1996)
TGF-α	Blastocyst	+ +	+	+	Schultz et al., (1996)
IL6	Morula	-	+	n.a.	Daniels et al., (2000)
FGF4	Morula, blastocyst	-	+	n.a.	Daniels et al., (2000)
FGFr2	Morula, blastocyst	(+)	+	n.a.	Daniels et al., (2000)
B. Quantitative differences:					
Hsp	Blastocyst	- / ↓	↑	∅	Wrenzycki et al., (2001b)
DNMT	Blastocyst	↑	↑ ↑	∅	Wrenzycki et al., (2001b)
Mash2	Blastocyst	↓	↓ ↓	∅	Wrenzycki et al., (2001b)
IF τ	Blastocyst	∅ / ↑	↑	∅	Wrenzycki et al., (2001b)

* + : transcript present
+ + : strong signal
(+) : weak signal
- : transcript absent
∅ : "normal" pattern
↑ : up regulation
↑ ↑ : strong up regulation
n.a. : not analysed
↓ : down regulated
↓ ↓ : strong down regulation

4. Alterations in gene expression patterns of preimplantation rodent embryos

As described for bovine embryos, the composition of the culture media that support preimplantation development of murine embryos from the zygote to the blastocyst stage is frequently based on that for somatic cells. A simplex optimisation approach was used to develop a totally defined medium that supports preimplantation development in vitro when compared to other media (Lawitts and Biggers, 1991; 1992). The SOM (simplex optimized medium) contains a significantly lower concentration of NaCl (85 mM) than typically found in most culture media, i.e. 125 mM. A modified version of SOM, KSOM, which contains increased concentrations of sodium (95 mM) and potassium chloride (2.5 mM vs 0.25 mM) enhanced development in vitro to the blastocyst stage even more than did SOM (Erbach et al., 1994).

4.1. Effects of culture systems on mRNA expression patterns of in vitro produced embryos

It has been shown that the transcriptional activity of early embryos is affected by manipulation and culture in vitro employing transgenic embryos (Vernet et al., 1993). Transgene expression (lacZ) was already induced at the 2-cell stage in embryos developed in vitro, while there was no constitutive expression in transgenic embryos developed in vivo. When the culture medium was supplemented with antioxidants, the transgene expression was markedly reduced. Employing a similar approach, expression of a reporter gene construct (Hsp70.1-luciferase) was fifteen times higher in in vitro cultured than in in vivo derived blastocysts (Christians et al., 1995).

In vitro produced blastocysts produced less than half Glut-1 mRNA compared to in vivo derived ones, suggesting that the glucose incorporation activity may be different in in vitro produced embryos (Morita et al., 1994). In addition, the level of Glut-1 expression at the protein level was greatest in blastocysts developed in vivo followed by those developed from fresh embryos in vitro. Only low expression was found in blastocysts developed from frozen-thawed 2-cell embryos suggesting that cryopreservation may affect embryonic development possibly through reduced expression of Glut-1 (Uechi et al., 1997).

The relative amounts of mRNAs for IGF-I and IGF-II and their cognate receptors were dramatically reduced in embryos cultured in SOM containing 125 mM NaCl when compared to embryos cultured in SOM with 85 mM NaCl, whereas only a mild reduction was observed in the relative abundance of actin mRNA (Ho et al., 1994). Culture of embryos in medium containing lower concentrations of NaCl seemed to result in higher rates of mRNA synthesis and greater mRNA stability.

The relative abundance of nine mRNAs [actin, glyceraldehydes-3-phosphate isomerase (G3PDH), Na/K-ATPase, Sp1, TATA box-binding protein TBP, IGF-I, IGF-II, IGF-IR, IGF-IIR] was determined in embryos grown in two different media containing low sodium chloride concentrations (KSOM vs. Whitten's) and compared with that of in vivo grown embryos (Ho et al., 1995). Expression of the genes analysed in embryos that developed in KSOM plus amino acids was very similar to that for embryos that developed in vivo and was, in general, higher than in embryos that developed in Whitten's medium (Ho et al., 1995). Furthermore, expression of growth arrest genes (gas5, gas6, CHOP-10/gadd 153) was not different in embryos cultured in KSOM medium compared with those developed in vivo (Fleming et al., 1997) indicating that these genes are expressed as part of normal early embryonic development. Recently, it has been shown that CHOP-10 is overexpressed in blastocysts following the inhibition of protein glycosylation indicating that its expression may be a marker for one of the pathways that lead to apoptosis in the blastocyst (Fontanier-Razzaq et al., 1999).

Analysis of the genetic expression of Hexokinase (HX) and glucose phosphate isomerase (GPI) in mouse embryos cultured either in glucose/phosphate-containing human tubal fluid (HTF) medium or glucose/phosphate-free preimplantation stage one (P1) medium showed more HX transcripts in the population of blastocysts compared with morulae. Furthermore, differences in transcript incidence between early blastocysts developed in HTF and P1 medium were observed (Johnson et al., 1997). HX transcripts were determined more frequently in early blastocysts cultured in HTF medium that in P1-cultured early blastocysts. In contrast, GPI transcripts were consistently present in all embryos analysed, and were thus not significantly affected by the culture environment.

Embryonic expression of laminin was sensitive to in vitro culture conditions indicated by a significantly higher level in in vivo derived embryos compared to in vitro produced counterparts. Thus the present results raise the possibility that laminin is one of the proteins whose expression may be under the influences of uterine environment (Shim et al., 1996). The pattern of relative expression of glutaminase in cultured embryos was not significantly different from embryos that developed in vivo (Chatot et al., 1997).

It has been shown that in vitro cultured embryos produced sevenfold less platelet-activating factor (PAF) than their in vivo counterparts (O'Neill, 1997). Fertilization of embryos in vitro caused the transcript of the PAF-receptor not to be expressed in the zygote. Culture of zygotes (irrespective of their method of fertilization, either ISF - in situ fertilized - or IVF - in vitro fertilized) caused expression from the zygotic genome to be retarded by more than 24 h. This retardation did not occur if culture commenced at the 2-cell stage (Stojanov and O'Neill, 1999). The reduced expression of PAF-receptor mRNA following IVF indicates a potential deficiency in autocrine stimulation of the receptor and may also suggest that supplementation of growth media with exogenous PAF would be only partially beneficial. In addition, IGF-II expression in embryos collected fresh from the reproductive tract was significantly greater than in embryos fertilized in the reproductive tract and cultured in vitro (ISF). Its expression was further reduced in IVF embryos. IGF-II expression was significantly lower when embryos were cultured individually compared with culture in groups. The addition of PAF to culture medium partially overcame the embryo density-dependent decline of expression. Culture of ISF and IVF zygotes also caused the onset of new IGF-II mRNA transcription from the zygotic genome to be significantly retarded, until at least the 8-cell stage of development. This effect was greater for IVF than for ISF embryos. Neither IVF nor culture had any obvious effects on IGF-IIR mRNA expression (Stojanov et al., 1999). A significant correlation between IGF-I gene expression and blastocyst formation in vitro was found. However, no association between IGF-IR expression and subsequent embryo development could be detected (Kowalik et al., 1999). Treatment of mouse blastocysts with either transforming growth factor (TGF)-α or TGF-β resulted in the increased expression of the b subunit of the F_0ATPase. TGF-β also stimulated the expression of the DNA polymerase α. TGF-α treatment resulted in the increase in expression of a gene homologous to the human HEPG2 (human liver HepG2 cell line) cDNA, as well as in a decrease in expression of fibronectin (Babalola et al., 1995).

Remarkably, culture (KSOM vs. Whitten's) also affected expression of the imprinted gene H19 (Doherty et al., 2000). Culture of murine 2-cell embryos to the blastocyst stage in Whitten's medium induced expression of the normally silent paternal H19 allele, whereas little paternal expression was observed following culture in KSOM containing amino acids. Furthermore, embryos cultured in Whitten's medium showed a reduction in the methylation of a CpG dinucleotide located in the upstream imprinting control region, whereas embryos cultured in KSOM approximated more closely the in vivo situation. Imprinted expression of Snrpn (small nuclear ribonucleoprotein N) was maintained following culture in either medium, indicating that not all imprinted genes are under the same stringent imprinting controls.

In preimplantation rabbit embryos cultured with 20% oxygen concentration, an increased expression of NADH-ubiquinone oxidoreductase chain 2 (ND2) was observed when compared with embryos cultured with 5% O_2 (Koerber et al., 1998). In addition, culture with moderate oxygen stress (20%) resulted in a 1.5 fold increase in cyclophilin18 (rabCyP18) transcription and an increased peptidyl-prolyl cis/trans isomerase (PPIase) activity compared that of blastocysts cultured with 5% O_2 (Navarrete Santos et al., 2000). These changes are suggested to be involved in oxygen defence mechanisms in rabbit preimplantation embryos.

The following table (Table 3) summarizes qualitative and quantitative differences in gene expression patterns of murine and rabbit preimplantation embryos due to different culture systems. If available, information of their in vivo counterparts is included.

Table 3: Gene expression in vivo or in vitro derived rodent embryos

Gene	Develop-mental stage	Expression patterns * in vitro	Expression patterns * in vivo	Reference
A. Murine embryos:				
lacZ	2-cell stage	+	-	Vernet et al., (1993)
Hsp70.1	Blastocyst	↑ (15 times)	∅	Christians et al., (1995)
Glut-1	Blastocyst	↓	∅	Morita et al., (1994)
IGF-I, IGF-II, IGF-IR, IGF-IIR	Blastocyst	↓↓ / ∅ (SOM with 125 mM / 85 mM NaCl)	n.a..	Ho et al., (1994)
TBP	Blastocyst	↑ / ↓ (KSOMaa / Whitten's)	∅	Ho et al., (1995)
G3PDH	Blastocyst	↓ (Whitten's)	∅	Ho et al., (1995)
IGF-I, IGF-II, IGF-IR, IGF-IIR		∅ / ↓ (KSOMaa / Whitten's)	∅	Ho et al., (1995)
HX	Early blastocyst	∅ / ↑ (P1 / HTF)	n.a.	Johnson et al., (1997)
Laminins	Blastocyst	↓	∅	Shim et al., (1996)
PAF-rec.	Blastocyst	↓ (retarded)	∅	Stojanov and O'Neill, (1999)
IGF-II	Blastocyst	↓ (retarded)	∅	Stojanov et al., (1999)
H19	Blastocyst	mono- / bi-allelic (KSOMaa / Whitten's)	mono-allelic	Doherty et al., (2000)
B. Rabbit embryos:				
ND2	Blastocyst	↑ / ∅ (20% / 5% oxygen)	n.a.	Koerber et al., (1998)
CyP18	Blastocyst	↑ / ∅ (20% / 5% oxygen)	n.a.	Navarrete Santos et al., (2000)

* + : transcript present
- : transcript absent
n.a. : not analysed
∅ : "normal" pattern
↑ : up regulation
↑↑ : strong up regulation
↓ : down regulation
↓↓ : strong down regulation

5. Gene expression patterns in human embryos

In human Assisted Reproductive Technologies (ART), the composition of an improved culture medium has recently gained great interest for the extended culture of human embryos to the blastocyst (Gardner and Lane, 1997; Bavister and Boatman, 1999; Ali et al., 2000).

Quantitative RT-PCR technology has been used to analyse mRNA expression patterns in single human blastomeres from in vitro generated preimplantation embryos, and to compare the mRNA pattern with the further development of the biopsied embryo in vitro. In one study, blastomeres from twelve embryos were examined: transcripts of β-actin and IL-1R type I were detected in all blastomeres, whereas IL-1β could be detected in only nine of the blastomeres. IL-1ra was expressed in only two of the blastomeres and those were simultaneously positive for IL-1β. Both IL-1ra positive embryos were arrested in development before reaching the blastocyst stage. From these findings it was postulated that embryos expressing IL-1ra mRNA in a detectable amount appear more likely to be arrested in early developmental stages (Krüssel et al., 1998).

In another study, the expression of three placental markers that are expressed prior to implantation, β-human chorionic gonadotropin (HCG), human leukocyte antigen (HLA)-G and pregnancy specific ß-1 glycoprotein (SP-1) was investigated (Jurisicova et al., 1999). No association could be found between the number of transcripts for these genes, total cell number or cell death rate. Interestingly, there was a positive correlation between the accumulation of β-HCG and HLA-G transcripts as well as between SP-1 concentrations (assessed in the culture medium of blastocysts using enzyme-linked immunosorbent assay) and blastocyst cell numbers. Moreover, synthetic oviduct medium enriched with potassium resulted in a SP-1 concentration twice as high as that observed using human tubal fluid medium. These data suggest that embryonic secretion of SP-1 may be used to select blastocysts with higher cell number, possibly resulting in higher pregnancy rates.

6. Summary and concluding remarks

Reproductive biotechnology in domestic animals have made great progress during the past fifteen years, especially in cattle. However, a high rate of embryonic, foetal and neonatal abnormalities has been consistently observed. Developmental abnormalities have also been reported in humans and mice.

The present data on expression pattern for preimplantation embryos clearly indicate that mRNA phenotypes obtained from IVP embryos do not necessarily represent those of in vivo derived embryos. Therefore, mRNA phenotyping is a useful tool to assess the normality of IVP derived and manipulated embryos. When combined with functional studies, the obtained results will help to gain a better insight into the molecular events directing early developmental processes such as cleavage, compaction, and blastocyst formation. Gene transcripts representing a large number of essential functions in preimplantation development will be selected by their importance in supporting normal development. This set of "marker genes" will provide insight in different physiological compartments of preimplantation embryos and will help to optimize in vitro culture conditions and embryo manipulation technologies, i.e. cloning and generation of transgenic animals. The results obtained from livestock species may also have strong implications for the application of ART in humans and serve as a model for studying the effects of culture conditions on the transcriptional activities in human embryos.

Acknowledgements

The authors gratefully acknowledge the financial support of the Deutsche Forschungsgemeinschaft (DFG Ni 256/12-1, -2), the European Commission (CT98-4339 and QLK3-CT-1999-00104) and the H.W. Schaumann Foundation, Hamburg.

References

Ali J, Shahata MAM, Al-Natsha SD. 2000. Formulation of a protein-free medium for human assisted reproduction. Hum Reprod 15:145-156.

Anderiesz C, Ferraretti A-P, Magli C, Fiorentino A, Fortini D, Gianaroli L, Jones GM, Trounson AO. 2000. Effect of recombinant human gonadotrophins on human, bovine and murine oocyte meiosis, fertilization and embryonic development in vitro. Hum Reprod 15:1140-1148.

Anderson JE, Matteri RL, Abeydeera LR, Day BN, Prather RS. 1999. Cyclin B1 transcript quantitation over the maternal to zygotic transition in both in vivo- and in vitro-derived 4-cell porcine embryos. Biol Reprod 61:1460-1467.

Babalola GO, Schultz RM. 1995. Modulation of gene expression in the preimplantation mouse embryo by TGF-α and TGF-β. Mol Reprod Dev 41:133-139.

Bavister BD. 1995. Culture of preimplantation embryos: facts and artefacts. Hum Reprod Update 1:91-148.

Bavister BD, Boatman D. 1999. The neglected human blastocyst revisited. Hum Reprod 12:1607-1609.

Blondin P, Farin PW, Crosier AE, Alexander JE, Farin CE. 2000. In vitro production of embryos alters levels of insulin-like growth factor-II messenger ribonucleic acid in bovine fetuses 63 days after transfer. Biol Reprod 62:384-389.

Bowmann P, McLaren A. 1970. Viability and growth of mouse embryos after in vitro culture and fusion. J Embryol Exp Morphol 23:693-704.

Campbell KH. 1999. Nuclear equivalence, nuclear transfer, and the cell cycle. Cloning 1:3-15.

Chatot CL, Lawry JR, Germain B, Ziomek CA. 1997. Analysis of glutaminase activity and RNA expression in preimplantation mouse embryos. Mol Reprod Dev 47:248-254.

Christians E, Campion E, Thompson EM, Renard J-P. 1995. Expression of the HSP 70.1 gene, a landmark of early zygotic activity in the mouse embryo is restricted to the first burst of transcription. Development 121:113-122.

Daniels R, Hall V, Trounson AO. 2000. Analysis of gene transcription in bovine nuclear transfer embryos reconstructed with granulosa cell nuclei. Biol Reprod 63:1034-1040.

Dean W, Bowden L, Aitchison A, Klose J, Moore T, Meneses JJ, Reik W, Feil R. 1998. Altered imprinted gene methylation and expression in completely ES cell-derived mouse fetuses: association with aberrant phenotypes. Development 125:2273-2282.

De Sousa PA, Watson AJ, Schultz GA, Bilodeau-Goeseels S. 1998a. Oogenetic and zygotic gene expression directing early bovine embryogenesis: a review. Mol Reprod Dev 51:112-121.

De Sousa PA, Watson AJ, Schultz RM. 1998b. Transient expression of a translation initiation factor is conservatively associated with embryonic gene activation in murine and bovine embryos. Biol Reprod 59:969-977.

Doherty AS, Temeles GL, Schultz RM. 1994. Temporal pattern of IGF-I expression during mouse preimplantation embryogenesis. Mol Reprod Dev 37:21-26.

Doherty AS, Mann MRW, Tremblay KD, Bartolomei MS, Schultz RM. 2000. Differential effects of culture on imprinted H19 expression in the preimplantation mouse embryo. Biol Reprod 62:1526-1535.

Dowsing AT, Gougoulidis T, Dowsing BJ, Draber P, Trounson, AO. 1998. The stage-specific expression of TEC-1, -2, -3, and -4 antigens on bovine preimplantation embryos. Mol Reprod Dev 49:19-28.

Eckert J, Niemann H. 1995. In vitro maturation, fertilization and culture to blastocysts of bovine oocytes in protein-free media. Theriogenology 43:1211-1225.

Eckert J, Niemann H. 1998. mRNA expression of leukemia inhibitory factor (LIF) and its receptor subunits glycoprotein 130 (gp130) and LIF-receptor β (LIF-β) in bovine embryos derived in vitro or in vivo. Mol Hum Reprod 4:957-965.

Erbach GT, Lawitts JA, Papaioannou VE, Biggers JD. 1994. Differential growth of the mouse preimplantation embryo in chemically defined media. Biol Reprod 50:1027-1033.

Fleming JV, Fontanier N, Harries DN, Rees WD. 1997. The growth arrest genes gas5, gas6, and CHOP-10 gadd153 are expressed in the mouse preimplantation embryo. Mol Reprod Dev 48:310-316.

Fontanier-Razzaq NC, Hay SM, Rees WD. 1999. Upregulation of CHOP-10 gadd153 expression in the mouse blastocyst as a response to stress. Mol Reprod Dev 54:326-332.

Fry RC. 1992. The effect of leukemia inhibitory factor LIF on embryogenesis. Reprod Fertil Dev 4:449-458.

Gardner DK, Lane M. 1993. Amino acids and ammonium regulate mouse embryo development in culture. Biol Reprod 48:377-385.

Gardner DK, Lane M. 1997. Culture and selection of viable blastocysts: a feasible proposition for human IVF. Hum Reprod Update 3:367-382.

Gurdon JB. 1974. The control of gene expression in animal development. Oxford University Press, Oxford.

Ho Y, Doherty AS, Schultz RM. 1994. Mouse preimplantation embryo development in vitro: effect of sodium concentration in culture media on RNA synthesis and accumulation and gene expression. Mol Reprod Dev 38:131-141.

Ho Y, Wigglesworth K, Eppig JJ, Schultz RM. 1995. Preimplantation development of mouse embryos in KSOM: augmentation by amino acids and analysis of gene expression. Mol Reprod Dev 41:232-238.

Holm P, Callesen H. 1998. In vivo versus in vitro produced bovine ova: similarities and differences relevant for practical application. Reprod Nutr Dev 38:579-594.

Johnson MD, Batey DW, Behr B, Barro J. 1997. Genetic expression of hexokinase and glucose phosphate isomerase in late-stage mouse preimplantation embryos: transcription activities in glucose/phosphate-containing HTF and glucose/phosphate-free P1 media. Mol Hum Reprod 3:351-357.

Jurisicova A, Antenos M, Kapasi K, Meriano J, Casper RF. 1999. Variability in the expression of trophectodermal markers β-human chorionic gonadotrophin, human leukocyte antigen-G and pregnancy specific β-1 glycoprotein by the human blastocyst. Hum Reprod 14:1852-1858.

Keskintepe L, Brackett BG. 1996. In vitro developmental competence of in vitro matured bovine oocytes fertilized and cultured in completely defined media. Biol Reprod 55:333-339.

Kidder GM. 1992. The genetic program for preimplantation development. Dev Genet 13:319-325.

Kirchhof N, Carnwath JW, Lemme E, Anastassiadis K, Schöler H, Niemann H. 2000. Expression pattern of Oct-4 in preimplantation embryos of different species. Biol Reprod 63:1698-1705.

Knijn HM, Wrenzycki C, Hendriksen PJM, Vos PLAM, van der Weijden GC, Niemann H, Dieleman SJ. 2001. Effects of in vitro versus in vivo maturation on gene expression in single bovine blastocysts. Theriogenology 55:238.

Koerber S, Navarrete Santos A, Tetens F, Küchenhoff A, Fischer B. 1998. Increased expression of NADH-Ubiquinone oxidoreductase chain 2 ND2 in preimplantation rabbit embryos cultured with 20% oxygen concentration. Mol Reprod Dev 49:394-399.

Kowalik A, Liu H-C, He Z-Y, Mele C, Barmat L, Rosenwaks Z. 1999. Expression of the insulin-like growth factor-1 gene and its receptor in preimplantation mouse embryos: is it a marker of embryo viability? Mol Hum Reprod 5:861-865.

Krüssel JS, Simón C, Rubio MC, Pape AR, Wen Y, Huang H-Y, Bielfeld P, Polan ML. 1998. Expression of interleukin-1 system mRNA in single blastomeres from human preimplantation embryos. Hum Reprod 13:2206-2211.

Lavoir M-C, Kelk D, Rumph N, Barnes F, Betteridge KJ, King WA. 1997. Transcription and translation in bovine nuclear transfer embryos. Biol Reprod 57:204-213.

Lawitts JA, Biggers JD. 1991. Optimization of mouse embryo culture media using simplex methods. J Reprod Fertil 91:543-556.

Lawitts JA, Biggers JD. 1992. Joint effects of sodium chloride, glutamine, and glucose in mouse preimplantation embryo culture media. Mol Reprod Dev 31:189-194.

Lee S, Gilula NB, Warner AE. 1987. Gap junctional communication and compaction during preimplantation stages of mouse development. Cell 51:851-860.

Lequarré A-S, Feugang J-M, Malhomme O, Donnay I, Massip A, Dessy F, Van Langendonckt A. 2001. Expression of Cu/Zn and Mn Superoxide dismutases during bovine embryo development: influence of in vitro culture. Mol Reprod Dev 58:45-53.

Lighten AD, Hardy K, Winston RML, Moore GE. 1997. Expression of mRNA for the insulin-like growth factors and their receptors in human preimplantation embryos. Mol Reprod Dev 47:134-139.

Lonergan P, Gutiérrez-Adán A, Pintado B, Fair T, Ward F, De La Fuente J, Boland M. 2000. Relationship between time of first cleavage and expression of IGF-I growth factor, its receptor, and two housekeeping genes in bovine two-cell embryos and blastocysts produced in vitro. Mol Reprod Dev 57:146-152.

Memili E, First NL. 2000. Zygotic and embryonic gene expression in cow: a review of timing and mechanisms of early gene expression as compared with other species. Zygote 8:87-96.

Menezo YJR, Veiga A, Pouly JL. 2000. Assisted reproductive technology ART in humans: facts and uncertainties. Theriogenology 53:599-610.

Morimoto RI, Santoro MG. 1998. Stress-inducible responses and heat shock proteins: New pharmacologic targets for cytoprotection. Nature Biotechnology 16:833-838.

Morita Y, Tsutsumi O, Oka Y, Taketani Y. 1994. Glucose transporter Glut1 mRNA expression in the ontogeny of glucose incorporation in mouse preimplantation embryos. Biochem Biophys Res Com 199:1525-1531.

Navarrete Santos A, Körber S, Küllertz G, Fischer G, Fischer B. 2000. Oxygen stress increases prolyl cis/trans isomerase activity and expression of cyclophilin 18 in rabbit blastocysts. Biol Reprod 62:1-7.

Neuber E, Powers RD. 2000. Is the mouse a clinically relevant model for human fertilization failures? Hum Reprod 15:171-174.

Niemann H, Wrenzycki C. 2000. Alterations of expression of developmentally important genes in preimplantation bovine embryos by in vitro culture conditions: implications for subsequent development. Theriogenology 53:21-34.

O'Neill C. 1997. Evidence for the requirement of autocrine growth factors for the development of mouse preimplantation embryos in vitro. Biol Reprod 56:229-237.

Pinyopummintr T, Bavister BD. 1991. In vitro-matured/in vitro-fertilized bovine oocytes can develop into morulae/blastocysts in chemically defined, protein-free culture media. Biol Reprod 45:736-742.

Prather RS, and First NL, 1993. Cell-to-cell coupling in early-stage bovine embryos: a preliminary report. Theriogenology 39:561-567

Rappolee DA, Sturm KS, Behrendtsen O, Schultz GA, Pedersen RA, Werb Z. 1992. Insulin-like growth factor II acts through an endogenous growth pathway regulated by imprinting in early mouse embryos. Genes Dev 6:939-952.

Reik W, Romer I, Barton SC, Surani MA, Howlett SK, Klose J. 1993. Adult phenotype in the mouse can be affected by epigenetic events in the early embryo. Development 119:933-942.

Schultz GA, Harvey MB, Watson AJ, Arcellana-Panlilio MY, Jones K, Westhusin ME. 1996. Regulation of early embryonic development by growth factors: growth factor gene expression in cloned bovine embryos. J Anim Sci 74 Suppl. 3:50-57.

Shehu D, Marsicano G, Flechon JE, Galli C. 1996. Developmentally regulated markers of in vitro produced preimplantation bovine embryos. Zygote 4:109-121.

Shim C, Kwon HB, Kim K. 1996. Differential expression of laminin chain-specific mRNA transcripts during mouse preimplantation embryo development. Mol Reprod Dev 44:44-55.

St Clair D, Oberley T, Muse K, St Clair W. 1994. Expression of manganese superoxide dismutase promotes cellular differentiation. Free Radical Biol Med 16:275-282

Stewart CL. 1994. Leukaemia inhibitory factor and the regulation of preimplantation development of the mammalian embryo. Mol Reprod Dev 39:233-238.

Stojanov T, O'Neill C. 1999. Ontogeny of expression of a receptor for platelet-activating factor in mouse preimplantation embryos and the effects of fertilization and culture in vitro on its expression. Biol Reprod 60:674-682.

Stojanov T, Alechna S, O'Neill C. 1999. In vitro fertilization and culture of mouse embryos in vitro significantly retards the onset of insulin-like growth factor-II expression from the zygotic genome. Mol Hum Reprod 5:116-124.

Tanbo T, Dale PO, Lunde O, Moe N, Abyholm T. 1995. Obstetric outcome in singleton pregnancies after assisted reproduction. Obstet Gynecol 86:188-192.

Telford NA, Watson AJ, Schultz GA. 1990. Transition from maternal to embryonic control in early mammalian development: a comparison of several species. Mol Reprod Dev 26:90-100.

Temeles GL, Schultz RM. 1997. Transient polyadenylation of a maternal mRNA follwing fertilization of mouse eggs. J Reprod Fertil 109:223-228.

Tervit HR, Whittingham DG, Rowson, LEA. 1972. Successful culture in vitro of sheep and cattle ova. J Reprod Fertil 30:493-497.

Thompson J, Partridge R, Houghton F, Cox C, Leese H. 1996 Oxygen uptake and carbohydrate metabolism by in vitro derived bovine embryos. J Reprod Fertil 106:299-306.

Thompson JG. 1997. Comparison between in vivo-derived and in vitro-produced preelongation embryos from domestic ruminants. Reprod Fertil Dev 9:341-354.

Uechi H, Tsutsumi O, Morita Y, Taketani Y. 1997. Cryopreservation of mouse embryos affects later embryonic development possibly through reduced expression of glucose transporter GLUT1. Mol Reprod Dev 48:496-500.

Van Stekelenburg-Hamers AEP, Rebel HG, Van Inzen WG, De Loos FAM, Drost M, Mummery CL, Weima SM, Trounson AO. 1994. Stage-specific appearance of the mouse antigen TEC-3 in normal and nuclear transfer bovine embryos: re-expression after nuclear transfer. Mol Reprod Dev 37:27-33.

Vernet M, Cavard C, Zider A, Fergelot P, Grimber G, Briand P. 1993. In vitro manipulation of early mouse embryos induces HIV1-LTRlacZ transgene expression. Development 119:1293-1300.

Walker SK, Hartwich KM, Seamark RF. 1996. The production of unusually large offspring following embryo manipulation: concepts and challanges. Theriogenology 45:111-120.

Wang JX, Clark AM, Kirby CA, Philipsen G, Petrucco O, Anderson G, Matthews CD. 1994. The obstetric outcome of singleton pregnancies following in vitro fertilization/gamete intra-Fallopian transfer. Hum Reprod 9:141-146.

Watson AJ, Hogan A, Hahnel A, Wiemer KE, Schultz GA. 1992. Expression of growth factor ligand and receptor genes in the preimplantation bovine embryo. Mol Reprod Dev 31:87-95.

Watson AJ, De Sousa P, Caveney A, Barcroft LC, Natale D, Urquhart J, Westhusin ME. 2000. Impact of bovine oocyte maturation media on oocyte transcript levels, blastocyst development, cell number, and apoptosis. Biol Reprod 62:355-364.

Winger QA, Hill JR, Shin T, Watson AJ, Kraemer DC, Westhusin ME. 2000. Genetic reprogramming of lactate dehydrogenase, citrate synthase, and phosphofructokinase mRNA in bovine nuclear transfer embryos produced using bovine fibroblast cell nuclei. Mol Reprod Dev 56:458-464.

Wrenzycki C, Herrmann D, Carnwath JW, Niemann H. 1996. Expression of the gap junction gene connexin43 Cx43 in preimplantation bovine embryos derived in vitro or in vivo. J Reprod Fertil 108:17-24.

Wrenzycki C, Herrmann D, Carnwath JW, Niemann H. 1998. Expression of RNA from developmentally important genes in preimplantation bovine embryos produced in TCM supplemented with bovine serum albumin BSA. J Reprod Fertil 112:387-398.

Wrenzycki C, Herrmann D, Carnwath JW, Niemann H. 1999. Alterations in the relative abundance of gene transcripts in preimplantation bovine embryos cultured in medium supplemented with either serum or PVA. Mol Reprod Dev 53:8-18.

Wrenzycki C, De Sousa P, Overström EW, Duby RT, Herrmann D, Watson AJ, Niemann H, O'Callaghan D, Boland MP. 2000. Effects of superovulated heifer diet type and quantity on relative mRNA abundances and pyruvate metabolism in recovered embryos. J Reprod Fertil 118:69-78.

Wrenzycki C, Herrmann D, Keskintepe L, Martins Jr A, Sirisathien, S, Brackett B, Niemann H. 2001a. Effects of culture system and protein supplementation on mRNA expression in preimplantation bovine embryos. Hum Reprod 16:893-901.

Wrenzycki C, Wells D, Herrmann D, Miller A, Oliver J, Tervit R, Niemann H. 2001b. The nuclear transfer protocol affects mRNA expression patterns in cloned bovine blastocysts. Biol Reprod 65:323-331.

Yaseen MA, Wrenzycki C, Herrmann D, Carnwath JW, Niemann H. 2001. Changes in the relative abundance of transcripts for insulin-like growth factor ligands IGF-I and IGF-II and their receptors IGF-IR/IGF-IIR in preimplantation bovine embryos derived from different in vitro systems. Reproduction 122:601-610.

Young LE, Sinclair KD, Wilmut I. 1998. Large offspring syndrome in cattle and sheep. Rev Reprod 3:155-163.

Young LE, Fernandes K, McEvoy TG, Butterwith SC, Gutierrez CG, Carolan C, Broadbent PJ, Robinson JJ, Wilmut I, Sinclair KD. 2001. Epigenetic change in IGF2R is associated with fetal overgrowth after sheep embryo culture. Nature Genetics 27:153-154.

Appendix I

THE ELECTRON MICROSCOPE

Poul Maddox-Hyttel
Department of Anatomy and Physiology, The Royal Veterinary and Agricultural University, Groennegaardsvej 7, DK-1870 Frederiksberg C, Denmark,

The transmission and scanning electron microscope

Basically, two types of electron microscopes are used for studying pre-implantation embryos: The scanning electron microscope and the transmission electron microscope. The scanning electron microscope is used for visualisation of cell and tissue surfaces by creating an image from the secondary electrons that are emitted from a thin gold coating of a specimen dried by e.g. critical point drying, when it is hit by a beam of primary electrons. The focus depth of the scanning electron microscope is relatively high and images of an embryo obtained by this technique are rather self explanatory: The embryo appears as a three dimensional spherical structure as a ball viewed by the naked eye. In contrast to its scenic appearance, the scientific information gained by scanning electron microscopy is rather limited. Thus in the following, transmission electron microscopy, which from a scientific point of view is of far greater importance, will be set in focus. At transmission electron microscopy, extremely thin sections of specimens are observed allowing for evaluation of the interior of e.g. the blastomeres of an embryo.

The transmission electron microscope basically consists of the following components: A high-voltage generator, an electron gun positioned at the end of a high-vacuum column that also comprises a set of electromagnetic lenses.

These components are connected to a vacuum pumping system and the necessary electronics for operating the entire system. The mode of operation can be compared with that of a traditional bright field microscope: The electron gun consists of a filament which generally is prepared from a thin wire of tungsten or more exotic material such as lanthanum hexaboride (LaB_6). When high voltage (generally in the range from 60 to 100 kV) is applied, the filament acts as the cathode (negatively charged) and below the filament the anode (positively charged) is positioned, which is a metal disk with a small hole in the centre. When the filament is heated, it acts as an electron emitter (as the bulb of the light microscope) and the negatively charged electrons are accelerated towards the positively charged anode. However, due to the inertia of the electrons they rush through the hole in the anode and form an electron beam passing down the high-vacuum column. The electron beam is controlled by the electron magnetic lenses: A first condenser lens determines the brightness of the beam (= the condenser of the light microscope), a next set of objective and so-called intermediate lenses, between which the specimen is introduced, determines the magnification (= the objective of the light microscope), and the last projection lens focuses the image onto a fluorescent screen for viewing (= the ocular of the light microscope), as the electron beam is not visible for the naked eye. The image can be recorded on either negative film or a CCD-camera.

Microscopical resolution

The resolution of a microscope can be defined as the minimum distance between two points that can be resolved by the equipment. The resolution can be calculated by the equation "$d = 0.621\lambda / n \sin \alpha$" defined by the mathematician Ernst Abbe in the late 19th century where d is the resolution, 0.621 is Abbe's constant, λ is the wavelength of the image-forming radiation, n is the index of refraction of the medium between the specimen and the lens, and α is the half-angle of the cone of radiation from the specimen plane accepted by the front surface of the lens. From this equation it is obvious that in order to get the best possible resolution one should have the smallest α and the largest possible $n \sin \alpha$ (also referred to as the numerical aperture of the lens). Because the half-angle of the cone of radiation from the specimen plane, α cannot exceed 90°, and media with a refractive index, n, greater than 1.6 (immersion oil has a refractive index of 1.55) are not known, it is nearly impossible to achieve a numerical aperture much greater than 1.4. Accordingly, using light radiation with a wavelength of 500 nm and an oil immersion objective with a numerical aperture of 1.4, the minimum resolution that can be obtained by the light microscope is 200 nm, i.e. 0.2 μm. Fortunately for microscopists, based on the work of Einstein and Planck, the French mathematician Louis de Broglie in 1924 advanced not only the theory that a beam of electrons has a wavelength but also an equation allowing for calculation of this wavelength depending on the velocity of the electrons. According to this equation, a 100,000 V potential electron beam would have a wavelength of approximately 0.0038 nm, which is almost 5 orders of magnitude shorter than that of visible light. According to Abbe's equation presented above, and where n is 1 (because the inside of the electron microscope is under high vacuum) and sin α is about equal to 1/100 (because the angle α is very small in the electron microscope) the predicted resolution of the electron microscope is about 0.24 nm, which is an improvement in resolution of about 3 orders of magnitude over the light microscope. This is a tremendous improvement sufficient of achieving atomic resolution! However, due to limitations in the preparation of biological specimens for transmission electron microscopy addressed in the following, the actual resolution that can be obtained is usually in the order of 2 nm, i.e. 0.002 μm.

Under practical circumstances studying pre-implantation embryos, the stereo microscope is generally used at magnifications up to about 150 times, the traditional bright field light microscope for sectioned specimens up to 1000 times, and the transmission electron microscope up to 100,000 times. Observing an embryo of about 150 μm in diameter, this small ball would in the stereo microscope correspond to performing observations with the naked eye on a sphere of 22,500 μm, i.e. 2.25 cm which is about the size of a hazelnut, in the traditional light microscope to observe a section of a sphere of 150,000 μm, i.e. 15 cm which is about the size of a grapefruit, and in the transmission electron microscope to observe a section of a sphere of 15,000,000 μm, i.e. 15 m which is about the size of a hot air balloon carrying people!

Specimen preparation for transmission electron microscopy

The outcome of transmission electron microscopy is dependent upon a proper fixation that preserve what can be described as a "controlled artifact" that reflects the image of the live cellular structure. For traditional transmission electron microscopy is generally used a primary fixation in 2.5 to 3% glutaraldehyde, which is a dialdehyde yielding a stable cross-linkage fixation of proteins, dissolved in a proper buffer of correct osmolarity and pH as e.g. 0.1M sodium phosphate or cacodylate buffer adjusted to a pH between 7.2 and 7.4. Paraformaldehyde, which is a monoaldehyde yielding a less stable cross-linkage fixation of proteins but having a quicker penetration due to a lower molecular weight, may be added to the fixative in a concentration of 2-4% and the glutaraldehyde concentration decreased to 1-2%. This so-called Karnowsky's fixative combines the advantage of stable cross-linkage by the glutaraldehyde with quick penetration of the paraformaldehyde. Fixation of embryos is generally performed for 1 h at 4°C. In order to facilitate the processing of the embryos, it is recommended to embed them in small (about 1 mm^3) agar blocks after the primary fixation allowing for handling without stereo microscopy and, thus, less exposure to the hazardous chemicals used for the subsequent processing (Hyttel and Madsen, 1987). After washing of the embryos in the agar blocks in the appropriate buffer, a secondary fixation of in particular lipids in 1-2% OsO_4 in the same buffer is, typically, applied for 1 h at 4°C. OsO_4 not only fixes the lipids, it also gives some staining contrast to, in particular, membranes. Subsequently, the specimens are dehydrated in graded series of ethanol, equilibrated in e.g. propylene oxide, which acts as an intermediate between ethanol and the epoxy resin in which the specimen is to be embedded, and finally equilibrated in the pure resin. Different types of epoxides, e.g. Epon, Araldite or Spurr's, can be used. Finally, the specimens are positioned in moulds in pure resin and polymerized. It should be kept in mind that if immunocytochemical detection of antigens in the embryos is to be applied, the primary fixative should be based on paraformaldehyde with up to 0.5% glutaraldehyde depending on the antigens in question, the secondary fixation should be omitted and the specimens should be embedded in acrylic resins, e.g. LR White, LR Gold or Lowicryl.

The subsequent sectioning of the specimens in the so-called ultra microtome aims at producing in the first round semi-thin sections of 1-2 μm thickness for light microscopical assessment, and in the second round ultra thin sections of about 70 nm thickness for electron microscopy. For the sectioning of mammalian oocytes and embryos a few strategic considerations should be applied. In most cases specific compartments of the oocyte or embryo, e.g. the germinal vesicle or the meiotic metaphase of the oocyte or the inner cell mass of the blastocyst, is of particular interest for the electron microscopical investigation. When the specimen is sectioned, only a limited number of sections will include these compartments of interest, and it is unknown at which stage of the sectioning procedure these areas will be cut. In order to avoid the need to stain and evaluate each semi-thin section in order to decide when to start cutting ultra thin sections from the block, we have elaborated a simple re-embedding procedure allowing for a rapid and precise evaluation procedure (Hyttel and Madsen, 1987): The entire oocyte or embryo is cut on glass knives into semi-thin sections of 2 μm thickness all of which are collected on glass slides, stained with 1% basic toluidine blue and examined by bright field light microscopy. The specimens undergo some shrinkage during dehydration and embedding, but a bovine oocyte or early embryo will, in general, result in approximately 50-60 2 μm semi-thin sections. Sections presenting features of particular interest for electron microscopy are then re-embedded by framing the sections with nail varnish and gluing a pre-polymerized epoxy resin block on to them with a droplet of un-polymerized epoxy resin. Following polymerization, the block with the semi-thin section at the end can be cracked off the slide by plunging into liquid nitrogen or heating to 70-80°C on a heating plate. Thus, the semi-thin section offers itself at the end of the resin block for ultra thin sectioning on a diamond knife. If the block is carefully aligned with the knife edge before starting the cutting, the 2 ῶm semi-thin section will allow for preparation of almost 30 ultra thin sections of 70 nm thickness that are collected on 2 or 3 electron microscopical grids. The sections are then stained, or rather contrasted as we refer to an electron beam with only one wave length and thus no colours, with uranyl acetate and lead citrate. The re-embedding technique is relatively rapid because it allows for semi-thin sectioning without interruptions for staining and assessment of the sections, and it is precise because it allows for electron microscopical assessment of exactly the same compartment as seen in the light microscope.

Appendix II

PROCESSING OF EGGS FOR TRANSMISSION ELECTRON MICROSCOPY

Poul Maddox-Hyttel
Department of Anatomy and Physiology, The Royal Veterinary and Agricultural University, Groennegaardsvej 7, DK-1870 Frederiksberg C, Denmark,

Practical protocol (example that can be modified)

Primary fixation

1. Fixation in 3% glutaraldehyde in 0.1M Na-phosphate buffer (PB) for 1 h at 4°C (pH 7.2-7.4).
2. Transfer to PB at 4°C.

Agar embedding

3. Embedding of eggs in 4% agar under stereo microscope (put egg in a drop of PB on a slide and put a drop of agar at about 45-55oC on top of it - build up with more agar on top in order to be able to cut out a small block (about 2 x 2 x 2mm) carrying the egg, with a scalpel). After embedding in agar the eggs can be handled without stereo microscope throughout the rest of the procedure.
4. Wash 2 x in PB for 5 min at 4°C.

Secondary, post-fixation

5. Post-fix in 1% OsO4 in PB for 1 h at RT.
6. Wash 2 x in PB for 5 min at RT.

Uranyl bloc staining

7. Wash 2 x in H_2O.
8. Stain in 0.5% uranyl acetate in H_2O for 60 min at RT in dark.
9. Wash 2 x in H_2O.

Dehydration and embedding (to be performed in rotor at RT):

10. 70% ethanol 10 min.
11. 96% ethanol 10 min.
12. 99% ethanol (pure MERCK) 3 x 20 min.
13. Propylene oxide 2 x 10 min.
14. Propylene oxide: Epon, 2:1 20 min.
15. Propylene oxide: Epon, 1:1 20 min.
16. Propylene oxide: Epon, 1:2 20 min.
17. Epon over night at RT without any stopper (cork) in glass.
18. Embedding in mulls with fresh Epon and hardening at 60°C for 2 days.

Semi-thin sectioning

19. Cut the entire egg into 2 micrometer semi-thin sections and stain with 1% basic toluidine blue for LM evaluation. Evaluate by LM and select sections presenting features of interest.

Re-embedding

20. Selected sections are framed with nail varnish and a pre-polymerized block of Epon is glued on top of them with a drop of fresh unpolymerized Epon put on the framed area (thc nail varnish will

prevent the unpolymerized Epon to stick to the glass, only to the section that is free of varnish). Polymerize at 60°C for 2 days.

21. The block is cracked off the slide with the semi-thin section on the end by warming the slide to about 80-90°C or plunging it into liquid nitrogen.

Ultra-thin sectioning

22. The block is carefully aligned and 70 nm sections (20-30 from one semi-thin section) are cut and collected on copper grids.
23. The sections are stained with uranyl acetate and lead citrate according top standard procedures.

Appendix III

HUMAN PREIMPLANTATION DIAGNOSIS PCR and RT-PCR

Deborah Taylor
Institute of Reproductive Biology, Faculty of Medicine, Imperial College, Hammersmith Hospital, Du Cane RoadW11 0NN, UK

Blastomere preparation and amplification of gDNA

Preparation of PCR buffers

Potassium [K+] free 10X PCR Buffer:
    25mM MgCl, pH8.3, 1mg/mL gelatine, 100mM Tris HCl, pH 8.3
Neutralising buffer (NB):
    900mM Tris HCl, pH8.3, 300mM KCl, 200mM HCl
Lysis buffer:
    200mM KOH, 50mM dithiothreitol (DTT)

Lysis of blastomere

1. Add 5 μl of the lysis to a 5ml microfuge tube.
2. After washing the isolated blastomere in PBS, place the blastomere in the lysis buffer using a pulled pasteur pipette.
3. Overlay the lysis buffer with 50 μl of silicone fluid.

4. Heat the sample for 10 minutes at 65 °C to ensure complete cell lysis.
5. Use immediately for PCR or store at –20 °C for up to 3 months.

gDNA amplification

1. For a 50 µl PCR reaction, add 5 µl of [K+] free 10X PCR Buffer and 5 µl of NB to the tube containing the lysed blastomere
2. Make up to 50 µl with 50 mM of each dNTP, 2.5 units taq polymerase (AmpliTaq, Perkin Elmer), and 0.4µM of each primer,
3. Denature at 94 °C for 2 minutes.
4. For nested PCR protocols, cycling conditions are the same for reactions with outer or inner primers: denature for 45 seconds at 94 °C, anneal at appropriate temperature for 60 seconds and extend at 72 °C for 90 seconds. Complete each PCR with a final extension step at 72 °C for 5 minutes.
5. To visualize the product on an ethidium-stained gel, run the outer primers for 25-30 cycles, and the nested primers for 30 cycles.
6. For the inner reaction, add 2µl of the first amplification product to freshly prepared PCR mix as above.

Reverse transcription of single blastomeres

Preparation of cDNA

1. Remove the zona pellucida from the embryo with acidified Tyrode's solution (pH 2.4)
2. Wash in phosphate buffered saline (PBS).
3. Lyse individual blastomeres in 5 µl of RT-lysis buffer (0.5% NP40, 10 mM Tris (pH 8.0), 10 mM NaCl and 3 mM MgC12) on ice (Gilliland et al., 1990).
4. Make up the lysate to 20 µl with reverse transcriptase buffer (50 mM Tris-HCl (pH 8.3), 75 mM KCl, 3 mM MgC12, 10 mM DDT (Pharmacia), 300 pmol oligo (dT)12-18 primer (Pharmacia) 0.5 mM dNTP (Pharmacia) and 100 units of M-MLV reverse transcriptase Incubate at 37 °C for 1 hour.
5. Terminate the reaction by heating at 95 °C for 5 minutes.

6. Spin the tubes for 1-2 minutes
7. Use the supernatant for PCR, or snap freeze the sample in liquid nitrogen and store at –70 °C for up to 3 months

RT-PCR cycling conditions

1. Add 10 or 20 µl of the cDNA product from the reverse transcription, to the PCR mix.
2. Make up to 30 µl with the PCR buffer (10mM Tris-HCl, pH 8.3, 50 mM of each dNTP, 2.5 units taq polymerase (AmpliTaq, Perkin Elmer), 0.4µM of each primer).
3. Cover with 50µl of silicone oil if necessary.
4. Denature at 94 °C for 2 minutes.
5. For nested PCR protocols, cycling conditions are the same for reactions with outer or inner primers: denature for 45 seconds at 94 °C anneal at appropriate temperature for 60 seconds and extend at 72 °C for 90 seconds. Complete each PCR with a final extension step at 72 °C for 5 minutes.
6. To visualize the product on a ethidium-stained gel, run the outer primers for 25-30 cycles, and the nested primers for 30 cycles.
7. For the inner reaction, add 2µl of the first amplification product to freshly prepared PCR mix as above.

Appendix IV

HUMAN PREIMPLANTATION DIAGNOSIS FLUORESCENCE IN SITU-HYBRIDIZATION (FISH) ON INDIVIDUAL BLASTOMERES

Deborah Taylor
Institute of Reproductive Biology, Faculty of Medicine, Imperial College, Hammersmith Hospital, Du Cane RoadW11 0NN, UK

Preparation of air-dried slides from single blastomeres

1. Draw a small circle on the reverse of a a poly-L-lysine slide (BDH)
2. Place a small drop of spreading solution (0.01M HCl; 0.1% Tween) on to the slide.
3. Lyse the blastomere by placing in the drop of spreading solution
4. Continually observe the cell down the stereo microscope, whilst agitating the drop to disperse the surrounding cytoplasm from the nucleus
5. Once the nucleus is clear of cytoplasm leave the slide to air dry
6. Dehydrate slides by passing through an ethanol series of 70%, 90% and then 100% for 5 minutes at room temperature
7. Use slides immediately for FISH analysis or store at room temperature for up to 6 months.

Hybridization of directly labelled fluorescent probes

1. Prior to hybridization, incubate slides in 0.1 mg/ml of pepsin for 20 minutes at 37oC to remove any protein attached to the nuclear material
2. Wash the air dried slides in PBS (Gibco) then incubate in 1% paraformaldehyde for 10 minutes at 4oC.

3. Wash slides again in PBS and then in distilled water.
4. Dehydrate slides by passing through an ethanol series of 70%, 90% and then 100% for 5 minutes at room temperature.
5. Hybridize the fluorescent probe according to the manufacturer's recommendations.

Triple colour DNA mixture (Chromosomes X, Y and 18) (Vysis™, Surrey, UK).

1. Co-denature both the nuclear and probe DNA by placing the slides on a hot block at 73oC for 5 minutes.
2. Place slides in a dark humidified container at 37°C for 45 minutes under a cover slip, to allow hybridization of the probe to nuclear DNA
3. Remove cover slips from the slides.
4. Wash slides in 0.4x SSC/0.3% NP-40 at 73°C for 2 minutes.
5. Then wash slides at room temperature in 2 xSSC/0.1% NP-40 for 1 minute.
6. Leave to air dry.
7. Mount slides with antifade Vectarshield (Vectar Laboratories, Burlingame, CA) containing 1.2 μg of 4', 6-diaminidino-2-phenyolindole (DAPI) to counter stain the nucleus.
8. Place a cover slip over the target area of each slide and remove air bubbles by applying pressure to the cover slip with a pair of blunt forceps
9. Score slides immediately using an Axophot fluorescent microscope (Zeiss), with a CCD camera controlled by SmartCaptureTM software (Vysis), or store for up to 3 months in the dark at –20°C.

Appendix V

FLUORESCENCE IN SITU-HYBRIDIZATION (FISH) ON METAPHASES IN BOVINE EMBRYOS

Dorthe Viuff
Molecular pharmacology, Novo Nordisk A/S, Novo Nordisk Park, 2760 Måløv, Denmark.

This protocol can be followed when two different labelled probes are used simultaneously. In our case the probes are a chromosome 6-specific bovine probe (p33E39) and a chromosome 7-specific bovine probe (cJAB8). These have been labelled by nick translation using biotin and digoxigenin, respectively. The biotinylated p33E39 DNA and the digoxigenated cJAB8 DNA were added to the hybridization solution (50% deionized formamide, 10% dextran sulphate, 2xSSC, 10μg salmon sperm DNA, and 10μg genomic DNA) at a final concentration of 40ng/μl and 30ng/μl, respectively.

Day 1

Solutions

- 100 ml Pepsin solution (0.1 mg/ml , 0.01N HCl:10 mg pepsin + 200μl 5N HCl ad 100 ml MilliQ H_2O)
- 200 ml Formaldehyde 1%:50 ml 4% formaldehyde ad 200 ml 0.15 M PBS (pH 7.4)
- 300 ml Formamide 70%:210 ml Formamide + 30 ml 20 x SSC ad 300 ml MilliQ H_2O
- 70 %, 90 %, 100 % ethanol.

Procedure

Slide preparation

Embryonic cells are normally incubated overnight with colchicine to arrest dividing cells at metaphase, and cells are then fixed in methanol:acetic acid and prepared by the air-drying method to obtain metaphase spreads (appendix IV).

Pepsin treatment

1. The slides are treated with pepsin solution at **37°C** from 3 to 20 min., depending on the number and quality of the nuclei and metaphase visualized by phase contrast microscope.
2. Wash twice at **room temperature (RT)** in PBS 0.15M, pH 7.4 for two min. each time.
3. Fix at **RT** in 1% formaldehyde for two min.
4. Wash twice at **RT** in PBS 0.15M, pH 7.4 for three min. each time.

Denaturation

1. Immerse the slides in 70% formamide at 71°C for two min. (Maximum two slides at a time)
2. Dehydrate in an ice-cold ascending ethanol series (70%, 90%, 100%) for two min. each.
3. Air dry.
4. Denature the probes at **75°C** for 5 min.
5. Leave to preanneal at **37°C** for 30-60 min.
6. Centrifuge at 13000 rpm/1 min.
7. Keep at **RT** until use.

Hybridization

1. Place 3-5μl of the probe on each slide.
2. Place a coverslip (9 mm in diameter) and seal with rubber cement.
3. Incubate overnight at **37°- 42°C.**

Day 2

Solutions

- 2000 ml 0.05 x SSC:5 ml 20x SSC ad 2000 ml MilliQ H_2O
- 2000 ml TN-buffer:400 ml 20x SSC + 2 ml Tween 20 ad 2000 ml MilliQ H_2O
- 100 ml TNS:5 g skim milk + 100 ml TN buffer
- Antibodies diluted in TNS
  1) C_y3-conjugated Streptavidin 1:100 +Anti-DIG-fluorescein 1:200
  2) Anti-Avidin D (biotinylated) 1:100 + Anti-mouse-Ig-DIG 1:200
  3) Same as nr. 1).
- Antifade:1 µl DAPI 1mg/ml + 999 µl DABCO antifade solution.

Procedure

Post hybridization

1. Remove the glue with a forceps and place the slides in 0.05 x SSC at **37°C**
2. for aprox.10 minutes or until the coverslip falls off.
3. Wash two times in 0.05 x SSC at **37°C** for 3 min.
4. Wash once in TNS for 10 min. at **37°C.**

Detection

1. On the wet slide place 100µl of antibody solution 1.
2. Cover with parafilm and incubate at **37°C** for 1 hr.
3. Wash 3 times in TN buffer at **RT** for 3 minutes each time
4. On the wet slide place 100 µl of antibody solution 2.
5. Repeat step 2.
6. Repeat step 3.
7. On the wet slide place 100 µl of antibody solution 1.
8. Repeat step 2.
9. Repeat step 3.

Visualisation

1. On the wet slide place 22 μl antifade and cover with a coverslip.
2. Analyse the specimens using a fluorescence microscope.

APPENDIX VI

DIFFERENTIAL STAINING

Ann Van Soom
Department of Reproduction, Obstetrics and Herd Health, Faculty of Veterinary Medicine, Salisburylaan 133, B-9820 Merelbeke, Belgium

This protocol is adapted for bovine in vitro produced embryos from Hardy et al. (1989). Different concentrations and/or incubation times may be applied if embryos from other species are used.

Solutions and Material

- Complement B (CL 5000, Cedarlane Laboratories Ltd): divide in 60 µl vials and store at -80°C.
- PBS/PVP: add 3 mg/ml polyvinylpyrrolidone (PVP) to PBS and store at 4°C
- Picryl sulphonic acid (Sigma P-2297) (5% (w/v)):
- Mix 9 ml PBS/PVP with 1 ml picryl sulphonic acid and check pH (7,4). Use NaOH to increase pH. Filter in 10 ml tube, label as TNBS and date-wrap in in aluminium foil and store at 4°C.
- Rabbit anti-DNP-BSA (ICN Biochemicals # 61-006-1): divide in 100 µl vials in freezer (-20°C is OK)
- Ethanol 100% (1115, UCB)
- Glycerol (1384, UCB)
- Hoechst stain # 33342, 10 mg/ml (Molecular Probes)
- Propidium iodide (PI), 1 mg/ml (P-4170, Sigma)
- TCM199 bicarbonate with Earle´s salts, w/o glutamine (M-2154, Sigma)
- TCM199 Hepes with Earle´s salts, w/o glutamine (M-7528, Sigma)
- 4-well plates, petridishes of 3 ml
- Microscopic slides, cover slips

- *Precautions!*
  Wear protective gloves
  Fluorescent stains are supposed to be mutagenic!

Protocol

If hatched blastocysts or zona free embryos (with acid tyrode and pronase; step 1) are available go directly to step 2.

Removal of the zona pellucida (ZP)

1. Hepes-TALP with 20% serum as the washing medium, 3 dishes, 3 ml -- washing medium per dish (warm to 38°C before ZP removal)
2. Culture medium (same as you used for experiment, warm to 38°C)
3. 0.5% pronase, warm to 38°C
4. Acid Tyrode (pH 2.1), warm to 38°C
5. Put embryos into 0.5% pronase under constant observation until ZP becomes wrinkly. It is better to treat embryos one by one.
6. Then transfer embryos to acid Tyrode. The ZP will blacken and close upon the embryo (this takes 1 max 2 min). When you cannot distinguish the ZP from the oolemma anymore: transfer the embryos immediately to Hepes-TALP with 20 % serum.
7. Wash the embryos in the first dish containing washing medium, then move to the second one, third one.
8. To recover leave the embryos one to two hours (or overnight) in adequate culture medium in CO_2 incubator after zona removal.

Prepare TNBS and antiserum

1. Prepare 0.5-1 ml TNBS (toxic) and store at 4°C.
2. Mix TCM199 bicarbonate (without serum) with anti-DNP-BSA to make antiserum (90 μl anti-DNP-BSA + 210 μl TCM199), and make 14 droplets of 20 μl under mineral oil, warm to 38°C in CO_2 incubator
3. Put 5 ml TCM199 Hepes at room temperature.

Incubation in TNBS (labelling of outer cells)

1. Take zona free embryos out of culture medium and wash them in TCM199 Hepes to get rid of oil.
2. Then place embryos in 0.5 ml TNBS for 10 min on ice.

Incubation in antiserum

1. Wash zona free embryos in some fresh TCM199 Hepes until yellow colour of TNBS is not visible anymore. Transfer with as little medium as possible to the DNP dilution (1 – 3 embryos per droplet), incubate at 38°C in CO_2 incubator for 30 min.
2. Meanwhile prepare your complement dilution (if you thaw complement too early it will loose its activity (heat labile): take 50 μl complement and dilute in 200 μl M199 (or 100 μl in 400 μl if necessary). Add 2.5 μl (or 5 μl) PI (be careful very toxic!!). Make 10 to 25 droplets of 25 à 30 μl under oil and incubate in CO_2 incubator

Incubation in complement

1. Take embryos out of the anti-DNP-BSA droplets and wash in the fresh TCM199 Hepes shortly, and put them into the complement PI droplets made in step 4 at 38°C for 30 min. (This step has to be checked for different species).
2. Meanwhile prepare everything for fixation.
3. Dilute 20 μl PI in 2 ml TCM199 Hepes and put on ice.
4. Take 5ml ethanol and put on ice also.
5. Prepare the microscopic slides. Indicate with pencil on the slide expt no., embryonic stage/treatment, hpi and number of embryos.

Fixation and staining

1. Take embryos out of the complement PI droplets and wash in icecold PI TCM199 Hepes made in step 5 for only a few seconds. (PI will be washed out if you wash at room temperature or longer than a few seconds)
2. Then fix embryos in ethanol for 5 min or overnight.

3. Transfer to Hoechst staining for 10 min at room temperature (ratio 1 µl Hoechst/ml ethanol), or keep 4 °C overnight and perform the Hoechst staining on the next day.
4. Then mount embryos on microscopic slides.

Mounting of embryos

1. Put a drop of glycerol 100% on a slide.
2. Put 5 embryos from Hoechst solution into glycerol
3. When ethanol has disappeared almost completely, put coverslip on top and locate the embryos with blue marker.

Index

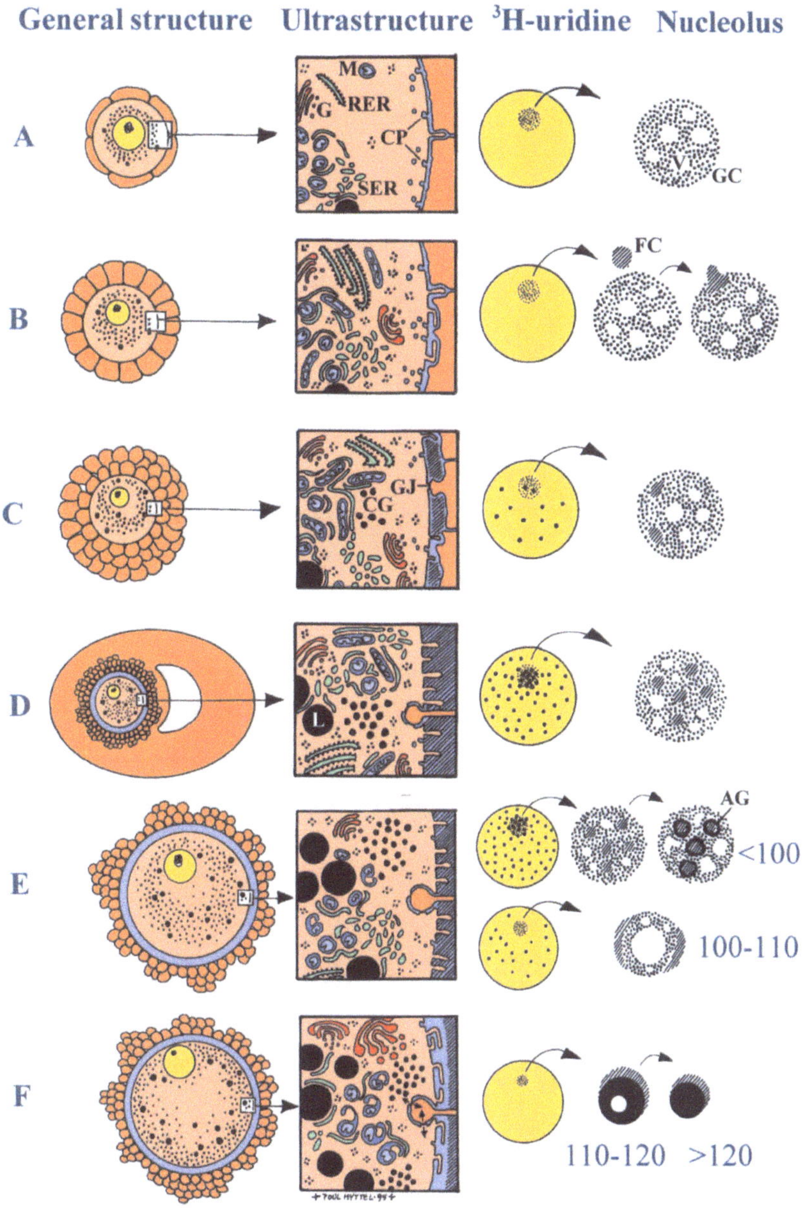

Figure 1 Chapter 13

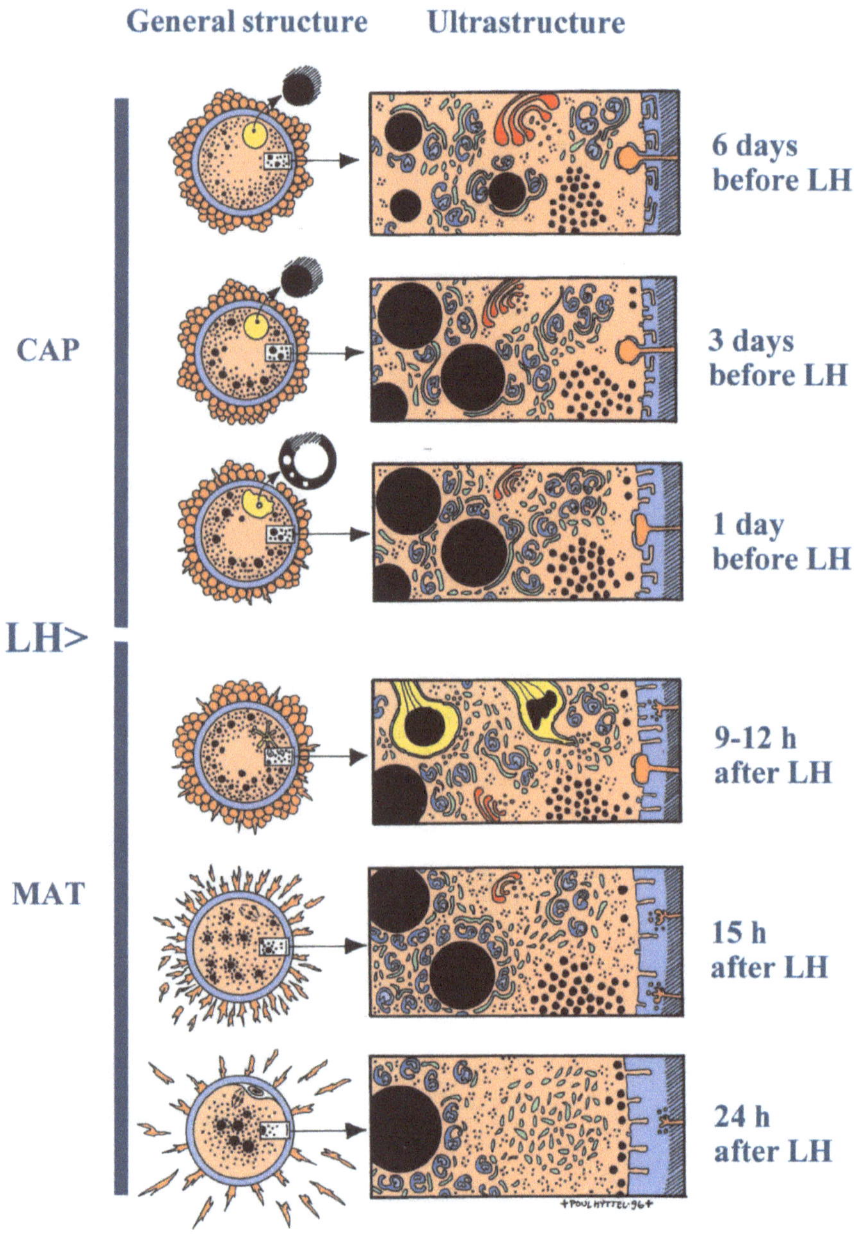

Figure 2 Chapter 13

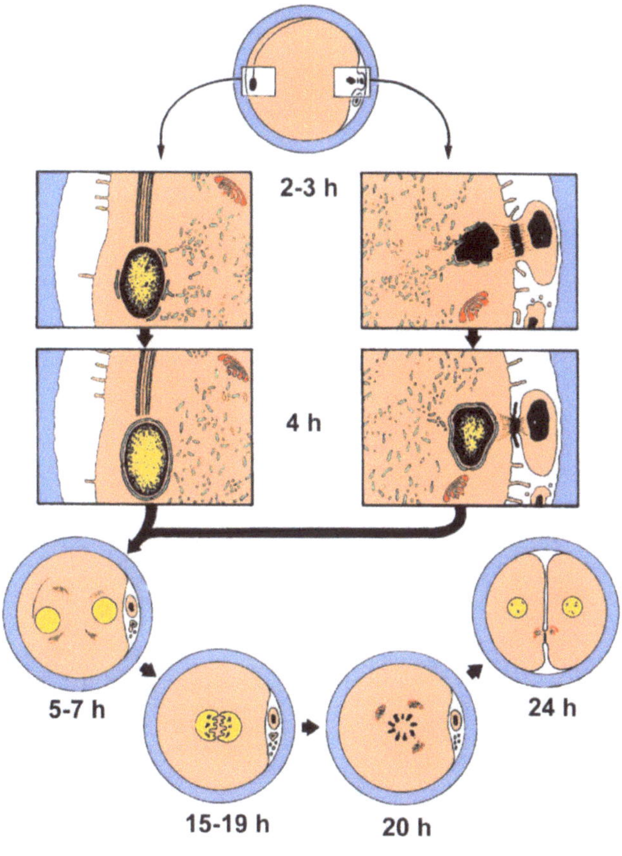

Figure 11 Chapter 13

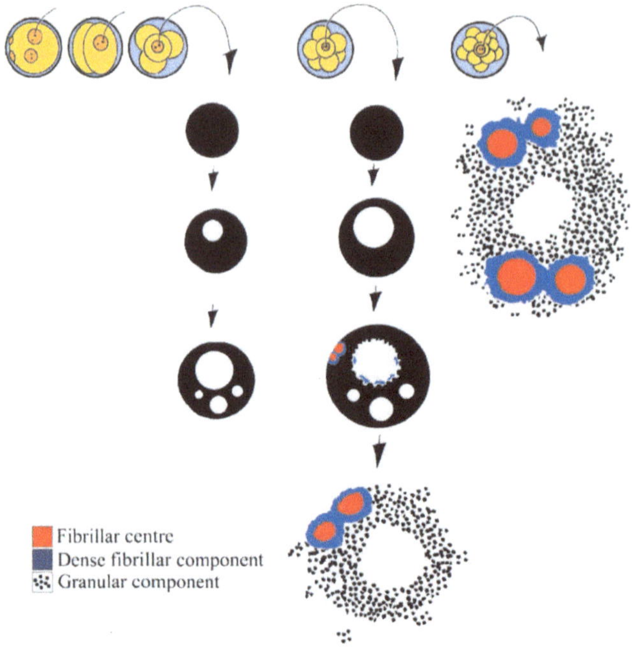

Figure 18 Chapter 13

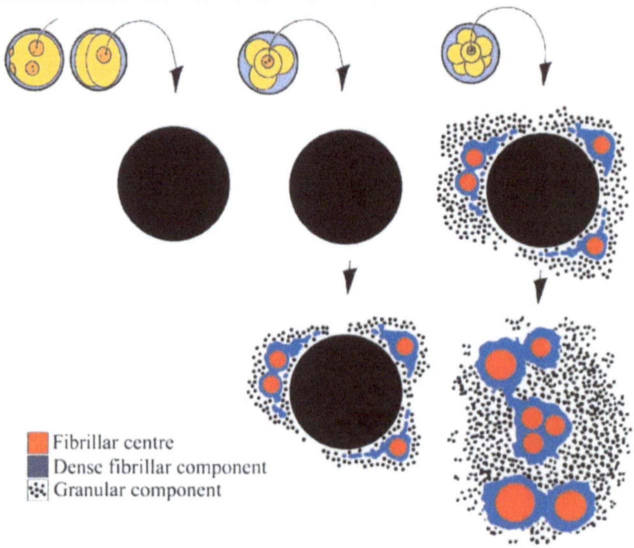

Figure 27 Chapter 13

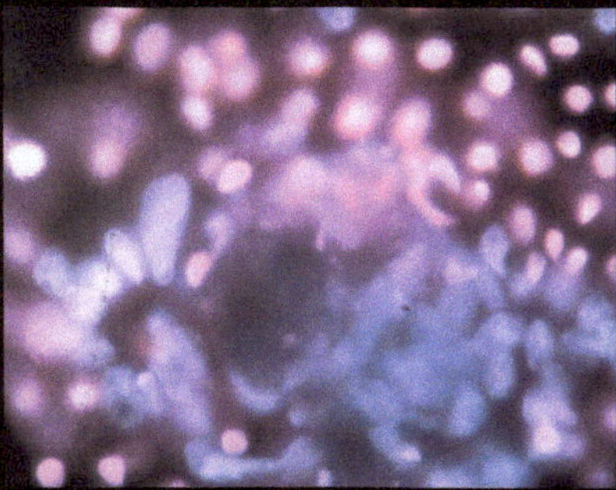

Figure 1a Chapter 11

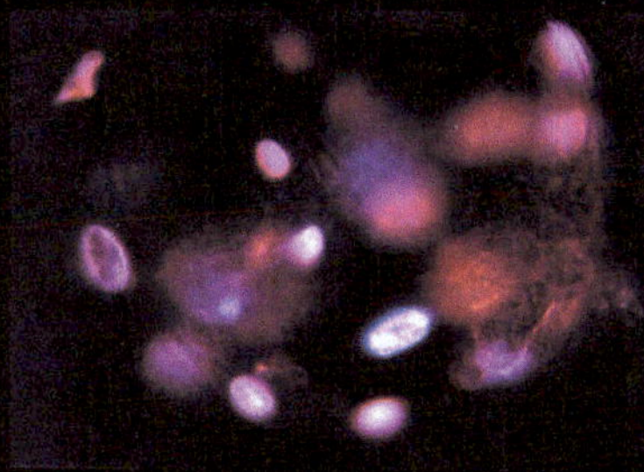

Figure 1b Chapter 11

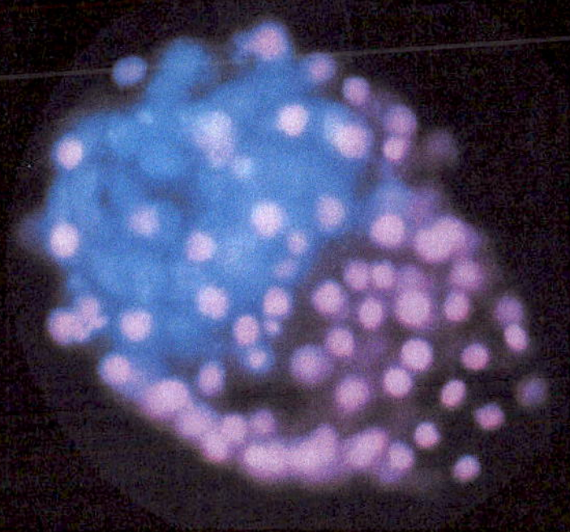

Figure 3a Chapter 11

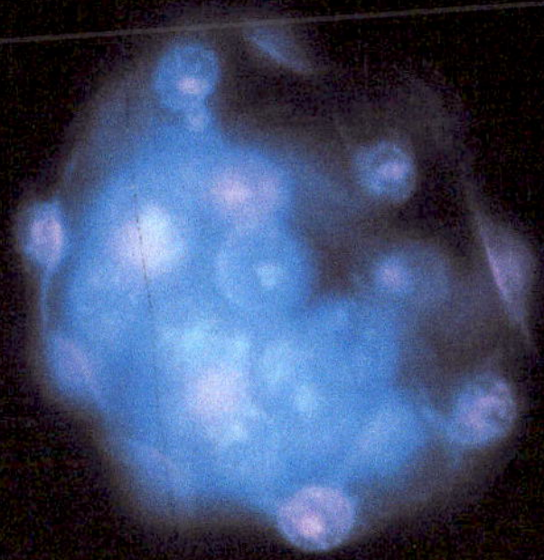

Figure 3b Chapter 11

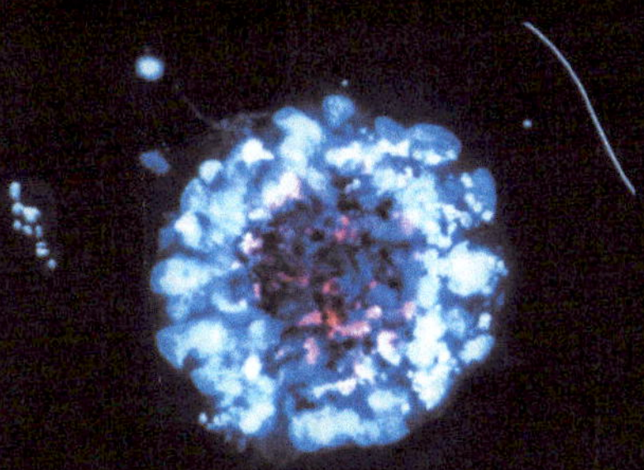

Figure 4a Chapter 11

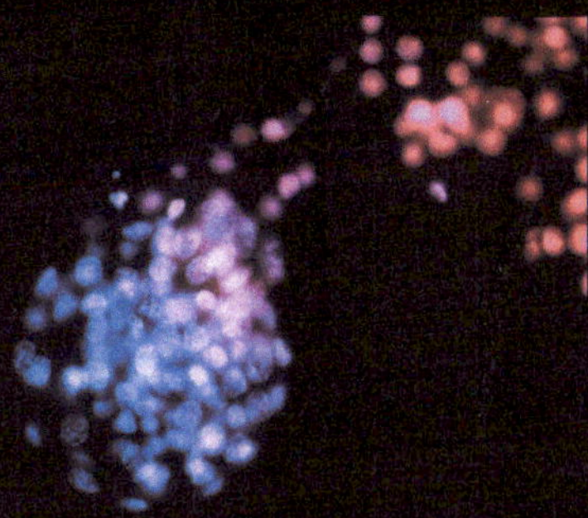

Figure 4b Chapter 11

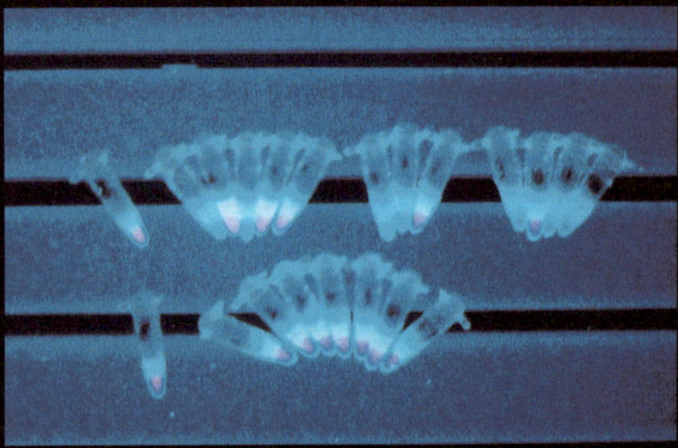

Figure 1 Chapter 8

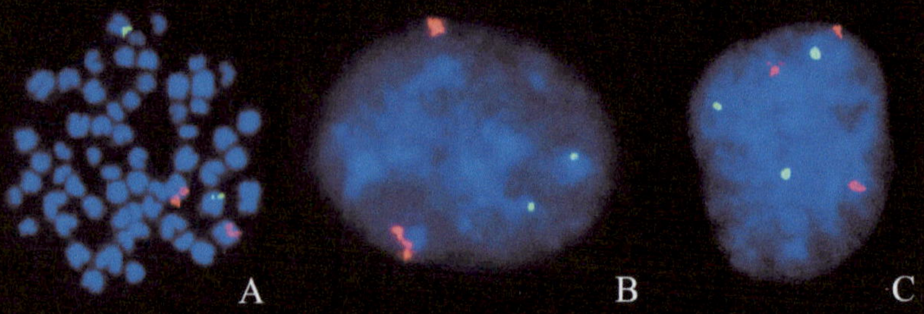

Figure 1 Chapter 10

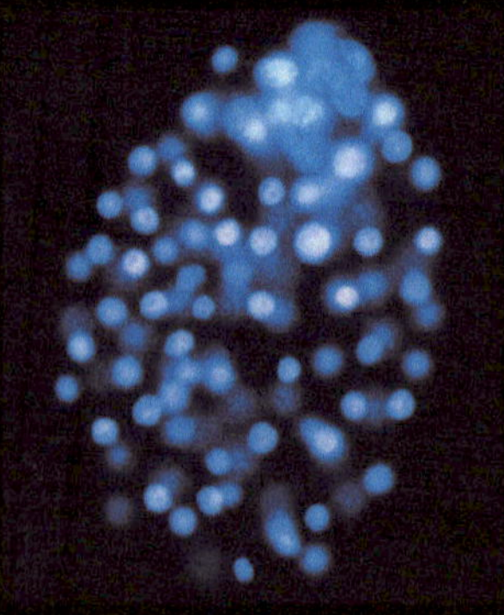

Figure 4c Chapter 11